Coherent and Polarization Optics

Coherent and Polarization Optics

Editors

Jiayi Yu
Yongtao Zhang

MDPI

Basel • Beijing • Wuhan • Barcelona • Belgrade • Novi Sad • Cluj • Manchester

Editors
Jiayi Yu
School of Physics and Electronics
Shandong Normal University
Jinan
China

Yongtao Zhang
School of Physics and Information Engineering
Minnan Normal University
Zhangzhou
China

Editorial Office
MDPI
St. Alban-Anlage 66
4052 Basel, Switzerland

This is a reprint of articles from the Special Issue published online in the open access journal *Photonics* (ISSN 2304-6732) (available at: www.mdpi.com/journal/photonics/special_issues/coherent_polarization_optics).

For citation purposes, cite each article independently as indicated on the article page online and as indicated below:

Lastname, A.A.; Lastname, B.B. Article Title. *Journal Name* **Year**, *Volume Number*, Page Range.

ISBN 978-3-7258-0190-9 (Hbk)
ISBN 978-3-7258-0189-3 (PDF)
doi.org/10.3390/books978-3-7258-0189-3

Contents

About the Editors

Jiayi Yu

Dr. Jiayi Yu received his Ph.D. from Soochow University and was a visiting scholar at University of North Carolina at Charlotte before he joined Shandong Normal University in 2019. His research expertise lies in the field of coherent optics, with a particular focus on the manipulation of coherence structures in light fields for applications in free space optical communications, optical tweezers and optical imaging. He has published about 60 papers with pioneering contributions in manipulation of nonuniform coherence structures and applications.

Yongtao Zhang

Prof. Yongtao Zhang received his PhD from Soochow University and was a Postdoctoral researcher at University of North Carolina at Charlotte before he joined Minnan Normal University in 2020. His research interests include optical coherence and singular optics. He has published more than 40 papers which focus on correlation singularities, orbital angular momentum, partially coherent beams, and scattering.

Preface

Coherence and polarization are two of the intrinsic properties of a light field. The theory of optical coherence and polarization was effectively developed over the past few decades. Extensive research shows that both of them play a crucial role in determining the light beam propagation and light–matter interaction. The manipulation of coherence and polarization was recognized as an available method to produce many peculiar physical phenomena, which can be superior in some of these applications, such as free-space optical communications, microdensitometry, optical information processing, and plasmonics.

However, there is no dedicated book or journal Special Issue published yet. We feel there is a need to edit a Special Issue on this topic to reflect the recent advances and developments as well as trends in the field. This Special Issue received many submissions and we selected 15 papers among these to be included. Our editorial in this reprint provides an overview and detailed analysis of each work in the Special Issue. We hope you will find this reprint useful and valuable for coherent and polarization optics research.

Jiayi Yu and Yongtao Zhang
Editors

photonics

MDPI

Editorial

Introduction to the Special Issue on Coherent and Polarization Optics

Yongtao Zhang [1,*] and Jiayi Yu [2,*]

1 Key Laboratory of Light Field Manipulation and System Integration Applications in Fujian Province, School of Physics and Information Engineering, Minnan Normal University, Zhangzhou 363000, China
2 Shandong Provincial Engineering and Technical Center of Light Manipulations & Shandong Provincial Key Laboratory of Optics and Photonic Devices, School of Physics and Electronics, Shandong Normal University, Jinan 250358, China
* Correspondence: yongtaozhang@mnnu.edu.cn (Y.Z.); jiayiyu0528@sdnu.edu.cn (J.Y.)

Both coherence and polarization are important inherent properties of light. Optical coherence, including spatial coherence and temporal coherence, demonstrates correlations between the components of a fluctuating electric field at two or more points. By adjusting the spatial and temporal coherence states of light sources, all fundamental properties of light beams can be changed, including the intensity, phase, color, polarization, and orbital angular momentum [1–10]. Polarization, on the other hand, arises from correlations between the components of a fluctuating electric field at a single point. The manipulation of polarization states of light beams also plays a crucial role in determining the properties of light beams [11–13]. Furthermore, though coherence and polarization have been treated independently from each other so far, in 2003, Wolf et al. found that they are also closely related aspects of statistical optics [14,15], which may help people to fully understand optical coherence and polarization. In this Special Issue, we highlight the recent progress on coherence and polarization optics, and show the properties of coherence and polarization and how both of them affect light beam propagation and light–matter interaction.

Coherence may be used to improve the light beam performance in random media. In Contribution 1, the authors study the propagation of partially coherent self-splitting structured beams in ocean turbulence. It was found that by reducing the coherence length and enhancing the order of the beam, the ocean-turbulence-induced negative effects could be reduced. The effect of ocean turbulence on the spectral degree of coherence on propagation is also discussed. The authors of Contribution 2 introduce a new class of non-uniformly correlated beam, termed the Lorentz non-uniformly correlated beam, and investigate its propagation both in free space and oceanic turbulence. The results show that such a beam exhibits self-focusing properties on propagation, and by increasing the beam width and decreasing the coherence length, the negative effects of the turbulence can be reduced. In Contribution 3, the authors propose a radially phase-locked Hermite–Gaussian-correlated beam array and study the propagation properties of such a beam array in ocean turbulence. The beam array with a smaller coherence length is shown to better retain the splitting properties during propagation in ocean turbulence.

The effect of coherence and polarization on the propagation of a light beam in anisotropic media is studied in Contribution 4. The authors of Contribution 4 investigated the propagation of a vector partially coherent twisted Laguerre–Gaussian pulsed beam through anisotropic atmospheric turbulence. Both the coherence and polarization of the beam are considered, and the advantage of the beam in reducing the atmospheric-turbulence-induced degeneration is discussed.

The tight focusing of optical fields can be used to control many properties of the fields, including the polarization, angular momentum, Poynting vector, etc. In Contribution 5, the authors investigate the properties of tightly focused Pearcey beams with a cross-phase,

Citation: Zhang, Y.; Yu, J. Introduction to the Special Issue on Coherent and Polarization Optics. *Photonics* **2024**, *11*, 50. https://doi.org/10.3390/photonics11010050

Received: 25 December 2023
Accepted: 26 December 2023
Published: 3 January 2024

and discuss the effect of intensity of the cross-phase on the intensity and gradient forces of these beams. The authors of Contribution 6 study the separation of polarization and orbital degrees of freedom, energy flow, and radial optical Hall effect of tightly focused Poincaré beams. In Contribution 7, the authors discuss the effect of the twisting phase on the conversion between the linear and circular polarizations of the tightly focused scalar and vector beams. The authors of Contribution 8 study the energy flow of the tightly focused azimuthally polarized beam with an x-type vortex, and show how to manipulate the longitudinal and transverse energy distributions by adjusting the anisotropic parameter of the noncanonical vortex. For an off-axis noncanonical vortex beam, the properties of energy flux are significantly influenced by the anisotropy parameter of such a tightly focused beam (Contribution 9). Partially coherent radially polarized Laguerre–Gaussian rotationally symmetrical power-exponent phase vortex beams with a Laguerre–Gaussian-correlated Schell model are introduced in Contribution 10. The intensity distribution, the degree of polarization and coherence, and the Stokes parameters of such tightly focused beams are demonstrated.

Asymmetric generalized Hermite–Gaussian and Laguerre–Gaussian beams are introduced in Contribution 11; they may have a high stability upon propagation in turbulent medium and could be used for optical trapping, rotating, and shifting of microparticles.

In Contribution 12, the authors study the possibility of obtaining powerful terahertz radiation with elliptical polarization by driving an orientated strong discharge current in a target with an elliptically shaped surface. The polarization properties of two-dimensional polarization holographic gratings in thin azopolymer films was studied in Contribution 13.

The effect of temporal coherence on interference contrast microscopy systems was studied for Contribution 14; the authors discuss the relationship between prism wedge angles and the interference color when the prism moves at multiple wavelengths.

In summary, the development of optical coherence, polarization, and their combination comprise an attractive research area, which may find a variety of potential applications for optical communication, imaging, sensing, and matter manipulation. I hope that this Special Issue will not only serve as a summary of the different research lines, but also encourage further study in this exciting field.

Author Contributions: Writing—original draft preparation, Y.Z.; writing—review and editing, J.Y. All authors have read and agreed to the published version of the manuscript.

Funding: National Natural Science Foundation of China (12174173, 12374276); Fujian Provincial Natural Science Foundation of China (2022J02047).

Conflicts of Interest: The authors declare no conflicts of interest.

List of Contributions

1. Liu, L.; Liu, Y.; Chang, H.; Huang, J.; Zhu, X.; Cai, Y.; Yu, J. Second-Order Statistics of Self-Splitting Structured Beams in Oceanic Turbulence. *Photonics* **2023**, *10*, 339.
2. Wei, D.; Wang, K.; Xu, Y.; Du, Q.; Liu, F.; Liu, J.; Dong, Y.; Zhang, L.; Yu, J.; Cai, Y.; Zhu, X. Propagation of a Lorentz Non-Uniformly Correlated Beam in a Turbulent Ocean. *Photonics* **2023**, *10*, 49.
3. Zhu, P.; Wang, G.; Yin, Y.; Zhong, H.; Wang, Y.; Liu, D. Radially Phased-Locked Hermite–Gaussian Correlated Beam Array and Its Properties in Oceanic Turbulence. *Photonics* **2023**, *10*, 551.
4. Xu Y.; Xu Y.; Wang T. Evolution properties of a partially coherent twisted laguerre-gaussian pulsed beam propagating through anisotropic atmospheric turbulence. *Photonics* **2023**, *9*, 707.
5. Yu, C.; He, Z.; Huang, C.; Chen, F.; Zeng, J.; Li, Y.; Zhang, Y.; Pu, J.; Lin, H. Tight Focusing Properties of Ring Pearcey Beams with a Cross Phase. *Photonics* **2022**, *9*, 964.
6. Kotlyar, V.V.; Stafeev, S.S.; Zaitsev, V.D.; Telegin, A.M. Poincaré Beams at the Tight Focus: Inseparability, Radial Spin Hall Effect, and Reverse Energy Flow. *Photonics* **2022**, *9*, 969.

7. Wu, S.D.; Chew, K.H.; Chen, R.P. Effect of Twisting Phases on Linear–Circular Polarization and Spin–Orbital Angular Momentum Conversions in Tightly Focused Vector and Scalar Beams. *Photonics* **2023**, *10*, 151.
8. Zhang, H.; Zhang, T.; Zhao, X; Pang, X. Manipulation of Energy Flow with X-Type Vortex. *Photonics* **2022**, *9*, 998.
9. Zhao, X.; Liang, H.; Wu, G; Pang, X. Influence of Off-Axis Noncanonical Vortex on the Dynamics of Energy Flux. *Photonics* **2023**, *10*, 346.
10. Ma, Z.; Pan, Y.; Dou, J.; Zhao, J.; Li, B; Hu, Y. Statistical Properties of Partially Coherent Higher-Order Laguerre-Gaussian Power-Exponent Phase Vortex Beams. *Photonics* **2023**, *10*, 461.
11. Abramochkin, E.G.; Kotlyar, V.V.; Kovalev, A.A; Stafeev, S.S. Generalized Asymmetric Hermite–Gaussian and Laguerre–Gaussian Beams. *Photonics* **2023**, *10*, 606.
12. Dmitriev, E.; Bukharskii, N.; Korneev, P. Powerful Elliptically Polarized Terahertz Radiation from Oscillating-Laser-Driven Discharge Surface Currents. *Photonics* **2023**, *10*, 803.
13. Mateev, G.; Nedelchev, L.; Nikolova, L.; Ivanov, B.; Strijkova, V.; Stoykova, E.; Choi, K.; Park, J.; Nazarova, D. Two-Dimensional Polarization Holographic Gratings in Azopolymer Thin Films: Polarization Properties in the Presence or Absence of Surface Relief. *Photonics* **2023**, *10*, 728.
14. Li, F; Zhao, T. Study of the Interference Color with Nomarski Prism Wedge Angle in a Differential Interference Contrast Microscopy System. *Photonics* **2023**, *10*, 678.

References

1. Wolf, E. Invariance of the spectrum of light on propagation. *Phys. Rev. Lett.* **1986**, *56*, 1370. [CrossRef] [PubMed]
2. James, D.F. Change of polarization of light beams on propagation in free space. *J. Opt. Soc. Am. A* **1994**, *11*, 1641. [CrossRef]
3. Zhang, Y.; Cai, Y.; Gbur, G. Partially coherent vortex beams of arbitrary radial order and a van Cittert–Zernike theorem for vortices. *Phys. Rev. A* **2020**, *101*, 043812. [CrossRef]
4. Yu, J.; Zhu, X.; Lin, S.; Wang, F.; Gbur, G.; Cai, Y. Vector partially coherent beams with prescribed non-uniform correlation structure. *Opt. Lett.* **2020**, *45*, 3824. [CrossRef] [PubMed]
5. Zhang, Y.; Cai, Y.; Gbur, G. Control of orbital angular momentum with partially coherent vortex beams. *Opt. Lett.* **2019**, *44*, 3617. [CrossRef] [PubMed]
6. Zhang, Y.; Korotkova, O.; Cai, Y.; Gbur, G. Correlation-induced orbital angular momentum changes. *Phys. Rev. A* **2020**, *102*, 063513. [CrossRef]
7. Peng, D.; Huang, Z.; Liu, Y.; Chen, Y.; Wang, F.; Ponomarenko, S.A.; Cai, Y. Optical coherence encryption with structured random light. *PhotoniX* **2021**, *2*, 1. [CrossRef]
8. Gbur, G.; Visser, T.D. The structure of partially coherent fields. *Prog. Opt.* **2010**, *55*, 285.
9. Korotkova, O.; Gbur, G. Applications of optical coherence theory. *Prog. Opt.* **2020**, *65*, 43.
10. Miao, W.; Zhang, Y.; Gbur, G. Deterministic vortices evolving from partially coherent fields. *Optica* **2023**, *10*, 1173. [CrossRef]
11. Zhan, Q. Cylindrical vector beams: From mathematical concepts to applications. *Adv. Opt. Photonics* **2009**, *1*, 1. [CrossRef]
12. Rosales-Guzmán, C.; Ndagano, B.; Forbes, A. A review of complex vector light fields and their applications. *J. Opt.* **2018**, *20*, 123001. [CrossRef]
13. Yang, Y.; Ren, Y.X.; Chen, M.; Arita, Y.; Rosales-Guzmán, C. Optical trapping with structured light: A review. *Adv. Photonics* **2021**, *3*, 034001. [CrossRef]
14. Wolf, E. Unified theory of coherence and polarization of random electromagnetic beams. *Phys. Lett. A* **2003**, *312*, 263. [CrossRef]
15. Wolf, E. *Introduction to the Theory of Coherence and Polarization of Light*; Cambridge University Press: Cambridge, UK, 2007.

photonics MDPI

Article

Powerful Elliptically Polarized Terahertz Radiation from Oscillating-Laser-Driven Discharge Surface Currents

Egor Dmitriev [1,2], Nikolai Bukharskii [1,2] and Philipp Korneev [1,2,*]

1 Department of Laser Plasma, Division of Quantum Radiophysics, P.N. Lebedev Physical Institute of the Russian Academy of Sciences, 53, Leninskiy Prospekt, 119991 Moscow, Russia
2 Institute for Laser and Plasma Technologies, National Research Nuclear University MEPHI, 31, Kashirskoe Shosse, 115409 Moscow, Russia
* Correspondence: ph.korneev@gmail.com

Abstract: This work presents a general concept of an intense laser-driven source of strong electromagnetic waves, which can be used for obtaining powerful terahertz radiation with controlled polarization. It is shown that the irradiation of a solid target surface by short relativistic laser pulses at small angles provides the excitation of strong compact relativistic discharge current pulses, propagating in a certain direction. For elliptical targets, this current emits elliptically polarized electromagnetic radiation at a given frequency with the ellipticity and the spectra defined by the target geometry. The proposed setup allows reaching extreme THz intensities and provides easy control of the radiation parameters, making it attractive for various scientific and technological applications.

Keywords: intense laser–matter interaction; discharge current; polarized THz radiation; intense THz radiation

1. Introduction

Radiation of the THz range possesses unique properties for fundamental material studies and high-end applications in medicine, chemistry, and communication [1–4]. The non-ionizing nature of the interaction makes possible non-destructive ultra-sensitive contamination [5] and chemical compound detection by terahertz spectroscopy [6,7]. In biology, with this kind of radiation, the spectroscopy of biomolecules [8], chemical imaging [9], and even the monitoring of bio-samples' content [10] become feasible. THz time domain spectroscopy [11,12] is used for protein and RNA dynamic polarization-dependence response [13] and DNA studies [14]. Terahertz technology is developing in intraoperative neurodiagnostics [15] and may be used for tissue detection [16]. The high potential of medical THz applications is defined by its safety at high intensities, which helps to obtain high image contrasts [17]. This radiation range is extremely interesting in the frames of fast material properties' modification [18] and terahertz wireless communication systems [19].

Among the different possible applications of THz radiation, many require certain controlled polarization properties, such as, e.g., molecular orientation and alignment [20–22], the detection of inclusions in matter [23], reflective time domain polarization spectroscopy [24], etc. Furthermore, some studies require a certain THz field strength, such as, e.g., the THz pump-probe technique for characterizing materials and studies aiming at understanding traditional nonlinear effects [25] at the high and ultra-high interaction level. There is a need for a good polarization control for strong-field terahertz time domain polarimetry and ellipsometry and nonlinear terahertz spectroscopy beyond the MV/cm electric field strength [3]. Along with the development of intense THz sources, intense THz magnetic fields have also been explored. The intense THz magnetic field can strongly couple to a spin subsystem and be used to coherently control the precession of macroscopic magnetization [25]. The intense THz radiation of different polarizations may be used for ionic motion control, lattice vibrations, coherent control of molecular orientation and rotation,

check for updates

Citation: Dmitriev, E.; Bukharskii, N.; Korneev, P. Powerful Elliptically Polarized Terahertz Radiation from Oscillating-Laser-Driven Discharge Surface Currents. *Photonics* **2023**, *10*, 803. https://doi.org/10.3390/photonics10070803

Received: 29 April 2023
Revised: 22 June 2023
Accepted: 5 July 2023
Published: 11 July 2023

spin manipulation in different materials, ionization at low frequencies, and induced phase transitions [26]. Many studies are devoted to particle acceleration with strong polarized THz radiation [27–30].

The polarization control of strong THz radiation may be obtained with further development of general schemes for THz generation, such as optical rectification in nonlinear crystals [31] and photoconductive emitters [32]. Higher intensities of polarized THz radiation are achievable using liquids [33] or gases under strong optical pumping. Some schemes involve additionally static external fields, such as strong magnetic fields of a few hundred Tesla [34] or electric fields of a special geometry [35]. Some other propositions deal with different combinations of pumping laser beams with different intensities, wavelengths, or polarizations [36,37]. However, these are usually schemes driven with TW-class fs lasers and with $\sim 10^{-3} \ldots 10^{-4}$ efficiency, which limits the expected THz power to the GW scale. The reachable power of polarized THz sources remains one of the most-important questions in the field. Its increase may allow stronger induced effects and lead to the development of new methods and applications.

Probably, plasma sources driven by high-intensity laser pulses are the best candidates for obtaining unprecedentedly strong THz radiation on the terawatt scale. Among the variety of plasma-based schemes, some of the most-interesting ones are emitters fed by strong laser-induced localized electric currents. They offer a promising potential as a source of powerful THz radiation and have been thoroughly investigated both numerically and experimentally for straight laser-driven metallic wires [38–42]. Note that, there, mainly the collective motion of fast laser-accelerated electrons was considered to be responsible for the THz emission, but the induced surface currents may provide a more-compact and dense radiation source [43]. One of the effective mechanisms for creating such currents involves driving an intense discharge pulse [44–46]. For smaller, e.g., sub-picosecond or even femtosecond laser pulses, the degree of localization of this discharge pulse may become sufficiently small on the target scale, and it may emit electromagnetic waves when propagating along a coil-shaped [45] or an undulating [43] wire. Here, we show that a simple solid target with an elliptical outer surface allows obtaining perfectly controlled polarized THz radiation along the normal to the target with the frequency defined by the target perimeter and the degree of ellipticity defined by the ratio of the minor and major axes of the target ellipse. Such a target may present either a disk with a mount connected to its center or a truncated cone with elliptical bases and the mount connected to the smaller base; see Figure 1a. In both cases, a short and fast current driven by an intense femtosecond laser pulse would propagate on the target's outermost surface because of the skin effect, i.e., along the disk outer surface or the cone larger base. Thus, for the considered effect, only the shape and size of the outer border should play a role, while other parts of the target may have arbitrary profiles as long as they do not carry any significant electric current. As we show below, for an interaction at small angles, the current propagation direction is defined by the irradiation direction, so that it is counterclockwise for the irradiation geometry in a possible experimental setup presented in Figure 1a and clockwise for the 3D simulation geometry in Figure 1b. As the target's outer surface presents a closed loop, no specific conditions are required to close the circuit, which enables the current pulse to make several round turns, as in the case of the coil with a plasma-connected slit [45]. The setup enables the production of multi-period strong waves with perfectly controlled polarization and spectra, which may provide unique possibilities in studies requiring ultra-strong polarized THz radiation. In the workflow of the previous experimental [39–42] and theoretical [35,37,38,43,45] studies aiming at the generation of strong controlled radiation in the THz domain, the work presents a design-theoretical description of the the powerful THz emitter with the possibility of enhanced polarization control, promoting further experimental research of THz sources driven by compact optically induced discharges.

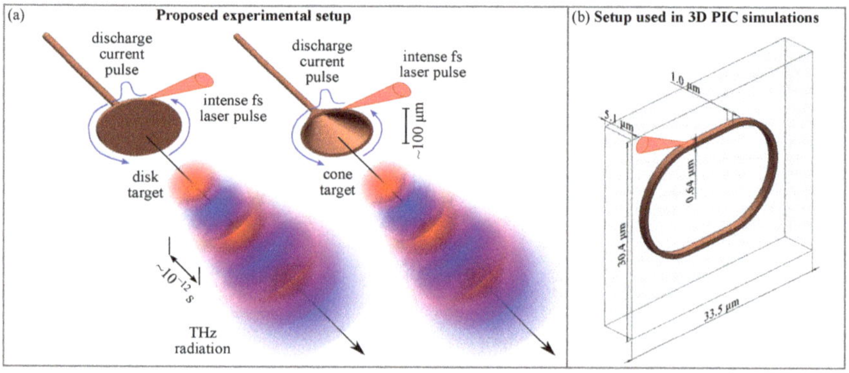

Figure 1. (**a**) Sketch of the proposed setup for the generation of intense THz radiation with elliptical polarization based on the interaction of an intense femtosecond laser pulse with elliptical disk (left) and cone (right) targets. The laser pulse hits the target's outermost surface at a grazing incidence; the created current pulse propagates along this surface counterclockwise, performing multiple oscillations and emitting THz waves. (**b**) Sketch of the setup used in the performed numerical simulations, showing the target profile, the simulation box, and the irradiation geometry.

2. Methods

2.1. Formation of the Strong Discharge Current Pulse

The process of the formation of a strong discharge current pulse oscillating on the target surface was demonstrated numerically with an exemplary small target, shown in Figure 1b, which presents an edge of either a disc or cone target, shown in Figure 1a. According to the general properties of fast surface waves [47] and the results of the related simulations, including those presented below, the discharge propagates mostly along the outer target part, so this simplification is not expected to qualitatively change the overall physical picture of the considered phenomenon. The process was modeled with the open-source 3D PIC code Smilei [48]. As other PIC algorithms, this code relies on the Vlasov–Maxwell system of equations. The distribution of a given species s of particles with charge q_s and mass m_s is described by the distribution function $f_s(t, \mathbf{x}, \mathbf{p})$ with $\mathbf{x}$ and $\mathbf{p}$ denoting the position and momentum of a phase-space element for a given time moment t. The function f_s is defined by

$$\left(\partial_t + \frac{\mathbf{p}}{m_s\gamma} \cdot \nabla + \mathbf{F}_L \cdot \nabla_{\mathbf{p}}\right) f_s = 0, \tag{1}$$

where $\gamma = \sqrt{1 + \mathbf{p}^2/m_s^2}$ is the Lorentz factor and $\mathbf{F}_L = q_s(\mathbf{E} + \mathbf{v} \times \mathbf{B})$ is the Lorentz force, while $\mathbf{E}(t, \mathbf{x})$ and $\mathbf{B}(t, \mathbf{x})$ are the collective electric and magnetic fields in a plasma satisfying the Maxwell equations with charge and current densities calculated from the distribution function f_s. The distribution function f_s is discretized as a sum of a finite number of quasi-particles. This allows describing the plasma dynamics by solving the equations of motion for individual quasi-particles, while the Maxwell equations are solved on a discrete spatial grid using the Finite Difference Time Domain (FDTD) approach. The described method allows obtaining both the phase-space distribution of particles in the plasma and the collective electromagnetic fields resulting from the presence and motion of these particles. The initial distribution of the plasma, i.e., its composition in terms of species, their positions and momenta, as well as the initial distribution of the electromagnetic fields in the simulation box served as the initial conditions. For more details on the specifics of the employed approach, see [48,49] and the references therein. Here, the Vlasov–Maxwell model was used to describe the three-dimensional dynamics of the initially cold plasma under the action of an intense laser pulse, which was introduced by defining a specific oscillating boundary condition for the fields on one of the surfaces of the rectangular 3D simulation box. This box had the sizes of $\approx$33.5 $\times$ 30.4 $\times$ 5.1 μm^3, or 2112 $\times$ 1920 $\times$ 320 cells. The spatial resolution was $\approx$16 nm, and the temporal resolution was 1.8×10^{-2} fs. The boundary conditions were generally absorbing for the electromagnetic fields, and the particles that reached the edges of the simulation box were removed from the simulation.

To obtain certain polarization properties of the emitted radiation, the target shape may differ, but in the interaction region, it should provide a good laser–matter coupling. For larger, experimentally feasible targets, the curvature of the ellipse is small enough to provide an effective excitation of the surface propagating discharge, but for the reduced target in the considered numerical setup, an additional flattening was introduced. This modified target profile is reminiscent of the "racetrack" or "stadium" geometry, which is two semicircles joined by two straight lines; see Figure 1b, where the 3D target profile used for the main simulation is presented. The radius of the semicircles was 8 μm, and the length of the straight segments was also set to 8 μm; the wire had a rectangular cross-section with the widths in the XY plane and along the Z axis of 0.64 μm and 1 μm, respectively. The target consisted of an electron–ion plasma with the degree of ionization $Z = 10$. The electron density was set to 10-times the critical density for the wavelength 0.8 μm, and each cell contained 10 quasi-particles. To save the calculation time, the ions were considered immobile, as for sufficiently heavy dense targets, i.e., copper or gold, ions exhibit very slow dynamics, which have an insignificant impact on the discharge pulse formation and propagation on the considered time scale of $\sim$1 ps. The laser pulse with linear p-polarization and the wavelength of 0.8 μm was introduced into the simulation box from the left boundary and was focused on the upper straight segment of the target at an angle of 20° relative to its surface, improving the absorption of the laser energy and creating a discharge current pulse with the desired propagation direction, which was in this case clockwise along the target surface. The duration of the laser pulse was set to 4 fs at Full Width at Half Maximum (FWHM), to accommodate a reduced target size in comparison to a more-realistic case. The size of the focal spot on the target surface was 2 μm ($1/e^2$ diameter), and the peak intensity at the focal spot was 10^{21} W/cm^2.

2.2. *Analysis of the Emitted Radiation*

The interesting region to consider is the wave zone, far from the target, which is beyond the numerical possibilities in the used 3D setup. The radiation properties were characterized with a standard analytical calculation assuming a short discharge pulse propagating with a given velocity v_0. Consider an ellipse circuit with semi-axes a and b. In elliptic coordinates, the ellipse may be described as

$$\begin{aligned} x &= \alpha \cosh u \cos \psi \\ y &= \alpha \sinh u \sin \psi \\ z &= 0 \end{aligned} \tag{2}$$

with $u = u_0 = \text{const}$ and $a = \alpha \cosh u_0$, $b = \alpha \sinh u_0$, and ψ changing in the range from 0 to 2π. A special case of a ring with radius r_0 can be obtained with substitution $\alpha = 2r_0e^{-u_0}$ and $u_0 \to \infty$. A linear target is achieved with $u_0 \to 0$.

Let a charged particle with charge e move at a constant velocity v_0 along the circuit. The current density of the particle:

$$\mathbf{j}(\mathbf{r},t) = e\mathbf{v}\delta(\mathbf{r} - \mathbf{r}_0(t)) = e\mathbf{v}\frac{\delta(u - u_0)\delta(\psi - \psi_0(t))\delta(z)}{\alpha^2 s^2}, \tag{3}$$

with $s(u,\psi) = \sqrt{\sinh^2 u + \sin^2 \psi}$. The function $\psi_0(t)$ represents the dependence of the angular position of the particle, moving at a constant velocity along the elliptical trajectory. The velocity of the particle in the elliptic coordinates reads

$$\mathbf{v} = \alpha s(u_0, \psi)\dot{\psi}_0 \mathbf{e}_\psi(u_0, \psi), \tag{4}$$

where

$$\mathbf{e}_\psi(u, \psi) = \frac{1}{s}(-\cosh u \sin \psi \mathbf{e}_x + \sinh u \cos \psi \mathbf{e}_y) \tag{5}$$

is the unitary tangent vector to the ellipse circuit. The particle velocity remains constant $v_0 = \text{Const}$; hence,

$$v_0 = \alpha s(u_0, \psi_0)\dot{\psi}_0, \tag{6}$$

which leads to the following implicit expression for $\psi_0(t)$:

$$\frac{v_0}{\alpha} t = \cosh u_0 \, E\left(\arcsin\left(\frac{\cosh u_0 \sin \psi_0}{s(u_0, \psi_0)}\right), \frac{1}{\cosh u_0}\right) - \frac{\sin 2\psi_0}{2s(u_0, \psi_0)} + \text{Const}, \tag{7}$$

with $E(\varphi, k)$ being the incomplete elliptic integral of the second kind. This expression may be solved, e.g., numerically to provide the full description of the particle motion. The current density reads

$$\mathbf{j}(\mathbf{r}, t) = e v_0 \frac{\delta(u - u_0)\delta(\psi - \psi_0(t))\delta(z)}{\alpha^2 s^2} \mathbf{e}_\psi, \tag{8}$$

and the vector potential in the far-field zone is

$$\mathbf{A} = \frac{1}{cr} \int \mathbf{j}\left(\mathbf{r}', t - \frac{r}{c} + \frac{\mathbf{n}\mathbf{r}'}{c}\right) d^3 r' = \frac{e v_0}{cr} \mathbf{e}_\psi(u_0, \psi'), \tag{9}$$

where $\mathbf{n} = \frac{\mathbf{r}}{r}$ and ψ' is the coordinate of the charge at a retarded time, i.e., the solution of the equation:

$$\psi' = \psi_0\left(t - \frac{r}{c} + \frac{\mathbf{n}\mathbf{r}'}{c}\right), \tag{10}$$

where $\mathbf{r}'$ is the position of the particle at the retarded time moment.

3. Results

The results of the performed 3D PIC modeling for the target profile shown in Figure 1b indicates that an ultra-short intense laser pulse hitting the upper straight segment of the target at a small angle, as shown in Figure 2a, creates a strong and short discharge current pulse; see Figure 2b. It propagates around the target clockwise (see Figure 2d–g) due to the chosen irradiation geometry, which defines the direction of the laser-driven discharge current pulse. During the simulated time, the discharge current pulse went a full round trip around the target with almost no observable dispersion and started another round; see Figure 2h,i. As no stretching of the pulse over the target perimeter occurred and it stayed well-localized on the coil scale, it preserved its capability of emitting electromagnetic waves. Although, it should be noted that its amplitude may decrease due to ionization [43] and collisionless damping [46], leading to slow variations of the envelope of the emitted radiation. Its main frequency is defined by the inverse time of a single round trip of the discharge current pulse, which depends on the target size and geometry, as well as on the propagation velocity of the discharge current pulse. The latter, according to the performed modeling, amounted to about 0.88 c, as the current pulse traveled 66 μm along the target perimeter in about 250 fs. The obtained propagation velocity was somewhat lower than the value reported in [43] for a straight wire target, but it should be noted that, here, due to the loop-like target geometry, the discharge current pulse propagated in the plasma formed by the weaker counter-clockwise discharge pulse, propagating in the opposite direction from the interaction region, in contrast to the case of a cold extended wire, where the plasma in front of the discharge current pulse was not excited. As was shown in [46], this near-surface plasma may modify the process of discharge current pulse propagation and lead to a decrease of the propagation velocity. The simulations over the times $\gtrsim 0.5$ ps in the considered 3D geometry for the dense plasma and small cell sizes were rather expensive and were not performed, as the results shown in Figure 2h,i and the results of other studies with similar setups [45] showed that the discharge pulse was able to make at least a few oscillations around the target before its amplitude reduced to some background level or before it stretched over the whole target perimeter.

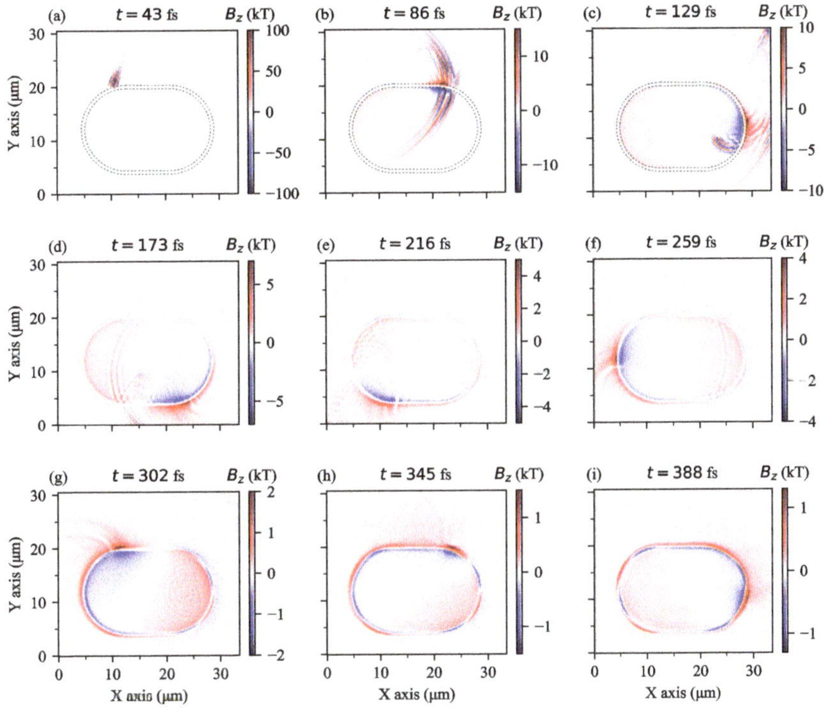

Figure 2. Results of 3D PIC modeling: B_z component of the electromagnetic field at time moments (**a**) $t = 43$ fs, (**b**) $t = 86$ fs, (**c**) $t = 129$ fs, (**d**) $t = 173$ fs, (**e**) $t = 216$ fs, (**f**) $t = 259$ fs, (**g**) $t = 302$ fs, (**h**) $t = 345$ fs, and (**i**) $t = 388$ fs, indicating annular propagation of the discharge current pulse along the target perimeter. The results correspond to the central cross-section of the target in the Z dimension and were averaged over a 1 μm-thick volume. The thin dashed lines in Panels (**a**–**c**) show a visual reference to the initial profile of the inner and outer surfaces of the "stadium" target.

For the two possible experimental realizations of the setup, shown in Figure 1a, it should be noted that, although the current propagated in the thin outermost layer of the target due to the skin effect, the presence of the inner conducting medium may affect qualitatively the process of the surface discharge current pulse formation and propagation. In order to demonstrate these changes, an additional 3D PIC simulation was performed with the same conditions as before, but for a target with the inner volume that was filled with plasma with the same density as for the external current-carrying layer. Due to the constraints imposed by the increased computational load required to model such a target, the simulation time was limited to half of the first oscillation period. The comparison of the resulting discharge current pulse profiles in the form of the B_z component at time moment $\approx$0.19 ps is shown in Figure 3a,b for the "hollow" and "filled" "stadium" targets, respectively; the corresponding electron density profiles are presented in Panels (c) and (d). From the presented plots (see Panels (a) and (b)), it can be seen that there were some differences in the process of discharge current pulse propagation along the coil, namely the profiles were slightly different, the amplitude appeared to be ≈2.5-times less than for a hollow target, and the propagation velocity was a few percent higher. However, qualitatively, the process did not change, supporting the main conclusions of the work, that an intense compact discharge current pulse was formed and propagated along the target's outer surface with no observable dispersion. This occurred regardless of the target's inner medium, and thus, both the disk and the cone target shown in Figure 1a enabled the charge oscillation along the given trajectory in a similar way.

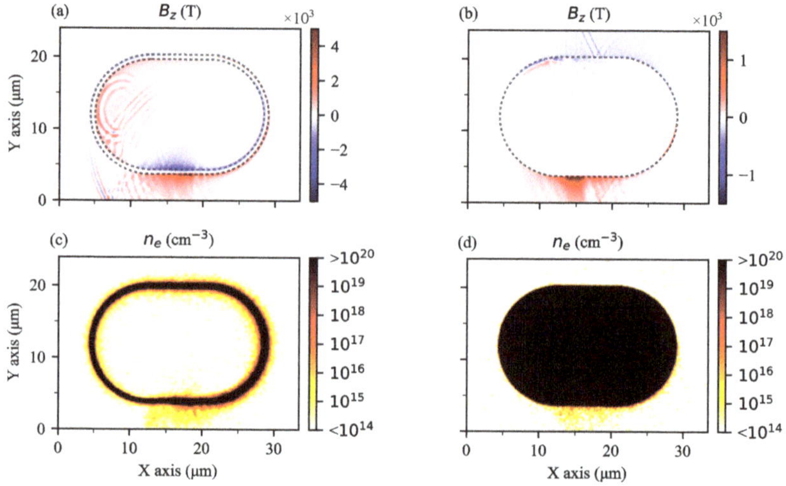

Figure 3. Comparison of the results of 3D PIC modeling for the hollow and filled "stadium" targets: Panels (**a**,**b**) show the B_z component at time moment ≈0.19 ps; Panels (**c**,**d**) show the corresponding electron density profiles. The thin dashed lines in Panels (**a**,**b**) show a visual reference to the initial profile of the inner and outer surfaces of the "stadium" target.

Further analysis was based on the results of the main simulation, i.e., that with the profile presented in Figure 1b. The examination of the spatial distributions of other physical quantities besides the axial component of the magnetic field may provide more insight into the interaction processes. The electric fields E_x and E_y, electron density n_e, and electric current densities J_x and J_y are presented in Figure 4, for two different time moments $t = 129$ fs (Panel (a)), when the discharge current pulse propagated along the right semicircle, and $t = 194$ fs (Panel (b)), when it propagated along the lower straight segment. From the presented distributions, it follows that the electric field also had a monopolar distribution, similar to that of the magnetic field; see Figure 4b1,c2. This was directed normally from the wire, confined fast electrons that escaped the target, and guided them along the surface of the wire; see Figure 4d1,d2. Because of that, another current component besides the return discharge current appeared. This can be seen in Figure 4f1,e2, where the two current pulses of opposite polarities can be identified. The one J_1 corresponded to the return flow of the bulk electrons and is shown with the blue arrow in Panel (f1). This flow was directed counter-clockwise, while the discharge current itself was directed clockwise. The other current, J_2, was created by the flow of electrons accelerated along the surface of the wire. It had the opposite direction to the return flow of the cold bulk electrons and is marked by the red arrow in Panel (f1). Here, the return flow dominated, which is the usual situation [50], and the polarity of the magnetic field created around the target was defined by the direction of this discharge current.

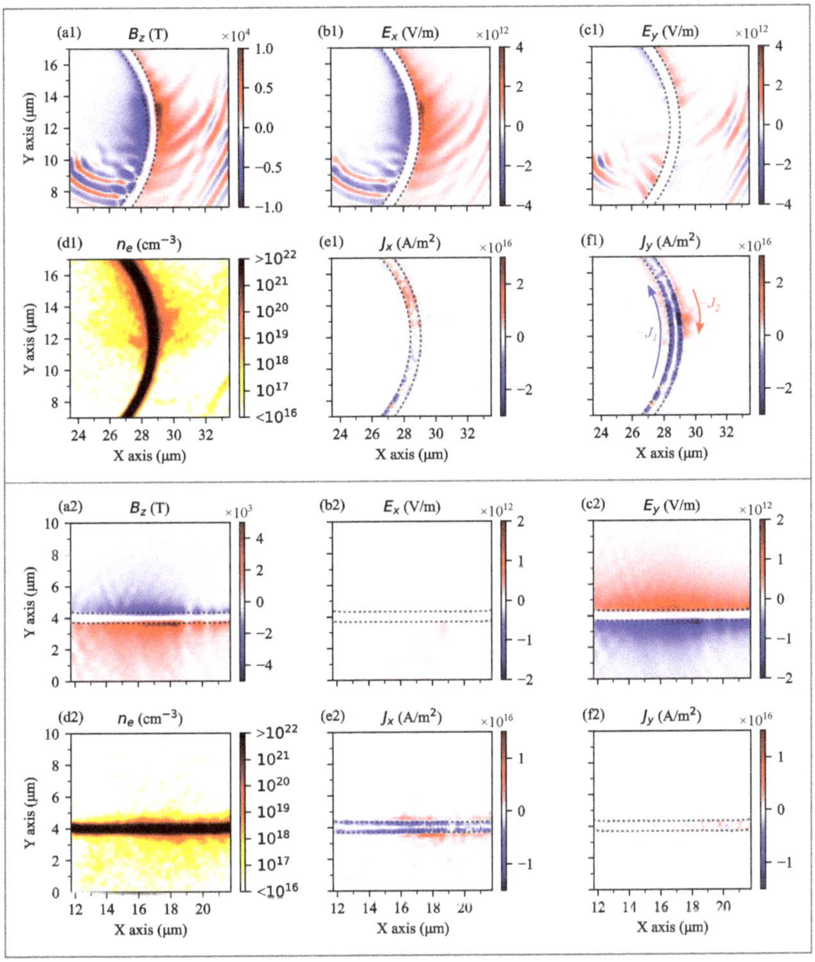

Figure 4. Results of 3D PIC modeling: spatial distribution of the electromagnetic field components (**a1**,**a2**) B_z, (**b1**,**b2**) E_x, and (**c1**,**c2**) E_y and (**d1**,**d2**) electron density n_e and electric current components (**e1**,**e2**) J_x and (**f1**,**f2**) J_y close to the wire at time moments $t = 129$ fs and $t = 194$ fs, respectively. The results correspond to the central cross-section of the target in the Z dimension and are averaged over a 1 μm-thick volume. The thin dashed lines show a visual reference to the initial profile of the inner and outer surfaces of the "stadium" target.

With the method, described above, the vector potential (9) may be represented using exponential forms of trigonometric functions

$$\mathbf{A} = \frac{ev_0}{2\alpha crs'}\mathrm{Re}\left\{e^{i(\psi'+\frac{\pi}{2})}\boldsymbol{\varkappa}_L\right\}, \tag{11}$$

where $s' = \sqrt{\sinh^2 u_0 + \sin^2\psi'}$ and $\boldsymbol{\varkappa}_L = a\mathbf{e}_x - ib\mathbf{e}_y$ is a polarization vector. Therefore, as probably could be expected, the electromagnetic field harmonics, radiated by a left-oriented current pulse propagating along an elliptic circuit in the direction normal to the ellipse, are left-handed elliptic polarization with the components ratio coinciding with the ratio of the ellipse semi-axes. As limiting cases, expressions for circular target $a = b$ or linear target $a = 0$ or $b = 0$ may be obtained from the general expression (11).

The electric and magnetic fields in a general elliptic case along an arbitrary direction **n** read

$$\begin{aligned} \mathbf{E} &= \frac{ev_0}{2\alpha c r s'}\mathrm{Re}\left\{\left(\frac{\sin 2\psi'}{2s'^2} - i\right)e^{i(\psi'+\frac{\pi}{2})}\frac{\partial\psi'}{\partial t}[[\mathbf{n}\boldsymbol{\varkappa}_L]\mathbf{n}]\right\}, \\ \mathbf{H} &= \frac{ev_0}{2\alpha c r s'}\mathrm{Re}\left\{\left(\frac{\sin 2\psi'}{2s'^2} - i\right)e^{i(\psi'+\frac{\pi}{2})}\frac{\partial\psi'}{\partial t}[\mathbf{n}\boldsymbol{\varkappa}_L]\right\}. \end{aligned} \tag{12}$$

Along the axis normal to the target plane ($\mathbf{n} = \mathbf{e}_z$, $r = z$), these expressions may be simplified as

$$\begin{aligned} \mathbf{E} &= \frac{ev_0}{2\alpha c r s'}\mathrm{Re}\left\{\left(\frac{\sin 2\psi'}{2s'^2} - i\right)e^{i(\psi'+\frac{\pi}{2})}\frac{\partial\psi'}{\partial t}\boldsymbol{\varkappa}_L\right\}, \\ \mathbf{H} &= \frac{ev_0}{2\alpha c r s'}\mathrm{Re}\left\{\left(\frac{\sin 2\psi'}{2s'^2} - i\right)e^{i(\psi'+\frac{\pi}{2})}\frac{\partial\psi'}{\partial t}[\mathbf{e}_z\boldsymbol{\varkappa}_L]\right\}, \end{aligned} \tag{13}$$

with $\psi' = \psi_0(t - \frac{r}{c})$, and in the target plane ($z = 0$), the far fields transform to

$$\begin{aligned} \mathbf{E} &= \frac{ev_0}{2\alpha c r s'}\mathrm{Re}\left\{\left(\frac{\sin 2\psi'}{2s'^2} - i\right)e^{i(\psi'+\frac{\pi}{2})}\frac{\partial\psi'}{\partial t}\left(-\frac{ay+ibx}{r}\mathbf{e}_\varphi\right)\right\}, \\ \mathbf{H} &= \frac{ev_0}{2\alpha c r s'}\mathrm{Re}\left\{\left(\frac{\sin 2\psi'}{2s'^2} - i\right)e^{i(\psi'+\frac{\pi}{2})}\frac{\partial\psi'}{\partial t}\left(-\frac{ay+ibx}{r}\mathbf{e}_z\right)\right\}, \end{aligned} \tag{14}$$

where $\mathbf{e}_\varphi$ it the azimuthal unit vector. In the later case, the electromagnetic wave is linearly polarized, the electric component is in the plane, and the magnetic one is normal to it.

A Fourier component of the electric field reads

$$\mathbf{E}_n = \frac{iev_0}{2\alpha c r T}[[\mathbf{n}\boldsymbol{\varkappa}_L]\mathbf{n}]e^{i\omega_0 n r}\int\limits_{-\frac{\pi}{2}}^{\frac{\pi}{2}}\frac{\frac{\sin 2\psi'}{2s'^2} - i}{s'}e^{i(\psi'+n\omega_0 f(\psi'))}\left(e^{-i\mathbf{n}\mathbf{r}'} - (-1)^n e^{i\mathbf{n}\mathbf{r}'}\right)d\psi', \tag{15}$$

where $T = \frac{4\alpha}{v_0}\cosh u_0 E\left(\frac{1}{\cosh u_0}\right)$ is the oscillation period, $\omega_0 = \frac{2\pi}{T}$ is the main radiation frequency, $E(k)$ is the full elliptic integral of the second kind, and

$$f(\psi') = \frac{\alpha}{v_0}\left(\cosh u_0\, E\left(\arcsin\left(\frac{\cosh u_0 \sin\psi'}{s(u_0,\psi')}\right), \frac{1}{\cosh u_0}\right) - \frac{\sin 2\psi'}{2s(u_0,\psi')}\right).$$

It follows from (15) that, on the target axis, where $\mathbf{n}\mathbf{r}' = 0$, only odd elliptically polarized harmonics are present. The absence of even harmonics is a general feature of the radiation related to the target symmetry [51,52].

Using the general expression, the effect of the finite pulse shape may be analyzed with the following generalization to the current (8):

$$\mathbf{j} = \int d\xi f(\xi) e\mathbf{v}\frac{\delta(u-u_0)\delta(\xi - (\psi - \psi_0(t)))\delta(z)}{\alpha^2 s^2}, \tag{16}$$

which introduces a current pulse profile $f(\xi)$. As we see, the polarization properties do not change for a finite pulse width. To model a damping of the discharge pulse during its propagation along the wire, a characteristic function $g(\psi)$ may be introduced, so that the current becomes:

$$\mathbf{j} = g(\psi)e\mathbf{v}\frac{\delta(u-u_0)\delta(\psi - \psi_0(t))\delta(z)}{\alpha^2 s^2} = g(\psi_0(t))e\mathbf{v}\frac{\delta(u-u_0)\delta(\psi - \psi_0(t))\delta(z)}{\alpha^2 s^2}. \tag{17}$$

Evidently, the polarization properties would not also change in this case, as the damping results just in multiplication by a time-dependent function.

4. Discussion

Spectral and polarization control of the radiation were provided by the target geometry as described above. In addition to the general analytical consideration, a detailed model analysis of these properties may be performed numerically, using the vector potential (9) for a given discharge current profile and the target geometry. Then, electromagnetic fields can be calculated as a function of time and their spectral and polarization characteristics can be studied. As an example, here, we presented the obtained results for three elliptical targets with the same perimeter, corresponding to an oscillation period of 1 ps, and different ratios of the major and minor semi-axes: $a/b = 1.1$, $a/b = 1.5$, and $a/b = 2$. This relatively large target was considered here for a more-realistic situation, which means also a longer laser beam, so the parameters of the interaction and those of the discharge current pulse were taken from [43]. The fields were calculated on the axis of the target in the wave zone, at a distance 10 cm from the center of the target, which was much larger than its characteristic size $\sim$100 μm. The obtained results are summarized in Figure 5. For an almost circular target, with $a/b = 1.1$, the power spectral density (PSD) distribution included a single pronounced peak, almost as it was for a rotating charge [51]; see Figure 5a1. The central frequency of this peak was $f = 1$ THz, which corresponded to the inverse time of one oscillation period and was consistent with previous results for the spectrum of the THz radiation produced by the short discharge current oscillating in annular targets [45]. The width of this peak is mainly defined by the finite duration of the THz emission resulting from the dissipation of the discharge pulse energy. To demonstrate the dependence of the spectral properties on the current pulse energy dissipation rate, consider a strong dissipation, so that the number of oscillation cycles the discharge current pulse makes before its amplitude reduced by e times was just $N_c = 0.5$. In this case, the spectral peaks widened significantly in comparison to the case of $N_c = 3$, and an additional enhancement attributed to the slow change of the emitted field intensity appeared at $f \lesssim 0.1$ THz. For larger a/b ratios, when the target substantially differed from a circle, the PSD distributions changed, so that additional frequency components appeared at the odd harmonics of the main frequency; see Figure 5a2,a3. This effect was more pronounced for $a/b = 2$, where the power emitted at the third harmonic amounted to about 25% of the power emitted at the main frequency; in addition, in this case, higher odd harmonics, i.e., the fifth one, became quite noticeable. If the electric fields at each of these harmonics were selected and filtered from the spectrum, it can be seen that the electric field vector drew an ellipse in time.

In addition to the ellipticity, the change of a/b also led to an increase of the higher harmonics. Each harmonics of the emitted radiation had a certain elliptical polarization defined by the target, though the total electric field drew a complicated line; see Figure 5c1–c3. There, dashed lines correspond to the case when there is no discharge dissipation, and the solid ones show the polarization curve in time for $N_c = 3$. For $a/b = 1.1$, the observed deviation from an ellipse was not that evident, but with the increase of a/b, when higher harmonics start to provide a significant contribution to the total emitted field, this effect became noticeable. The inclusion of the dissipation into the calculations led to the gradual decrease of the field amplitude with no changes in the overall polarization, which remained the same during the whole process of the THz wave emission. As was already noted above, the polarization of each harmonic was elliptical, but it was also slowly affected by dissipation; see Figure 5b1–b3. There, the dashed line for the main frequency presented the energy conservation limit, while the real polarizations of the radiation at odd frequencies, emitted by a dissipating current pulse, were elliptic spirals, shown in Figure 5b1–b3 for a relatively strong dissipation rate with $N_c = 3$. The phase difference between the harmonics resulted in a complex non-elliptical polarization, shown in Figure 5c1–c3. Filtering frequencies and retaining only the necessary one may be applied to obtain almost elliptically polarized THz radiation in the wave zone on the target axis. Note that, for different angles to the target axis, the polarization of each harmonic changed its ellipticity from linear (in the plane of the target) to a more-general one with the parameters of ellipticity defined by the target and the angle. This spatial polarization sweeping may in principle be converted to the temporal one for use with a single sample. This is just one example of how this setup may provide the unique possibility to reveal the properties of matter and study the material modification and solid-state excitation under strong controlled radiation in a given temporal and spatial domain.

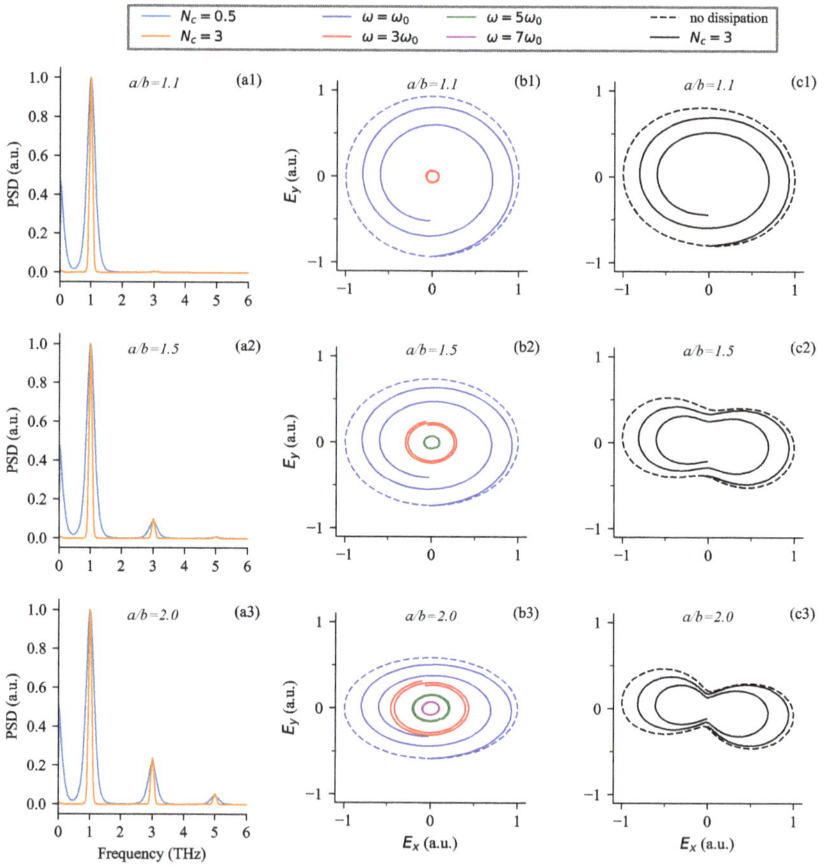

Figure 5. (**a1–a3**) Power spectral density (PSD) for elliptical targets with the same perimeter, corresponding to the oscillation period of 1 ps, and different ratios of the major and minor semi-axes $a/b = 1.1$, $a/b = 1.5$, and $a/b = 2.0$, respectively; for each target, PSD plots are given for two different exponential decay lengths of the discharge current, corresponding to different numbers of oscillation cycles N_c made by the discharge current pulse before its amplitude reduces by e-times: $N_c = 0.5$ and $N_c = 3$. (**b1–b3**) Curves drawn by different odd frequency components of the electric field vector from the first to the seventh one, with the solid curves showing the results for $N_c = 3$ and the dashed one showing the results for the main frequency with no dissipation taken into account. (**c1–c3**) Curves drawn by the total electric field vector, with the dashed curve corresponding to the case without dissipation and the solid curves corresponding to the case when $N_c = 3$. The fields were calculated on the axis of the target in the wave zone, at a distance of 10 cm from the target.

Another merit of the considered scheme is that the spectral and polarization control is coupled with the capability of obtaining extremely intense terahertz radiation when using modern laser systems. Estimate the expected THz radiation power in this case. Consider as a simple example the elliptical target with the relation of the major and minor axes close to unity. The emitted power density per unit solid angle $d\Omega$ can be calculated using the expression for an annular coil target on the axis [45]:

$$\frac{dP_\omega}{d\Omega d\omega} = \frac{a^2k^2I_\omega^2}{4\pi c} \tag{18}$$

where a is the radius of the coil and k is the wave vector. Assuming the main frequency is defined by the perimeter of the coil ($ak \simeq 1$), a monochromatic spectrum and the effective emission solid angle of $\Delta\Omega \approx 1$ sr, the total emitted power:

$$P = \frac{I_0^2}{4\pi c} \tag{19}$$

and depends only on the amplitude of the electric current in the coil. In the reduced numerical simulations performed here with a total laser energy of $\approx$130 mJ, the total electric current amplitude was about 10^4 A. Converting to the Gaussian units and using (19), one obtains $\sim$0.3 GW of emitted power in the THz range. For a realistic laser and target parameters, the energy invested into the discharge current pulse excitation may be several orders of magnitude greater, which increases the estimate, respectively. The analysis performed in [43] showed that, for intensities in the range $(10^{21}–10^{22})$ W/cm^2, achievable with modern petawatt and multi-petawatt laser facilities and realistic focusing parameters, electric currents up to $(10^5–10^6)$ A may be obtained. This corresponds to the peak power of the 10 GW-1 TW level, suitable for a wide range of studies of intense and ultra-intense terahertz radiation interaction with matter.

A more careful estimate of the emitted power should take into account its angular distribution, as well as a more-realistic target geometry and the profile of the discharge current pulse. This can be achieved by the numerical calculation of the retarded vector potential given by (9) and the corresponding electromagnetic fields **E** and **B**. These fields may be used to find the Poynting vector $\mathbf{S} = \mathbf{E} \times \mathbf{B}$ and its integration over time to find the energy fluence. Such calculations were performed for elliptical targets with different ratios of the major and minor semi-axes a and b and the same perimeter corresponding to the oscillation period $T = 1$ ps. The used discharge current velocity $v_0 = 0.88c$ was obtained in the PIC simulations; the discharge current profile was taken from [43]. In addition, two separate calculations of the energy fluence were performed for the "stadium" target considered in the PIC simulations and the elliptical target with the same a/b and the same perimeter as for the "stadium" target. The results are presented in Figure 6.

For a circular target with $a/b = 1$, most of the power was radiated in the plane of the target, i.e., the XY plane (see Figure 6a), resulting in bright and relatively wide maxima in the $x = 1.28$ mm and $y = -1.28$ mm planes in the presented fluence distribution. The difference between this distribution and that for the circular target with a maximum along the target axis presented in [45] came from the wider discharge current profile. Indeed, a rotating point relativistic charge emitted mostly in the rotation plane, and the spectrum had higher frequencies, the higher the charge velocity was [51]. For a finite charge profile, the higher frequencies were suppressed, and for the profile with a characteristic width of the order of the target perimeter, only the main harmonic in the radiation spectrum with a maximum along the target axis survived [45]; see also the results for the small target and the same charge profile width, shown in Figure 6e,f.

As the relation a/b increased, the radiation patterns changed substantially; see Figure 6b–d. The intensity of the radiation along the target axis grew and became comparable in magnitude to the intensity in the plane of the target. At the same time, due to the introduced asymmetry of the target sides, patterns in the $x = 1.28$ mm and $y = -1.28$ mm planes began to differ from one another. The one in $y = -1.28$ mm mostly retained its uniform shape, while for the pattern in $x = 1.28$ mm, a noticeable drop in the intensity appeared in the direction of the x axis, i.e., along the longer side of the target. At $a/b = 10$, the distribution in $x = 1.28$ mm plane attained a doughnut-like shape with a hole in the center; see Figure 6d. This result was consistent with the angular distributions of the power emitted by a compact current oscillating in a straight wire; see the Supplementary Material in [43] for details. For a smaller "stadium" target with the sizes from the simulation above and the discharge current with the same parameters as those for the larger elliptical targets, the distribution had a more pronounced maximum along the axis of the target, in the $z = -1.28$ mm plane, rather than in the $x = 1.28$ mm and $y = -1.28$ mm planes; see Figure 6e. For a comparison, an elliptical target with the same perimeter, corresponding to the oscillation period of 0.25 ps and with the same a/b relation, was considered. The calculations yielded similar results as for the "stadium" target (see Figure 6f), confirming that the degree of localization of the discharge current on the coil perimeter played an important role and defined the angular distribution of the emitted power.

Integrating fluence distributions over the boundaries of the simulation box allowed the estimation of the total emitted energy and the average emitted power. For the electric current of 10^4 A driven in the small "stadium" target, as in the PIC simulations presented here, the total emitted energy was $\approx 7 \times 10^{-4}$ J per one period and the average emitted power was $\approx$2.8 GW, somewhat larger than predicted by a simplified treatment using (19), probably resulting from a larger effective solid angle of the emission than the initially anticipated 1 sr. The resulting conversion efficiency with the invested laser energy of 130 mJ appeared to be about 0.5% per period. For the electric current of 10^5 A, which can be driven [43] by a laser pulse with intensity $\gtrsim 10^{21}$ W/cm^2, the emitted energy per period may reach $\approx 7 \times 10^{-2}$ J and the average emitted power per period increased to $\approx$0.3 TW. For larger elliptical targets and the same electric current of 10^5 A, the integration of the fluence distributions (Figure 6a–d), per all emission directions, yielded somewhat larger total emitted energies of $\approx$0.1 J per period, though due to the greater period, the average emitted power was less than for smaller targets and amounted to $\approx$0.1 TW for all the considered a/b ratios. The estimated laser energy required to drive the electric current of 10^5 A was about 25 J, assuming a tight focusing of a 25 fs-long laser pulse to obtain the intensity of $\sim 10^{21}$ W/cm^2. The conversion efficiency then also amounted to about 0.5%, similar to the estimate provided for the "stadium" target and the laser energy used in the simulations. If the discharge pulse was capable of traveling several target periods without a significant decrease of its amplitude, the efficiency of 0.5% per period would result in the total conversion efficiency of about a few percent, which was comparable or surpassed the efficiency of other setups for the generation of intense THz radiation [53].

An important question that defines the possible range of applications and the efficiency of the proposed scheme is the decay rate of the discharge current pulse during its propagation along the wire surface. According to the results of the performed numerical simulations, for the thin wire considered here, the decay rate appeared to be considerable, similar to that reported for a 1 µm-thick wire in [43]. However, for thicker wires, much lower decay rates have been observed experimentally; see [46], where no dissipation was noticed on the scale of a few millimeters, and [54], where the electron motion guided by the laser-induced fields traveling along the wire was observed over distances of up to 1 m. These experimental results suggest that, under appropriate conditions, the damping of the discharge wave can be lowered and become negligible on the spatial scale of an oscillation period $\sim$(100–1000) µm. In this situation, it should be possible for the discharge pulse to make several full turns, emitting multi-period polarized THz waves. Although, a careful treatment of discharge propagation along a wire with realistic sizes is rather complicated and requires further research.

Another point that has to be addressed in the context of the further realization of the proposed scheme is the relation between the reduced simulated geometry and that experimentally accessible. In a more-feasible setup, laser pulses are expected to have a duration of a few tens of femtoseconds rather than of the 4 fs considered here, with the focal spot on the target surface being about a few microns. The process of the discharge current pulse formation under the laser irradiation of a wire target with such laser pulses at different laser intensity levels was investigated numerically in [43], where it was shown that the discharge current pulse remained well-localized on the target scale, with a linear relation between its duration and the duration of the laser driver. The target size should meet the requirements of a particular experiment. Thus, for example, in order to obtain a spectral maximum at ≈1 THz, the target perimeter should be about 300 μm so that the oscillation period is ≈1 ps. Producing such targets is feasible, e.g., with laser cutting of thin metallic foils [55]. In order to shift the spectral maximum to a higher-frequency range, the single oscillation time should be reduced by decreasing the size of the target. For the target considered here in the simulations, this time was about 250 fs and corresponded to 4 THz. The required target size in this case was a few tens of microns in the target plane. Although the manufacturing of such targets may be complicated, it is not impossible and can be performed with modern techniques, e.g., laser 3D microlithography or nanolithography [56] using commercially available high-resolution 3D printers. Thus, in principle, with the appropriate miniaturization of the target, the proposed setup can be useful if the required spectral maximum is in the range from a few tenths up to several THz.

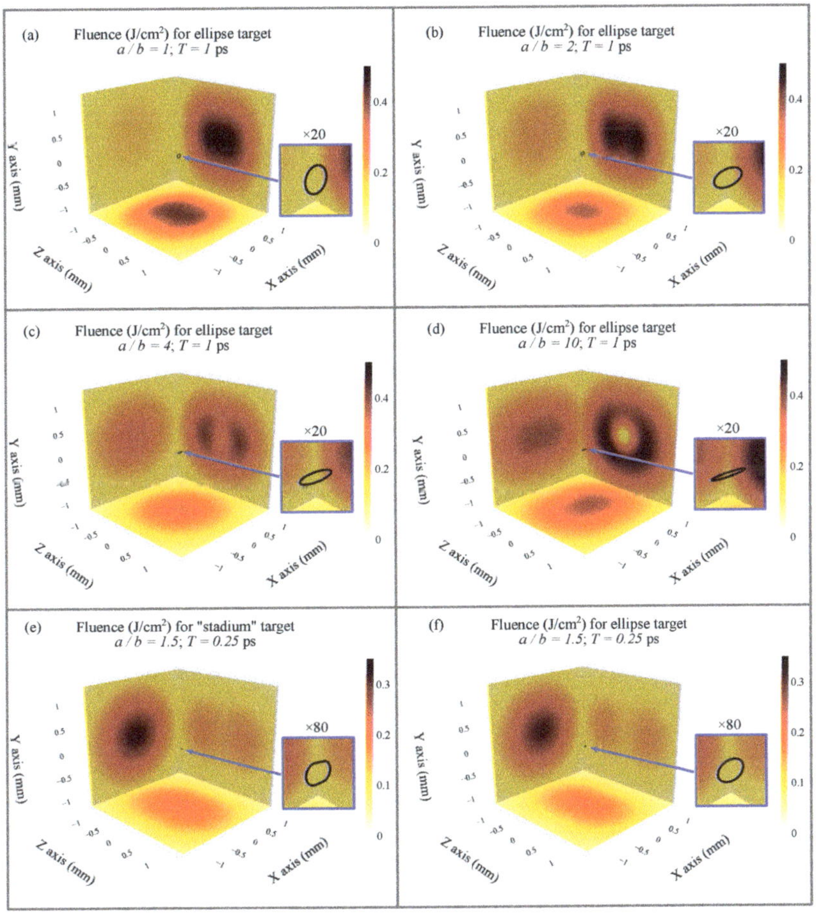

Figure 6. **(a–f)** Energy fluence through the edges of the $2.56 \times 2.56 \times 2.56$ mm^3 simulation box for elliptical targets with the same oscillation period $T = 1$ ps and different ratios of the major and minor semi-axes: $a/b = 1$ (circle), $a/b = 2$, $a/b = 4$, and $a/b = 10$, respectively. The target is positioned at the center of the simulation box in the XY plane; see the inset panels with the magnified images of the target. The fluences correspond to the total electric current of 10^5 A and are normalized per one oscillation period.

The considered setup was a single-shot emitter, implying that the target was destroyed during the shot. Therefore, it is important that the target heat expansion does not develop until the discharge current pulse travels several times along its perimeter. This is possible as on the picosecond time scale, heavy ions exhibit rather slow dynamics, e.g., for copper ions with characteristic energies up to a few MeV, the characteristic expansion velocity is less than a few microns per picosecond. In this case, the target remained almost unchanged for several turns of the discharge current pulse. For multi-petawatt laser facilities with higher energies delivered to the target, the latter may be further stabilized by using heavier metals, e.g., gold instead of copper, but regardless of the target material and the laser energy, if the latter extends to a range of a few mJ, the target never survives after a single shot and has to be replaced. It actually mostly removes the limitations on the laser energy delivered to the target, so that petawatt-class laser facilities may be employed for boosting the THz output power and extending it to the multi-terawatt range. Note that modern PW laser facilities allow operation at relatively high repetition rates, up to 1 Hz at the 1 PW level of the output power and up to 10 Hz at the 100 TW level of the output power [57], enabling tens of shots per day with an appropriate target revolver system. The comparison with other state-of-the-art high-power THz sources (see, e.g., [53,58] and the references there) indicates that techniques that are not based on ultra-intense laser–matter interaction, such as linear accelerators, gyrotrons, and THz free electron lasers, or schemes involving optical rectification or photoconductive emitters, normally allow for peak powers in the MW-GW range, depending on a particular scheme. Nevertheless, the use of petawatt laser facilities is not necessary for the scheme proposed here to work as a strong THz source. Ultrashort discharge currents can be already excited at laser intensities as low as $\sim 10^{19}$ W/cm^2 (see [43], which in the femtosecond regime can be achieved with more-modest commercial terawatt laser systems, making the proposed setup less expensive and readily available.

An important connection to the previous results on the generation of strong radiation in the THz domain by irradiating dense extended targets is that the considered short relativistic discharge current pulses represent a compact moving charge bunch surrounded by a hot electron cloud; see, e.g., the figures here and in [46] for a demonstration. In previous studies, related to the THz emission using straight wires, the confined hot electrons from the interaction region were shown to perform a helical motion and were considered as an emitter of THz waves [40,41]. However, a more-confined and perfectly geometrically controlled discharge in the wire may probably be a more-efficient emitter, though it has the opposite electrical charge. Recently, several propositions of powerful THz emitters based on these discharges were presented in [43,45], where the wire shape was a ring or an undulator. It was also noticed that the discharge propagated along the wire surface, not penetrating inside the material further than the corresponding skin depth. This makes it possible to use as a THz emitter just a surface of, e.g., an elliptical shape, as we considered in this work. Indeed, as seen in Figure 4d1,d2, there was an electron cloud around the discharge, which was reminiscent of a projection of those considered in [40], but the current of the positive charge on the surface dominated; see Figure 4f1. The electron cloud actually slightly suppressed the radiation according to Le Chatelier's principle, and without it, the emission's effectiveness would probably increase. Another important question is the excitation of the considered surface discharge wave and forcing its propagation in a certain direction. For that, the electron cloud seems important, as laser-accelerated electrons moving in the direction of the laser beam's propagation form an electromagnetic response on the nearby surface. The problem of the excitation of a strong relativistic discharge pulse moving in a certain direction is in general rather nontrivial, but as was shown here, such a directed discharge excitation is possible under certain conditions, e.g., it may require a driver with a relativistic intensity. The demonstration of the possibility to excite the directed surface discharge pulse on a closed circuit, along with the description of the related THz emission, was the main point of this work.

5. Conclusions

In conclusion, we demonstrated the possibility of obtaining powerful terahertz radiation with arbitrary elliptic polarization at a certain frequency by driving an oriented strong discharge current in a target with an elliptically shaped surface. The use of intense femtosecond laser pulses for exciting such a current made it take the form of a short compact pulse propagating over the target surface and performing multiple oscillations along its perimeter. During these oscillations, the current emitted electromagnetic waves with the frequency defined by the perimeter of the circuit. The polarization of these waves for elliptic targets, as the performed analysis showed, in the most-general case at a given harmonic was elliptical and in more-specific cases can be turned into either linear or circular by adjusting the target shape and the propagation direction. This feature, along with the dependence of the main frequency on the size of the target, makes the proposed scheme attractive for various applications where powerful polarized THz radiation with a good control of its spectral distribution is required. The proposed method of strong polarized THz wave generation may be readily applied in laboratories with powerful table-top lasers, which makes it a convenient and versatile tool for atomic, molecular, and material studies.

Author Contributions: Conceptualization, supervision, project administration, funding acquisition P.K.; methodology, software, validation, formal analysis, investigation, resources, data curation, writing—original draft preparation, writing—review and editing, visualization, P.K, E.D. and N.B. All authors have read and agreed to the published version of the manuscript.

Funding: The work was supported by the Foundation for the Advancement of Theoretical Physics and Mathematics "BASIS".

Data Availability Statement: Not applicable.

Conflicts of Interest: The authors declare no conflict of interest.

References

1. Mittleman, D.M. Perspective: Terahertz science and technology. *J. Appl. Phys.* **2017**, *122*, 230901. [CrossRef]
2. Dhillon, S.S.; Vitiello, M.S.; Linfield, E.H.; Davies, A.G.; Hoffmann, M.C.; Booske, J.; Paoloni, C.; Gensch, M.; Weightman, P.; Williams, G.P.; et al. The 2017 terahertz science and technology roadmap. *J. Phys. D Appl. Phys.* **2017**, *50*, 043001. [CrossRef]
3. Leitenstorfer, A.; Moskalenko, A.S.; Kampfrath, T.; Kono, J.; Castro-Camus, E.; Peng, K.; Qureshi, N.; Turchinovich, D.; Tanaka, K.; Markelz, A.; et al. The 2023 Terahertz Science and Technology Roadmap. *J. Phys. D Appl. Phys.* **2023**, *56*, 223001. [CrossRef]
4. Sun, Q.; He, Y.; Liu, K.; Fan, S.; Parrott, E.P.J.; Pickwell-MacPherson, E. Recent advances in terahertz technology for biomedical applications. *Quant. Imaging Med. Surg.* **2017**, *7*, 345–355. [CrossRef]
5. Stelmaszczyk, K.; Karpierz-Marczewska, E.; Mikhnev, V.; Cywinski, G.; Skotnicki, T.; Knap, W. Ultrafast Time-of-Flight Method of Gasoline Contamination Detection Down to ppm Levels by Means of Terahertz Time-Domain Spectroscopy. *Appl. Sci.* **2022**, *12*, 1629. [CrossRef]
6. Yang, L.; Guo, T.; Zhang, X.; Cao, S.; Ding, X. Toxic chemical compound detection by terahertz spectroscopy: A review. *Rev. Anal. Chem.* **2017**, *37*, 1–10. [CrossRef]
7. Davies, A.G.; Burnett, A.D.; Fan, W.; Linfield, E.H.; Cunningham, J.E. Terahertz spectroscopy of explosives and drugs. *Mater. Today* **2008**, *11*, 18–26. [CrossRef]
8. Liu, W.; Liu, Y.; Huang, J.; Huang, G.; Zhang, Y.; Fu, W. Application of terahertz spectroscopy in biomolecule detection. *Front. Lab. Med.* **2018**, *2*, 127–133. [CrossRef]
9. Ajito, K.; Ueno, Y. THz Chemical Imaging for Biological Applications. *IEEE Trans. Terahertz Sci. Technol.* **2011**, *1*, 293–300. [CrossRef]
10. Gente, R.; Koch, M. Monitoring leaf water content with THz and sub-THz waves. *Plant Methods* **2015**, *11*, 1–9. [CrossRef] [PubMed]
11. Grischkowsky, D.; Keiding, S.; van Exter, M.; Fattinger, C. Far-infrared time-domain spectroscopy with terahertz beams of dielectrics and semiconductors. *J. Opt. Soc. Am. B* **1990**, *7*, 2006–2015. [CrossRef]
12. Neu, J.; Schmuttenmaer, C.A. Tutorial: An introduction to terahertz time domain spectroscopy (THz-TDS). *J. Appl. Phys.* **2018**, *124*, 231101. [CrossRef]
13. Niessen, K.A.; Xu, M.; George, D.K.; Chen, M.C.; Ferré-D'Amaré, A.R.; Snell, E.H.; Cody, V.; Pace, J.; Schmidt, M.; Markelz, A.G. Protein and RNA dynamical fingerprinting. *Nat. Commun.* **2019**, *10*, 1026. [CrossRef]
14. Brucherseifer, M.; Nagel, M.; Bolivar, P.H. Label-free probing of the binding state of DNA by time-domain terahertz sensing. *Appl. Phys. Lett.* **2000**, *77*, 4049–4051. [CrossRef]

15. Chernomyrdin, N.V.; Musina, G.R.; Nikitin, P.V.; Dolganova, I.N.; Kucheryavenko, A.S.; Alekseeva, A.I.; Wang, Y.; Xu, D.; Shi, Q.; Tuchin, V.V.; et al. Terahertz technology in intraoperative neurodiagnostics: A review. *Opto-Electron. Adv.* **2023**, *6*, 220071. [CrossRef]
16. Martins, I.S.; Silva, H.F.; Lazareva, E.N.; Chernomyrdin, N.V.; Zaytsev, K.I.; Oliveira, L.M.; Tuchin, V.V. Measurement of tissue optical properties in a wide spectral range: A review [Invited]. *Biomed. Opt. Express* **2023**, *14*, 249. [CrossRef]
17. Yan, Z.; Zhu, L.G.; Meng, K.; Huang, W.; Shi, Q. THz medical imaging: From in vitro to in vivo. *Trends Biotechnol.* **2022**, *40*, 816–830. [CrossRef]
18. Vicario, C.; Shalaby, M.; Hauri, C.P. Subcycle Extreme Nonlinearities in GaP Induced by an Ultrastrong Terahertz Field. *Phys. Rev. Lett.* **2017**, *118*, 083901. [CrossRef]
19. Huang, K.C.; Wang, Z. Terahertz terabit wireless communication. *IEEE Microw. Mag.* **2011**, *12*, 108–116. [CrossRef]
20. Fleischer, S.; Zhou, Y.; Field, R.W.; Nelson, K.A. Molecular Orientation and Alignment by Intense Single-Cycle THz Pulses. *Phys. Rev. Lett.* **2011**, *107*, 163603. [CrossRef]
21. Zhang, X.M.; Li, J.; Yu, J.; Cong, S.L. Field-Free Molecular Orientation Induced by a Single-Cycle THz Laser Pulse Train. *Commun. Comput. Phys.* **2016**, *20*, 689–702. [CrossRef]
22. Lapert, M.; Sugny, D. Field-free molecular orientation by terahertz laser pulses at high temperature. *Phys. Rev. A* **2012**, *85*, 063418. [CrossRef]
23. Yu, Z.; Cui, Y.; Wang, X.; Zhang, Y. Identification of ores and gems using THz polarization information. In *Infrared, Millimeter Wave, and Terahertz Technologies*; SPIE: Bellingham, WA, USA, 2010; Volume 7854, pp. 413–425. [CrossRef]
24. Zhang, Z.; Zhong, C.; Fan, F.; Liu, G.; Chang, S. Terahertz polarization and chirality sensing for amino acid solution based on chiral metasurface sensor. *Sens. Actuators B Chem.* **2021**, *330*, 129315. [CrossRef]
25. Zhang, Y.; Li, K.; Zhao, H. Intense terahertz radiation: Generation and application. *Front. Optoelectron.* **2020**, *14*, 4–36. [CrossRef] [PubMed]
26. Kampfrath, T.; Tanaka, K.; Nelson, K.A. Resonant and nonresonant control over matter and light by intense terahertz transients. *Nat. Photon.* **2013**, *7*, 680–690. [CrossRef]
27. Zhang, D.; Fakhari, M.; Cankaya, H.; Calendron, A.L.; Matlis, N.H.; Kärtner, F.X. Cascaded Multicycle Terahertz-Driven Ultrafast Electron Acceleration and Manipulation. *Phys. Rev. X* **2020**, *10*, 011067. [CrossRef]
28. Zhang, D.; Zeng, Y.; Fakhari, M.; He, X.; Matlis, N.H.; Kärtner, F.X. Long range terahertz driven electron acceleration using phase shifters. *Appl. Phys. Rev.* **2022**, *9*, 031407. [CrossRef]
29. Tang, H.; Zhao, L.; Zou, X. Stable and scalable two-stage terahertz-driven particle accelerator. In Proceedings of the 2021 46th International Conference on Infrared, Millimeter and Terahertz Waves (IRMMW-THz), Chengdu, China, 29 August–3 September 2021. [CrossRef]
30. Salamin, Y.I. Electron acceleration in vacuum by a linearly-polarized ultra-short tightly-focused THz pulse. *Phys. Lett. A* **2017**, *381*, 3010–3013. [CrossRef]
31. Amer, N.; Hurlbut, W.C.; Norton, B.J.; Lee, Y.S.; Norris, T.B. Generation of Terahertz Pulses with Arbitrary Elliptical Polarization. *Appl. Phys. Lett.* **2005**, *87*, 221111. [CrossRef]
32. Mosley, C.D.W. *Enhanced Polarisation Control and Extreme Electric Fields: Advances in Terahertz Spectroscopy Applied to Anisotropic Materials and Magnetic Phase Transitions*; Springer Theses, Springer International Publishing: Cham, Switzerland, 2021. [CrossRef]
33. Huang, H.h.; Juodkazis, S.; Gamaly, E.G.; Nagashima, T.; Yonezawa, T.; Hatanaka, K. Spatio-Temporal Control of THz Emission. *Commun. Phys.* **2022**, *5*, 134. [CrossRef]
34. Wang, W.M.; Gibbon, P.; Sheng, Z.M.; Li, Y.T. Tunable Circularly Polarized Terahertz Radiation from Magnetized Gas Plasma. *Phys. Rev. Lett.* **2015**, *114*, 253901. [CrossRef] [PubMed]
35. Lu, X.; Zhang, X.C. Generation of Elliptically Polarized Terahertz Waves from Laser-Induced Plasma with Double Helix Electrodes. *Phys. Rev. Lett.* **2012**, *108*, 123903. [CrossRef] [PubMed]
36. Song, Q.; Yuan, X.; Hu, S.; Huang, J.; Zhong, H.; Lin, Q.; Wang, H.; Lu, X.; Zheng, M.; Cai, Y.; et al. Enhanced terahertz radiation with controllable polarization from two filaments pumped by two-color femtosecond laser fields. *Opt. Express* **2021**, *29*, 22659–22666. [CrossRef]
37. Wang, G.L.; Qi, H.X.; Li, Y.N.; Jiao, Z.H.; Zhao, S.F.; Zhang, L. Polarization-Controlled Terahertz Generation by Bicircular Longer-Wavelength Laser Fields. *J. Opt. Soc. Am.* **2022**, *39*, 1370. [CrossRef]
38. Li, Z.; Zheng, J. Terahertz radiation from a wire target irradiated by an ultra-intense laser pulse. *Phys. Plasmas* **2007**, *14*, 054505. [CrossRef]
39. Zhuo, H.B.; Zhang, S.J.; Li, X.H.; Zhou, H.Y.; Li, X.Z.; Zou, D.B.; Yu, M.Y.; Wu, H.C.; Sheng, Z.M.; Zhou, C.T. Terahertz generation from laser-driven ultrafast current propagation along a wire target. *Phys. Rev. E* **2017**, *95*, 013201. [CrossRef]
40. Tian, Y.; Liu, J.; Bai, Y.; Zhou, S.; Sun, H.; Liu, W.; Zhao, J.; Li, R.; Xu, Z. Femtosecond-laser-driven wire-guided helical undulator for intense terahertz radiation. *Nat. Photonics* **2017**, *11*, 242–246 . [CrossRef]
41. Zeng, Y.; Zhou, C.; Song, L.; Lu, X.; Li, Z.; Ding, Y.; Bai, Y.; Xu, Y.; Leng, Y.; Leng, Y.; et al. Guiding and emission of milijoule single-cycle THz pulse from laser-driven wire-like targets. *Opt. Express* **2020**, *28*, 15258–15267 . [CrossRef]
42. Zhang, D.; Bai, Y.; Zeng, Y.; Ding, Y.; Li, Z.; Zhou, C.; Leng, Y.; Song, L.; Tian, Y.; Li, R. Towards High-Repetition-Rate Intense Terahertz Source with Metal Wire-Based Plasma. *IEEE Photonics J.* **2022**, *14*, 1–5. [CrossRef]

43. Bukharskii, N.; Korneev, P. Intense widely controlled terahertz radiation from laser-driven wires. *Matter Radiat. Extrem.* **2023**, *8*, 044401. [CrossRef]
44. Quinn, K.; Wilson, P.A.; Cecchetti, C.A.; Ramakrishna, B.; Romagnani, L.; Sarri, G.; Lancia, L.; Fuchs, J.; Pipahl, A.; Toncian, T.; et al. Laser-Driven Ultrafast Field Propagation on Solid Surfaces. *Phys. Rev. Lett.* **2009**, *102*, 194801. [CrossRef]
45. Bukharskii, N.; Kochetkov, I.; Korneev, P. Terahertz annular antenna driven with a short intense laser pulse. *Appl. Phys. Lett.* **2022**, *120*, 014102. [CrossRef]
46. Ehret, M.; Bailly-Grandvaux, M.; Korneev, P.; Apiñaniz, J.I.; Brabetz, C.; Morace, A.; Bradford, P.; d'Humières, E.; Schaumann, G.; Bagnoud, V.; et al. Guided electromagnetic discharge pulses driven by short intense laser pulses: Characterization and modeling. *Phys. Plasmas* **2023**, *30*, 013105. [CrossRef]
47. Landau, L.D.; Bell, J.S.; Kearsley, M.J.; Pitaevskii, L.P.; Lifshitz, E.M.; Sykes, J.B. *Electrodynamics of Continuous Media*; Elsevier Science: Amsterdam, Netherlands, 2013.
48. Derouillat, J.; Beck, A.; Pérez, F.; Vinci, T.; Chiaramello, M.; Grassi, A.; Flé, M.; Bouchard, G.; Plotnikov, I.; Aunai, N.; et al. Smilei: A collaborative, open-source, multi-purpose particle-in-cell code for plasma simulation. *Comput. Phys. Commun.* **2018**, *222*, 351–373. [CrossRef]
49. PIC Algorithms. Available online: https://smileipic.github.io/Smilei/Understand/algorithms.html (accessed on 21 May 2023).
50. Korneev, P.; d'Humières, E.; Tikhonchuk, V. Gigagauss-scale quasistatic magnetic field generation in a snail-shaped target. *Phys. Rev. E* **2015**, *91*, 043107. [CrossRef]
51. Landau, L.D.; Lifshitz, E.M. *The Classical Theory of Fields*; Butterworth Heinemann: Oxford, UK, 1980; Volume 2.
52. Boyd, R.W. *Nonlinear Optics*, 3rd ed.; Elsevier Science: Amsterdam, The Netherlands, 2007.
53. Liao, G.Q.; Liu, H.; Scott, G.G.; Zhang, Y.H.; Zhu, B.J.; Zhang, Z.; Li, Y.T.; Armstrong, C.; Zemaityte, E.; Bradford, P.; et al. Towards Terawatt-Scale Spectrally Tunable Terahertz Pulses via Relativistic Laser-Foil Interactions. *Phys. Rev. X* **2020**, *10*, 031062. [CrossRef]
54. Nakajima, H.; Tokita, S.; Inoue, S.; Hashida, M.; Sakabe, S. Divergence-Free Transport of Laser-Produced Fast Electrons Along a Meter-Long Wire Target. *Phys. Rev. Lett.* **2013**, *110*, 155001. [CrossRef] [PubMed]
55. Kochetkov, I.V.; Bukharskii, N.D.; Ehret, M.; Abe, Y.; Law, K.F.F.; Ospina-Bohorquez, V.; Santos, J.J.; Fujioka, S.; Schaumann, G.; Zielbauer, B.; et al. Neural network analysis of quasistationary magnetic fields in microcoils driven by short laser pulses. *Sci. Rep.* **2022**, *12*, 13734. [CrossRef]
56. Gonzalez-Hernandez, D.; Varapnickas, S.; Bertoncini, A.; Liberale, C.; Malinauskas, M. Micro-Optics 3D Printed via Multi-Photon Laser Lithography. *Adv. Opt. Mater.* **2023**, *11*, 2201701. [CrossRef]
57. Lureau, F.; Matras, G.; Chalus, O.; Derycke, C.; Morbieu, T.; Radier, C.; Casagrande, O.; Laux, S.; Ricaud, S.; Rey, G.; et al. High-energy hybrid femtosecond laser system demonstrating 2 × 10 PW capability. *High Power Laser Sci. Eng.* **2020**, *8*, e43. [CrossRef]
58. Thumm, M. State-of-the-Art of High-Power Gyro-Devices and Free Electron Masers. *J. Infrared Millim. Terahertz Waves* **2020**, *41*, 1–140. [CrossRef]

photonics

MDPI

Article

Two-Dimensional Polarization Holographic Gratings in Azopolymer Thin Films: Polarization Properties in the Presence or Absence of Surface Relief

Georgi Mateev [1,2], Lian Nedelchev [1,2,*], Ludmila Nikolova [1], Branimir Ivanov [1], Velichka Strijkova [1], Elena Stoykova [1], Kihong Choi [3], Joongki Park [3] and Dimana Nazarova [1,2]

1 Institute of Optical Materials and Technologies, Bulgarian Academy of Sciences, 1113 Sofia, Bulgaria; georgi@iomt.bas.bg (G.M.); lnikolova@iomt.bas.bg (L.N.); branimirivanov@iomt.bas.bg (B.I.); vily@iomt.bas.bg (V.S.); estoykova@iomt.bas.bg (E.S.); dimana@iomt.bas.bg (D.N.)
2 University of Chemical Technology and Metallurgy, 8 St. Kliment Ohridski Blvd, 1756 Sofia, Bulgaria
3 Digital Holography Research Section, Electronics and Telecommunications Research Institute, Daejeon 34129, Republic of Korea; kihong08@etri.re.kr (K.C.); jkp@etri.re.kr (J.P.)
* Correspondence: lian@iomt.bas.bg

Abstract: During polarization holographic recording in azopolymer thin films, usually together with the volume anisotropic grating, a surface relief grating (SRG) is also formed. By using two consecutive exposures, it is possible to obtain a two-dimensional (2D) grating. To the best of our knowledge, the polarization properties of such gratings have not been studied yet. To determine the influence of the surface relief on the polarization selectivity of the 2D gratings, we propose two methods to suppress the SRG formation: by varying the recording conditions or varying the sample structure. In these experiments we have used the commercially available azopolymer PAZO, poly[1-4-(3-carboxy-4-hydrophenylazo) benzene sulfonamido]-1,2-ethanediyl, sodium salt] to perform the polarization holographic recording using a 442 nm He-Cd laser. As indicated by our results, when the surface relief is present, it strongly dominates the response of the 2D grating and it behaves almost as a scalar polarization insensitive grating. Conversely, when the SRG formation is suppressed, the polarization properties of the 2D grating in all four diffracted orders are very well pronounced. In this way, we demonstrate that we can easily control SRG formation and, if desired, obtain 2D grating with high surface relief modulation, or alternatively record polarization-selective 2D gratings with virtually no surface relief.

Keywords: polarization holography; 2D polarization gratings; surface relief gratings; AFM; azopolymer PAZO

check for updates

Citation: Mateev, G.; Nedelchev, L.; Nikolova, L.; Ivanov, B.; Strijkova, V.; Stoykova, E.; Choi, K.; Park, J.; Nazarova, D. Two-Dimensional Polarization Holographic Gratings in Azopolymer Thin Films: Polarization Properties in the Presence or Absence of Surface Relief. *Photonics* **2023**, *10*, 728. https://doi.org/10.3390/photonics10070728

Received: 30 May 2023
Revised: 20 June 2023
Accepted: 22 June 2023
Published: 26 June 2023

1. Introduction

One of the challenges in modern photonics is the creation of increasingly complex optical elements combining different unique properties and performing multiple tasks. One example is the polarization holographic gratings (PHG) recorded in azobenzene-containing materials. A detailed overview of the properties and applications of polarization gratings and azopolymer materials can be found in [1–3]. An important property of azobenzenes is their polarization sensitivity, discovered as early as 1984 [4]. Following a series of *trans-cis-trans* isomerizations, the azobenzene chromophores reorient perpendicularly to the polarization of the recording light. During polarization holographic recording, this leads to the formation of significant spatially modulated photoinduced birefringence, which can be used to create polarization-selective elements. Additionally, it was discovered in 1995 that it is possible to induce a surface relief grating (SRG) of considerable height in azopolymers when recording polarization diffraction gratings [5,6]. In the following years, considerable efforts were made to establish the mechanism of this process [7–11]. The process occurs at laser powers insufficient for ablation and it is believed to be based again on *trans-cis-trans* isomerization, which, in combination with a force gradient formed in the layer, redistributes the azo molecules, forming topographic structures with a sinusoidal profile. Thus, the polarization sensitivity of the material is the basis of the mass movement in azobenzene. The formed relief has a scalar nature. This is because, in a relief structure, we have a grating caused by the difference in refractive indices of the azopolymer film and air, which is much higher than the photoinduced birefringence in the azobenzene.

Various applications of the SRGs formed in azo materials have been suggested [12–16]. Using computer-generated holograms, Oscurato et al. have obtained complex surface relief structures with possible applications in the fields of surface engineering, biology, and photonics [12]. Another example we find in [13], where the effect of metal nanostructures on azopolymer after irradiation with polarized light was investigated using atomic force microscopy (AFM). Thus, SRGs are used as a non-invasive, easy-to-use microscopic method to characterize metallic nanostructures. Surface relief gratings with high diffraction efficiency have also been studied in azopolymer-based nanocomposite media with different types of nanoparticles [14–16].

Additionally, more complex, two-dimensional surface relief structures realized in azo materials have also been reported [17–22]. Yi et al. [17] show how controlling the relief height in the created two-dimensional structures can affect the formation of 3D patterns of colloids on the symmetrically and asymmetrically patterned 2D SRG substrates. The gratings were fabricated on azopolymer films by exposure to an interference pattern of Ar+ laser beams. The spacing of the gratings was one and the same in both directions; however, the asymmetric 2D grating had a larger modulation depth in the vertical direction. In [18], Stiller et al. analyzed how the relief in 1D and 2D structures changes the elastic and mechanical properties of the used azopolymer (pDR1M). Additionally, using the tip of a scanning near-field optical microscope (SNOM), structures with a size of about 100 nm were inscribed onto the polymer surface. In [19], it is demonstrated how the relief formed on the azopolymer poly(Disperse orange 3) thin films can be used to easily fabricate various TiO_2 nanostructures for applications in photonics and electronics. Varying the angle of rotation of the substrate between the exposures from 90° to 60°, both square and hexagonal 2D SRG patterns were obtained. Huang et al. studied 2D structures created with two-step holographic recording and the dependence of surface modulation of the 2D SRG on the rotational angle between the 1D gratings for different azobenzene media [20]. Using nematic liquid crystal as an interface, it was demonstrated that the depth of the SRG groove could be enhanced about 2–3 times. The main material investigated in [21] is the azopolymer PAZO, showing the possibility to fabricate 1D and 2D SRG with an ultraviolet laser at 355 nm. Due to the short wavelength of the recording laser, gratings with a period on the order of 200 nm were obtained. The ability to reversibly generate relief structures is the main topic of research in [22]. Linearly polarized light or heat were successfully used as the stimuli for the reversible reshaping of periodic structures, imprinted in a blend of low-molecular azo glass and polystyrene.

It is very important to note, however, that the surface relief gratings (1D or 2D) in azopolymer materials coexist with an underlying volume polarization grating. In fact, for some applications, the properties of the polarization grating are more important, and the SRG has an adverse effect, deteriorating the polarization characteristics of the optical element. Examples of such applications are optical switches [23], a circular polarization beam splitter [24], bifocal polarization lenses [25], and a reconfigurable fiber Bragg filter [26], all based on azo-containing materials. In these cases, it is essential to find an efficient method to suppress the formation of the surface relief grating or to evaluate its influence on the polarization properties of the obtained holographic optical element. To the best of our knowledge however, no detailed analysis of the polarization properties of 2D gratings in the presence or absence of SRG has yet been reported.

Thus, the main goal of the present work is to investigate the polarization properties of two-dimensional polarization holographic gratings, comparing the case when an

SRG with significant height is present and the case when SRG formation is prevented. The experiments were performed using the well-known commercially available azopolymer PAZO, poly[1-4-(3-carboxy-4-hydrophenylazo) benzene sulfonamido]-1,2-ethanediyl, sodium salt] [27–29]. Two methods of SRG suppression were employed: recording from the back (or substrate) side of the sample, for which the azopolymer layer acts as a filter reducing the intensity of light before it reaches its free surface, and enclosing the azopolymer film between two glasses when no free surface is present.

2. Materials and Methods

2.1. Samples Preparation

Thin film samples for polarization holographic recording were prepared from the commercially available azopolymer PAZO—poly[1-4-(3-carboxy-4-hydrophenylazo) benzensulfonamido]-1,2-ethanediyl, sodium salt] (Sigma-Aldrich, St. Louis, Missouri, United States, #346411). The absorbance spectrum of PAZO is measured earlier [30] and is presented in Figure 1. Its chemical structure is shown in the inset.

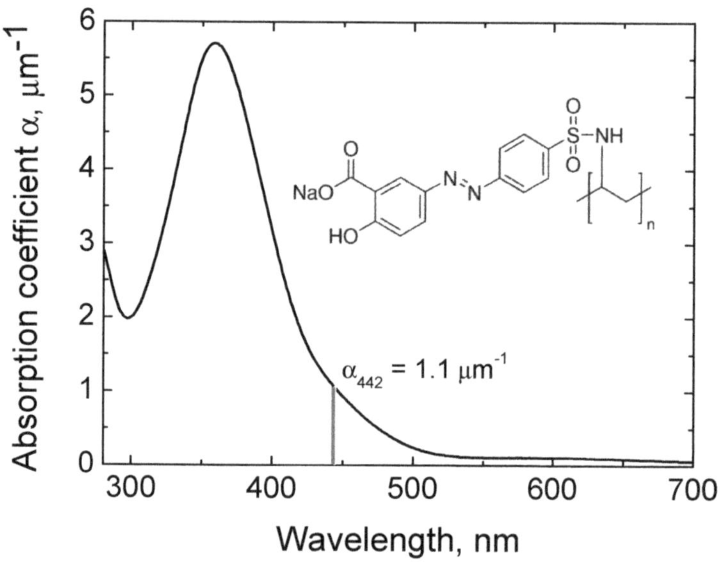

Figure 1. UV–Vis spectrum of the absorption coefficient of the azopolymer PAZO (adapted from [30]). The inset shows the chemical structure of PAZO.

The samples were deposited via spin-coating of PAZO/methanol solution with 100 mg/mL concentration on high-quality optical glass substrates (BK7) at 1000 rpm. The substrates are rectangular in shape and their dimensions are 24 × 36 mm. The thickness of the azopolymer layer is approximately 1800 nm, as measured by F20 Optical Thin-Film Analyzer (Filmetrics). The film thickness variation across the samples' surface is about 1%, or around 20 nm.

Additionally, we have prepared samples with a double-layer or "sandwich" structure, as shown in Figure 2c. They consist of two thin film samples bonded together with the optically transparent glue ethyl 2-cyanoacrylate (CAS No. 7085-85-0).

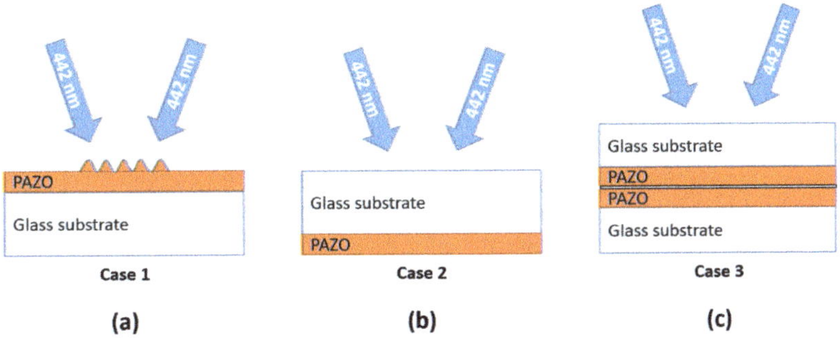

Figure 2. (**a**) Case 1—single-layer sample, recording from the azopolymer film side; (**b**) Case 2—single-layer sample, recording from the substrate side; (**c**) Case 3—double-layer "sandwich" sample.

The azopolymer films are from the inside of this sample, and the glass substrates are from the outside. In this case, two important advantages are achieved: (i) the glass slides protect the sample from mechanical damage or scratches; and (ii) there is no free surface of the azopolymer film, so the formation of surface relief grating can be effectively suppressed.

2.2. Polarization Holographic Setup

The polarization holographic recording of diffraction gratings requires the interference of two coherent light beams with wavelengths within the absorbance band of the azopolymer. In our experimental setup, illustrated in Figure 3, the coherent light source is He-Cd gas laser (IK4171I G, Kimmon Koha, Tokyo, Japan) with λ = 442 nm. Using half-wave plate (HWP) and polarizing beam splitter (PBS), the laser light is split into two beams with equal intensities and by means of two quarter-wave plates (QWP), their polarizations are set to left and right circular (LCP and RCP). The resulting interference field has constant intensity, and its polarization is linear with periodically varying azimuth. These polarizations of the recording beams were selected, as it is known that, in this case, polarization holographic gratings with highest diffraction efficiency and highest surface relief modulation are obtained in azopolymers [2].

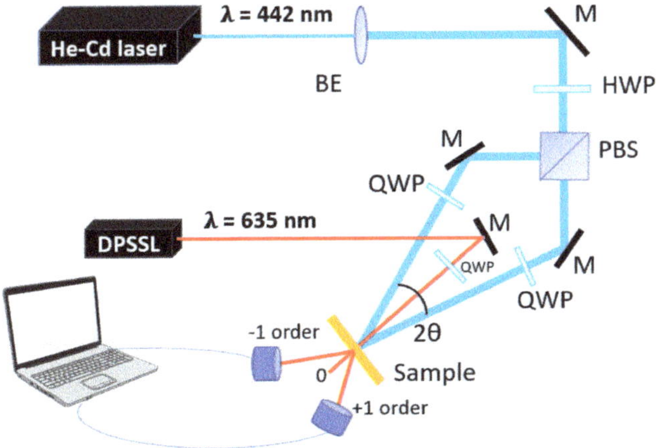

Figure 3. Optical setup for inscription of polarization holographic gratings. BE—beam expander, M—mirror, HWP—half-wave plate, QWP—quarter-wave plate, PBS—polarizing beam splitter, DPSSL—diode-pumped solid-state laser.

After use of the beam expander (BE), the diameter of the recording beams was 7 mm. The total light intensity on the sample was approximately 160 mW/cm^2, and the recording angle (the angle between the two recording beams, as shown in Figure 3) was $2\theta = 20°$, which corresponds to grating period of about 1.3 μm. The angle, θ, is defined as the angle between each of the recording beams and the line perpendicular to the surface of the sample. The diffraction efficiency (DE) is measured in real time during the inscription of the polarization holographic gratings, simultaneously in both +1 and –1, diffracted order by two computer-operated power meters (PM100D, Thorlabs). The probe laser beam used to determine the DE is with left circular polarization and wavelength of 635 nm from a DPSS laser (B&W Tek Inc., Newark, Delaware, United States).

We employ three modes of recording, denoted as case 1, case 2, and case 3, as illustrated in Figure 2. In case 1, the recording is performed with the azopolymer layer facing the recording beams (Figure 2a). In this case, the intensity on the free surface of the film is maximal and this results in formation of surface relief grating. In case 2, the sample is turned back, so the substrate side is facing the recording beams (Figure 2b). The light first encounters the substrate/film interface where, due to the adhesion of the azopolymer layer to the substrate, SRG cannot form. Then, during its propagation through the azopolymer film, the intensity of light is significantly reduced due to the absorption (about 100 times for the given wavelength and sample thickness), and when it reaches the free surface of the layer, it is not sufficient to induce SRG. Finally, in case 3 (Figure 2c), there is no free azopolymer/air surface where the SRG can form.

The inscription of the 2D polarization holographic gratings is performed in two steps. First, a 1D grating is recorded until reaching saturation of the diffraction efficiency. Afterward, the sample is rotated at 90° counterclockwise, as shown in Figure 4a, and the recording of the second superimposed grating starts. The indices for the diffracted orders of the first grating are denoted as $\pm1, 0$, while for the second grating, they are $0, \pm1$ (Figure 4b).

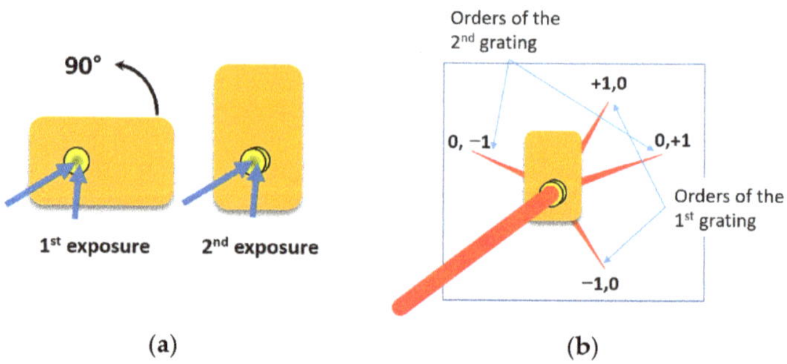

Figure 4. (**a**) Rotation of the sample between the two exposures to obtain the 2D PHG; (**b**) Illustration of the notations of the diffracted orders of the 2D gratings.

The inscription of the second grating reduces the diffraction efficiency of the first, so our aim is to terminate the second recording when the diffraction efficiencies of the two gratings are approximately equal. This task is even more complex because we have to take into account that the surface relief grating is more stable than the volume polarization grating.

2.3. Characterization of the Recorded 2D Gratings

To determine the polarization state and the diffraction efficiencies of the four diffracted orders of the 2D polarization holographic gratings, we used probe beam from a DPSS laser at 635 nm with polarization state, controlled via polarizer and quarter-wave plate (Figure 4b). The polarization incident on the sample was consecutively set to left circular (LCP), right circular (RCP), and linear, with azimuth varying from 0° to 180°. The polarization state of the ±1, 0 and 0, ±1 diffracted orders were measured by a polarimeter (Thorlabs PAX 5710).

In order to analyze the surface topography of the gratings, we employed an atomic force microscope (Asylum Research MFP-3D, Oxford Instruments, Abingdon, UK) supplied with standard silicon probes AC160TS-R3 operating at 300 kHz with spring constant 26 N/m. For optimal visualization of the SRG features, the dimensions of the measurement area were set to 10 μm × 10 μm.

3. Results and Discussion

This section consists of three parts. In the first part, we present the kinetics of the DE for the ±1 diffracted orders obtained during the inscription of the first and the second 1D grating. Next, we present the measured ellipticities and DE for all four diffracted orders of the 2D PHGs for probe light with LCP, RCP, and linear polarizations. In the last case, we also vary the input azimuth from 0° to 180°. Finally, in the third part, we use AFM analysis to obtain the parameters of the relief structures for case 1 and case 2.

3.1. Diffraction Efficiency Kinetics

The diffraction efficiency kinetics for cases 1, 2, and 3 are shown in Figure 5. The main graphs correspond to the inscription of the first 1D gratings, while the insets correspond to the recording of the second 1D gratings. The polarization of the probe beam is set up to LCP. In all figures diffraction efficiency kinetics of both the +1 order (DE_{+1}) and the −1 order (DE_{-1}) are plotted.

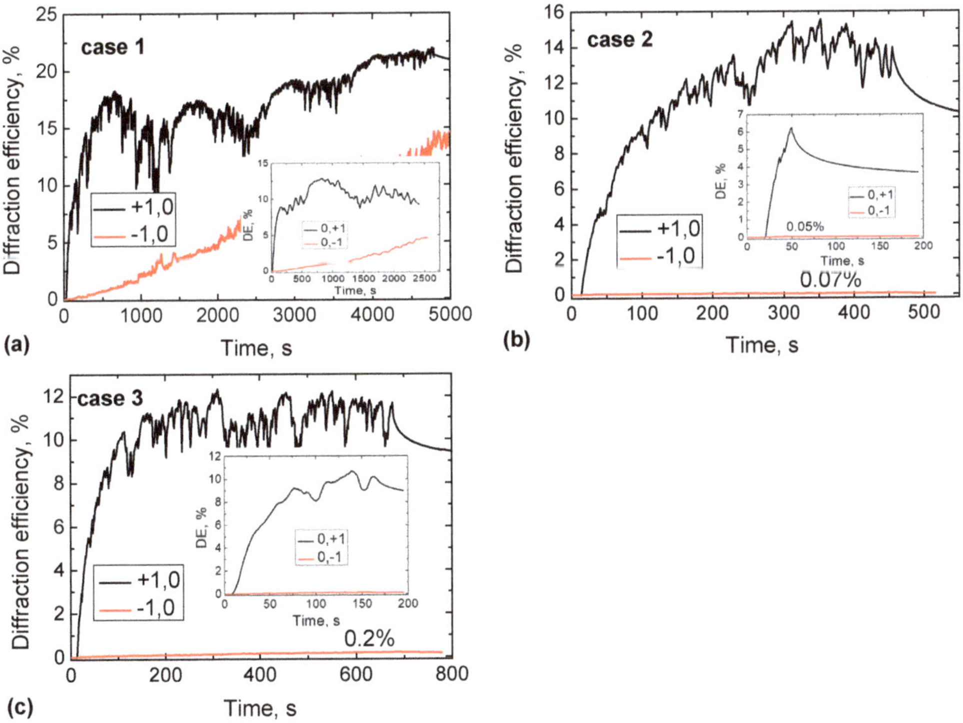

Figure 5. Kinetics of the DE for the first 1D grating in both +1 and −1 diffracted orders for case 1 (**a**), case 2 (**b**), and case 3 (**c**). The insets present the DE kinetics for the second grating recording.

Here, we analyze in detail the data presented in Figure 5. It is well known from the theory of polarization holography [2], that when a polarization grating, recorded with left and right circularly polarized beams, is illuminated with circularly polarized light, the light will be directed only in the +1 order of diffraction. In case 1, however, we observe significant diffraction also in the −1 order, both for the first and second grating. This clearly indicates the formation of surface relief grating. Because of its scalar nature, the SRG distributes light equally in +1 and −1 orders. On the other hand, the fast increase in DE_{+1} at the beginning of the recording and the fact that the value of DE_{+1} is higher than DE_{-1} prove that a volume polarization grating is also present. Still, the ratio of the maximal diffraction efficiencies for the ±1 orders, DE_{-1max}/DE_{+1max} = 65%, indicates that the SRG has a more significant contribution to the total DE in comparison with the polarization grating. The noticeable fluctuations of the kinetic curve $DE_{+1}(t)$ are due not only to external factors but also to the change in the polarizations of the recording beams caused by the surface grating. This issue will be discussed in more detail in our forthcoming article.

In contrast, in case 2 and case 3, the ratio DE_{-1max}/DE_{+1max} is negligible (about or even below 1%). Hence, we can conclude that in these two cases, the formation of the surface relief grating was efficiently suppressed. The absence of SRG leads to a lower value of the maximal DE compared to case 1, but at the same time significantly improves the polarization properties of the PHG, as will be demonstrated in the next section.

We should also note that the recording time of the second grating is shorter than the recording time of the first grating. The first grating is recorded until saturation of DE_{+1} is reached, while the recording of the second grating is stopped when its DE is approximately the same as that of the first one.

3.2. Polarization Properties of the 2D Gratings

After the inscription of the 2D polarization holographic gratings, their polarization properties were evaluated using the method described in Section 2.3. Generally, we could expect the behavior of each couple of diffracted orders of the 2D polarization holographic grating (+1, 0 and −1, 0; 0, +1 and 0, −1) to be similar to that of the orders of the 1D grating. The dependencies of the diffraction efficiency and ellipticity for all four diffracted orders on the azimuth, φ, of a linearly polarized probe beam are presented in Figure 6.

If we consider the ellipticity vs. azimuth dependences, we can see that the presence of the surface relief in case 1 significantly influences the polarization properties of the 2D grating. The average value of the ellipticity is approximately zero (or linear, as is the polarization of the probe beam) due to the dominance of the SRG, with a modulation of about 0.3. On the other hand, in case 2 and case 3 where the surface relief grating is absent, we observe a very clear distinction between the ellipticity of the diffracted orders: two of them are close to RCP, and the other two are close to LCP. These characteristics are typical for polarization-selective grating.

The variation of the diffraction efficiency on the input azimuth in case 1 is most pronounced for the 0, +1 order, and it is in the range 7–11.5%. The DE values in case 1 are higher than in case 2 and case 3, and this can be ascribed to the presence of SRG, which contributes to the DE. We should also note that in case 2 and case 3, there is no significant modulation for three of the diffracted orders.

Table 1 summarizes the experimental data on the diffraction efficiency and the output ellipticity (e_{out}) of the four diffracted orders when the probe laser beam has left circular polarization ($e_{in} = -1$) or right circular polarization ($e_{in} = +1$).

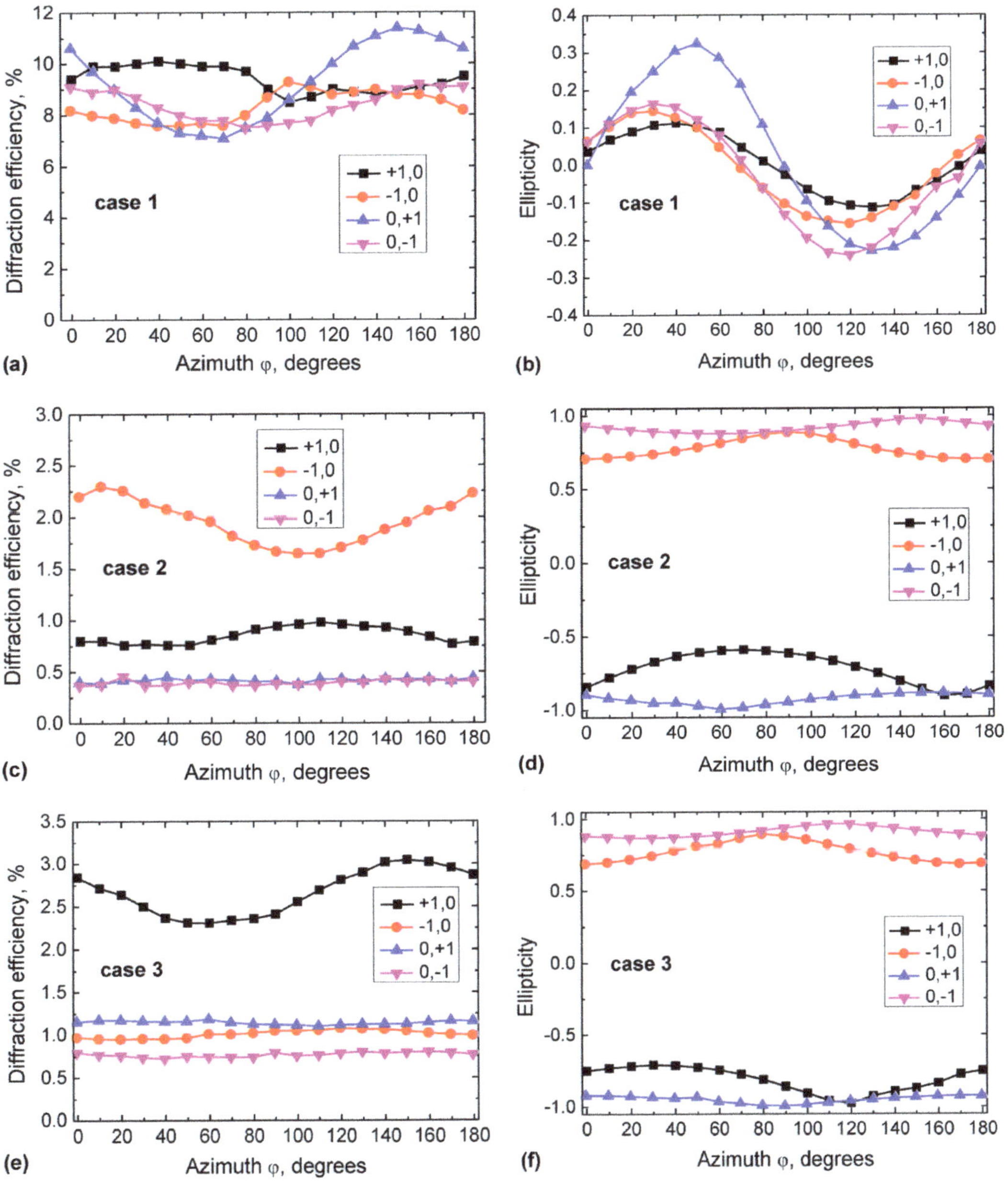

Figure 6. Dependencies of the diffraction efficiency (**a**,**c**,**e**) and ellipticity (**b**,**d**,**f**) for all four diffracted orders of the 2D grating on the azimuth, φ, of a linearly polarized probe beam. (**a**,**b**) Correspond to case 1; (**c**,**d**) to case 2; (**e**,**f**) to case 3.

Table 1. Experimental data for DE and e_{out} for cases 1, 2, and 3 for LCP and RCP probe light (e_{in} = ±1).

		Case I		Case II		Case III	
e_{in}	Diffr. Order	e_{out}	DE, %	e_{out}	DE, %	e_{out}	DE, %
−1	+1,0	−0.72	10.4	0.84	4.0	0.94	2.5
−1	−1,0	−0.77	5.5	-	0.2	0.68	0.5
−1	0, +1	−0.67	10.7	0.90	1.6	0.95	2.6
−1	0, −1	−0.62	6.8	-	0.02	-	0.2
+1	+1, 0	0.86	12.2	-	0.3	−0.94	0.1
+1	−1, 0	0.87	7.1	−0.83	2.7	−0.78	2.0
+1	0, +1	0.90	12.1	-	0.01	-	0.1
+1	0, −1	0.83	10.5	−0.97	1.2	−0.88	4.7

As seen from the data in Table 1, in case 1, the output ellipticity for all orders has the same sign as the input ellipticity. This fact again confirms the presence of the scalar surface relief grating. The total diffraction efficiency (the sum of the DE for all four orders), however, is relatively high: for the RCP probe light, it is DE_{total} = 41.9%.

In contrast, for case 2 and case 3, we observe the typical behavior of a polarization-selective grating: the LCP probe light (e_{in} = −1) is diffracted mainly in the +1, 0 and 0, +1 orders, while the RCP probe light (e_{in} = +1) is diffracted in the −1, 0 and 0, −1 orders At the same time, the ellipticity of the diffracted orders (e_{out}) is with reverse sign compared to the input ellipticity. The high polarization selectivity of the gratings in cases 2 and 3 is also the reason for the missing values of the output ellipticity for some of the diffracted orders: the light intensity is so low that it is not possible to obtain reliable data using the polarimeter. We should also note that for case 2 and case 3, the ratio between the diffraction efficiency in the +1 and −1 order (DE + 1/DE − 1) is higher for the second gratings than for the first ones. A possible explanation is that the inscription of the second polarization grating reduces the polarization selectivity of the first.

3.3. Atomic Force Microscopy Analysis

Using AFM, we have analyzed the surface topography of the 2D holographic gratings, recorded in case 1 and case 2. In case 3, the azopolymer layer is covered from both sides with glass slides and there is no free film surface to be studied by AFM. The results are shown in Figure 7.

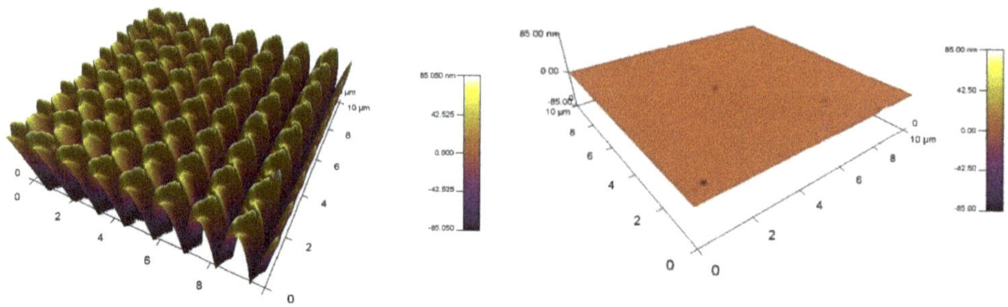

(a) (b)

Figure 7. (**a**) AFM image of the 2D grating in case 1, where a significant surface relief is present; (**b**) AFM image of the 2D grating in case 2, where the SRG formation is suppressed. The scale of the color bars is from −85 nm to +85 nm.

In case 1 (Figure 7a), a periodic two-dimensional surface relief structure is clearly observed with a period of 1.3 × 1.3 μm and a height of about 170 nm. This fully corresponds to the already presented data about the diffraction efficiency and polarization properties of this type of grating. Figure 7b is plotted on the same scale and depicts case 2. No periodic modulation of the surface can be seen, and the maximal height is below 5 nm, which corresponds to the typical surface roughness of a non-recorded azopolymer film. This demonstrates that for case 2, SRG formation has been successfully prevented.

4. Conclusions

We have recorded two-dimensional polarization holographic gratings in thin azopolymer films and studied their polarization properties and surface topography. Three different modes of recording were employed. In case 1, the azopolymer film was facing the recording beams. As a result, 2D surface relief grating was formed with a height of about 170 nm. The total diffraction efficiency in this case reached 42%. However, the polarization selectivity of the grating was low, as the scalar SRG had a more significant contribution than the volume polarization grating. On the other hand, by recording from the substrate side of the sample (case 2) or using a "sandwich" sample, covered with glass from both sides (case 3), we successfully suppressed the formation of SRG and obtained gratings with high polarization selectivity. Therefore, using the same optical setup and simply flipping the sample, we can easily obtain 2D polarization holographic gratings either with significant surface relief or with high polarization selectivity.

Author Contributions: Conceptualization, L.N. (Lian Nedelchev), L.N. (Ludmila Nikolova) and D.N.; methodology, G.M.; validation, L.N. (Lian Nedelchev) and D.N.; formal analysis, K.C. and J.P.; investigation, G.M., B.I. and V.S.; writing—original draft preparation, G.M.; writing—review and editing, L.N. (Lian Nedelchev); visualization, G.M.; supervision, L.N. (Lian Nedelchev) and D.N.; funding acquisition, L.N. (Lian Nedelchev) and E.S. All authors have read and agreed to the published version of the manuscript.

Funding: This work was supported by the Institute of Information & Communications Technology Planning & Evaluation (IITP), grant No. 2019-0-00001. G. Mateev acknowledges the financial support from the Bulgarian Ministry of Education and Science under the National Research Programme "Young scientists and postdoctoral students-2", approved by DCM 206/07.04.2022. L. Nedelchev and D. Nazarova are grateful for the funding from the European Union–NextGenerationEU, through the National Recovery and Resilience Plan of the Republic of Bulgaria, project № BG-RRP-2.004-0002, "BiOrgaMCT".

Institutional Review Board Statement: Not applicable.

Informed Consent Statement: Not applicable.

Data Availability Statement: All data generated or analyzed during this study are included in this published article.

Conflicts of Interest: The authors declare no conflict of interest. The funders had no role in the design of the study; in the collection, analyses, or interpretation of data; in the writing of the manuscript; or in the decision to publish the results.

References

1. Natansohn, A.; Rochon, P. Photoinduced Motions in Azo-Containing Polymers. *Chem. Rev.* **2002**, *102*, 4139–4175. [CrossRef]
2. Nikolova, L.; Ramanujam, P.S. *Polarization Holography*; Cambridge University Press: Cambridge, UK, 2009. [CrossRef]
3. Wang, X. *Azo Polymers: Synthesis, Functions and Applications*; Springer: Berlin/Heidelberg, Germany, 2017. [CrossRef]
4. Todorov, T.; Nikolova, L.; Tomova, N. Polarization holography. 1: A new high-efficiency organic material with reversible photoinduced birefringence. *Appl. Opt.* **1984**, *23*, 4309–4312. [CrossRef]
5. Rochon, P.; Batalla, E.; Natansohn, A. Optically induced surface gratings on azoaromatic polymer films. *Appl. Phys. Lett.* **1995**, *66*, 136–138. [CrossRef]
6. Kim, D.Y.; Tripathy, S.K.; Li, L.; Kumar, J. Laser-induced holographic surface relief gratings on nonlinear optical polymer films. *Appl. Phys. Lett.* **1995**, *66*, 1166–1168. [CrossRef]

7. Kumar, J.; Li, L.; Jiang, X.L.; Kim, D.Y.; Lee, T.S.; Tripathy, S. Gradient force: The mechanism for surface relief grating formation in azobenzene functionalized polymers. *Appl. Phys. Lett.* **1998**, *72*, 2096–2098. [CrossRef]
8. Barrett, C.J.; Rochon, P.; Natansohn, A. Model of laser-driven mass transport in thin films of dye-functionalized polymers. *J. Chem. Phys.* **1998**, *109*, 1505–1516. [CrossRef]
9. Baldus, O.; Leopold, A.; Hagen, R.; Bieringer, T.; Zilker, S.J. Surface relief gratings generated by pulsed holography: A simple way to polymer nanostructures without isomerizing side-chains. *J. Chem. Phys.* **2001**, *114*, 1344–1349. [CrossRef]
10. Bublitz, D.; Fleck, B.; Wenke, L. A model for surface-relief formation in azobenzene polymers. *Appl. Phys. B* **2001**, *72*, 931–936. [CrossRef]
11. Pedersen, T.G.; Johansen, P.M.; Holme, N.C.R.; Ramanujam, P.S.; Hvilsted, S. Mean-field theory of photoinduced formation of surface reliefs in side-chain azobenzene polymers. *Phys. Rev. Lett.* **1998**, *80*, 89–92. [CrossRef]
12. Oscurato, S.L.; Salvatore, M.; Borbone, F.; Maddalena, P.; Ambrosio, A. Computer-generated holograms for complex surface reliefs on azopolymer films. *Sci. Rep.* **2019**, *9*, 6775. [CrossRef]
13. Hubert, C.; Rumyantseva, A.; Lerondel, G.; Grand, J.; Kostcheev, S.; Billot, L.; Vial, A.; Bachelot, R.; Royer, P.; Chang, S.H.; et al. Near-field photochemical imaging of noble metal nanostructures. *Nano Lett.* **2005**, *5*, 615–619. [CrossRef]
14. Nedelchev, L.; Mateev, G.; Strijkova, V.; Salgueiriño, V.; Schmool, D.S.; Berberova-Buhova, N.; Stoykova, E.; Nazarova, D. Tunable Polarization and Surface Relief Holographic Gratings in Azopolymer Nanocomposites with Incorporated Goethite (α-FeOOH) Nanorods. *Photonics* **2021**, *8*, 306. [CrossRef]
15. Stoilova, A.; Mateev, G.; Nazarova, D.; Nedelchev, L.; Stoykova, E.; Blagoeva, B.; Berberova, N.; Georgieva, S.; Todorov, P. Polarization holographic gratings in PAZO polymer films doped with particles of biometals. *J. Photochem. Photobiol. A Chem.* **2021**, *411*, 113196. [CrossRef]
16. Berberova-Buhova, N.; Nedelchev, L.; Mateev, G.; Stoykova, E.; Strijkova, V.; Nazarova, D. Influence of the size of Au nanoparticles on the photoinduced birefringence and diffraction efficiency of polarization holographic gratings in thin films of azopolymer nanocomposites. *Opt. Mater.* **2021**, *121*, 111560. [CrossRef]
17. Yi, D.K.; Seo, E.M.; Kim, D.Y. Surface-modulation-controlled three-dimensional colloidal crystals. *Appl. Phys. Lett.* **2002**, *80*, 225–227. [CrossRef]
18. Stiller, B.; Geue, T.; Morawetz, K.; Saphiannikova, M. Optical patterning in azobenzene polymer films. *J. Microsc.* **2005**, *219*, 109–114. [CrossRef]
19. Kim, S.S.; Chun, C.; Hong, J.C.; Kim, D.Y. Well-ordered TiO_2 nanostructures fabricated using surface relief gratings on polymer films. *J. Mater. Chem.* **2006**, *16*, 370–375. [CrossRef]
20. Huang, B.Y.; Yu, K.Y.; Huang, S.Y.; Kuo, C.T. The investigation of the two-dimensional surface relief grating on dye-doped polymer film. *Opt. Mater. Express* **2014**, *4*, 308–314. [CrossRef]
21. Sakhno, O.; Goldenberg, L.M.; Wegener, M.; Dreyer, C.; Berdin, A.; Stumpe, J. Generation of sub-micrometer surface relief gratings in an azobenzene-containing material using a 355 nm laser. *Opt. Mater.* **2022**, *128*, 112457. [CrossRef]
22. Kaban, B.; Bagatur, S.; Soter, M.; Hillmer, H.; Fuhrmann-Lieker, T. Reversible Photo-Induced Reshaping of Imprinted Microstructures Using a Low Molecular Azo Dye. *Polymers* **2022**, *14*, 586. [CrossRef]
23. Ikeda, T.; Tsutsumi, O. Optical switching and image storage by means of azobenzene liquid-crystal films. *Science* **1995**, *268*, 1873–1875. [CrossRef]
24. Nedelchev, L.; Todorov, T.; Nikolova, L.; Petrova, T.; Tomova, N.; Dragostinova, V. Characteristics of high-efficient polarization holographic gratings. *Proc. SPIE* **2001**, *4397*, 338–342. [CrossRef]
25. Nedelchev, L.; Mateev, G.; Nikolova, L.; Nazarova, D.; Ivanov, B.; Strijkova, V.; Stoykova, E.; Choi, K.; Park, J. In-line and off-axis polarization-selective holographic lenses recorded in azopolymer thin films via polarization holography and polarization multiplexing. *Appl. Opt.* **2023**, *62*, D1–D7. [CrossRef]
26. Lausten, R.; Rochon, P.; Ivanov, M.; Cheben, P.; Janz, S.; Desjardins, P.; Ripmeester, J.; Siebert, T.; Stolow, A. Optically reconfigurable azobenzene polymer-based fiber Bragg filter. *Appl. Opt.* **2005**, *44*, 7039–7042. [CrossRef]
27. Minkov, D.; Nedelchev, L.; Angelov, G.; Marquez, E.; Blagoeva, B.; Mateev, G.; Nazarova, D. Hybrid dispersion model characterization of PAZO polymer thin films over the entire transmittance spectrum measured in the UV/VIS/NIR spectral region. *Materials* **2022**, *15*, 8617. [CrossRef]
28. Nedelchev, L.; Nazarova, D.; Mateev, G.; Berberova, N. Birefringence induced in azopolymer (PAZO) films with different thickness. *Proc. SPIE* **2015**, *9447*, 94471I. [CrossRef]
29. Yadavalli, N.S.; Santer, S. In-situ atomic force microscopy study of the mechanism of surface relief grating formation in photosensitive polymer films. *J. Appl. Phys.* **2013**, *113*, 224304. [CrossRef]
30. Nedelchev, L.; Ivanov, D.; Berberova, N.; Strijkova, V.; Nazarova, D. Polarization holographic gratings with high diffraction efficiency recorded in azopolymer PAZO. *Opt. Quantum Electron.* **2018**, *50*, 212. [CrossRef]

photonics

MDPI

Article

Study of the Interference Color with Nomarski Prism Wedge Angle in a Differential Interference Contrast Microscopy System

Fei Li and Tingyu Zhao *

Department of Physics, Key Laboratory of Optical Field Manipulation of Zhejiang Province, Zhejiang Sci-Tech University, 928 2nd Street, Hangzhou 310018, China
* Correspondence: zhaotingyu@zstu.edu.cn

Abstract: Differential interference contrast microscopy systems demonstrate the phase (optical path) rather than the amplitude of a sample. Previous studies have usually approximated the optical path difference produced by the Normarski prism. We derive the mathematical expression for the optical path difference produced by the incident light at any position of the Nomarski prism. As a result, the optical path difference introduced by the differential interference contrast microscope can be calculated. Moreover, the optical path difference is linearly related to the position of the incident light. In addition, the differential interference contrast microscopy system uses a composite light source, while previous studies were basically performed at a single wavelength. The standard wavelengths d (656 nm), F (587 nm), and C (486 nm) are taken as examples to analyze the relationship between the interference color change of the observation surface under the superposition of different wavelengths and the prism wedge angle when the prism moves. When the prism moves the same distance, the larger the prism wedge Angle is, the faster the interference color change is. The analysis based on practical considerations in this paper is believed to provide a method for studying Nomarski prisms.

Keywords: differential interference contrast; Nomarski prism; plane of apparent splitting; splitting angle

check for updates

Citation: Li, F.; Zhao, T. Study of the Interference Color with Nomarski Prism Wedge Angle in a Differential Interference Contrast Microscopy System. *Photonics* **2023**, *10*, 678. https://doi.org/10.3390/photonics10060678

Received: 2 May 2023
Revised: 4 June 2023
Accepted: 8 June 2023
Published: 11 June 2023

1. Introduction

In recent years, reflective differential interference contrast (DIC) digital microscopy system has been widely used in precision processing and inspection fields, such as the observation of conductive particles in thin-film transistor liquid crystal displays [1], the measurement of martensite microstructures, and the classification of individual dielectric nanoparticles [2,3]. DIC microscopy system can be used for qualitative or quantitative measurements of samples [4–7], as well as for reflective surface assessment [8] and optical profiling [9].

As one of the main 3D morphological tools available for microsurface inspection, differential interference contrast microscopy [10] can transform small phase or optical path length variations on the object surface into light intensity, i.e., the phase or optical path length variations on the object surface are transformed into intensity variations due to the interference of light from two closely separated points on the object surface. It shows the subtle bumps on the sample surface with strong light intensity and interference color variations. The images captured by the DIC microscopy system have a distinctive shadow effect and a strong three-dimensional relief.

Studying the Nomarski prism is of utmost importance as it serves a critical function in differential interference contrast microscopy. Numerous researchers have conducted meticulous studies on the design theory of Nomarski prisms and engaged in DIC microscopy [11–20]. Early studies proposed simplified mathematical theoretical models for DIC imaging, such as the DIC imaging model constructed based on differential methods [21,22]. These studies provided a good research basis and reference value for accurate models in later research. However, these imaging models did not take into account the phase difference produced by the light in the Nomarski prism itself, so they were only suitable for qualitative observation of the sample and could not be used for quantitative measurements.

In order to use the differential interference contrast microscopy for quantitative measurements, the phase change due to the Nomarski prism must be calculated. The above theoretical study only calculated the phase change for the half of the light path, i.e., the linearly polarized light incident to the prism and out to the intersection of the two linearly polarized beams, and did not take into account the light incident to the Nomarski prism, reflected from the sample surface and then returned to the Nomarski prism, and the outgoing process. Hartman [23] makes an approximation to the phase change caused by the Nomarski prism and does not obtain an exact solution. In this paper, the theoretical calculation of the phase difference arising during the passage of the incident light through the prism to the horizontal viewing plane and back to the prism is investigated based on the incident light perpendicular to the plane of apparent splitting, covering the entire light path and obtaining an algebraic solution.

In addition, previous theoretical studies of differential interference contrast microscopy systems have been dominated by single wavelength analysis. Practical differential interference contrast microscopy systems usually use a composite light source containing multiple wavelengths, such as white light, rather than a monochromatic source. The standard wavelengths d (656 nm), F (587 nm), and C (486 nm) are taken as examples to analyze the relationship between the interference color change of the observation surface under the superposition of different wavelengths and the prism wedge angle when the prism moves

The research content of this paper includes three parts. In Section 2, we derive an expression between the prism parameters based on a geometric model of incident light perpendicular to the plane of apparent splitting and then derive the relationship between the optical path difference produced by the light in the prism as it is incident to the prism and reflected the prism through the image plane and the position of the incident light. In Section 3, we discuss the effect of different prism wedge angles on the interference color when the prism is moved at the same distance under a superposition of light intensities of multiple wavelengths. We briefly draw a conclusion in Section 4.

2. Methods

Figure 1 shows a diagram of the incidence plane of a Nomarski prism with a wedge angle of γ, the thickness of the prism d, δ the angle between the crystal axis of the first crystal and the lower surface (axially orthogonal), and the optical axis of the second crystal perpendicular to the incidence plane, which divides the optical system into zones 1, 2, 3 and 4. Assuming that the linearly polarized light is incident on the prism at an angle of incidence α, ε is the angle of separation between the ordinary(o) and extraordinary(e) rays, where the principal refractive index of the ordinary ray is n_o, and the principal refractive index of the extraordinary ray is n_e.

First, we discuss the component of light whose vibrational direction is perpendicular to the incident surface—ray o in zone 2 and ray e in zone 3, and we can then obtain the exit angle β_{oe}, and the subscript oe indicates that the component of polarized light changes from ray o (zone 2) to ray e (zone 3). We call it ray oe.

$$\beta_{oe} = \sin^{-1}\left[n_e \sin\left[\sin^{-1}\left\{\frac{n_o}{n_e}\sin\left[\sin^{-1}\left[\frac{\sin\alpha}{n_o}\right]+\gamma\right]\right\}-\gamma\right]\right] \tag{1}$$

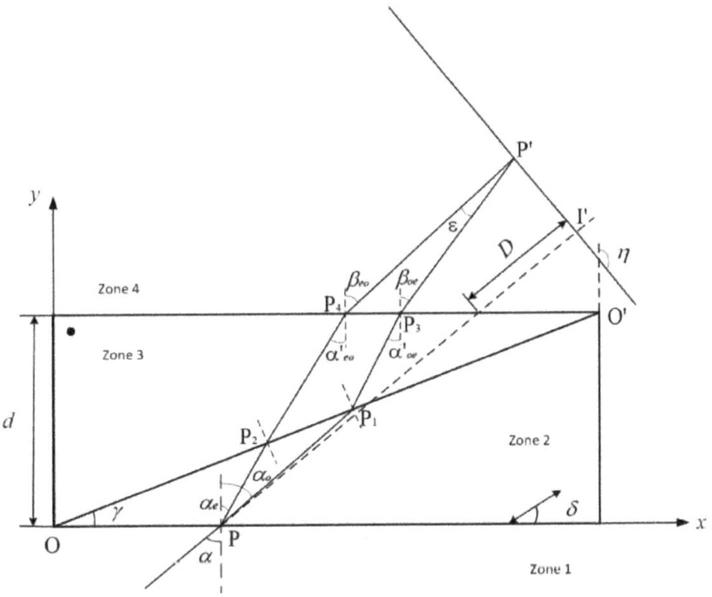

Figure 1. Optical schematic diagram of the Nomarski prism.

Then, we focus on the light component, whose vibrational direction is parallel to the incident surface, ray *e* in zone 2 and ray *o* in zone 3, and we can then obtain the exit angle β_{eo}, and the subscript *eo* indicates that the component of polarized light changes from ray *e* (zone 2) to ray *o* (zone 3). We call it ray *eo*.

$$\beta_{eo} = \sin^{-1}\left[n_o \sin\left[\sin^{-1}\left\{\frac{n_\theta}{n_o}\sin\left[\sin^{-1}\left[\frac{\sin\alpha}{n_\theta}\right]+\gamma\right]\right\}-\gamma\right]\right] \tag{2}$$

where, in Equation (2)

$$n_\theta = \frac{n_o n_e}{\left(n_o^2 \sin^2\theta + n_e^2 \cos^2\theta\right)^{1/2}} \tag{3}$$

where θ is the angle between the direction normal to the ray *e* light in zone 2 and the crystal optical axis.

The detailed derivation process of Equations (1) and (2) is shown in Appendix A.

Therefore, the beam splitting angle ε after the two beams exit through the prism is

$$\varepsilon = \beta_{oe} - \beta_{eo} \tag{4}$$

The right-angle coordinate system is shown in Figure 1, and η represents the angle between the ray intersection plane and the positive direction of the y-axis. Let the coordinates of the incident point be $P(X_0, 0)$, the coordinates of rays *o* and *e* on the second dividing plane are $P_1(X_o, Y_o)$ and $P_2(X_e, Y_e)$, respectively; the coordinates of rays *oe* and *eo* on the third dividing plane are $P_3(X_{oe}, Y_{oe})$ and $P_4(X_{eo}, Y_{eo})$, respectively; the two beams intersect at the point $P'(X_0', Y_0')$.

The geometric path of light in zone 2 is

$$t_o = \overline{PP_1} = \frac{X_0 \sin\gamma}{\cos(\alpha_o + \gamma)} \tag{5}$$

where α_o is the refraction angle of ray o in zone 2; then, the coordinates of point P_1 as

$$\begin{aligned} X_o &= X_0 + t_o \sin\alpha_o \\ Y_o &= t_o \cos\alpha_o \end{aligned} \tag{6}$$

Similarly, the coordinates of point P_3 can be found as

$$\begin{aligned} X_{oe} &= X_0 + t_o \sin\alpha_o + (d - Y_o)\tan\alpha'_{oe} \\ Y_{oe} &= d \end{aligned} \tag{7}$$

where α'_{oe} is the angle of incidence at the junction of zone 3 and zone 4, and the geometric path of ray e in zone 2 is

$$t_e = \overline{PP_2} = \frac{X_0 \sin\gamma}{\cos(\alpha_e + \gamma)} \tag{8}$$

where α_e is the refraction angle of ray e in zone 2; then, the coordinates of point P_2 as

$$\begin{aligned} X_e &= X_0 + t_e \sin\alpha_e \\ Y_e &= t_e \cos\alpha_e \end{aligned} \tag{9}$$

Similarly, the coordinates of point P_4 can be found as

$$\begin{aligned} X_{eo} &= X_0 + t_e \sin\alpha_e + (d - Y_e)\tan\alpha'_{eo} \\ Y_{eo} &= d \end{aligned} \tag{10}$$

where α'_{eo} is the angle of incidence at the junction of zone 3 and zone 4, and the final two outgoing beams are determined by the following set of equations:

$$\begin{aligned} Y'_0 &= d + \tfrac{X'_0 - X_{oe}}{\tan\beta_{oe}} \\ Y'_0 &= d + \tfrac{X'_0 - X_{eo}}{\tan\beta_{eo}} \end{aligned} \tag{11}$$

Solving the coordinates of the intersection point of the two beams in Equation (11), we can obtain the following:

$$\begin{aligned} X'_0 &= \tfrac{X_{oe}\tan\beta_{eo} - X_{eo}\tan\beta_{oe}}{\tan\beta_{eo} - \tan\beta_{oe}} \\ Y'_0 &= d + \tfrac{X'_0 - X_{eo}}{\tan\beta_{eo}} \end{aligned} \tag{12}$$

The distance between the upper surface of the prism and the intersecting surface of the outgoing ray is $D = (Y'_0 - \frac{X'_0 - X_0}{\tan(90+\alpha)})\cos\alpha - \frac{d}{\cos\alpha}$.

The two polarized beams of light pass through the Nomarski prism and produce an optical path difference. The optical path of ray oe is

$$\begin{aligned} OPL_{oe} &= n_o\sqrt{(X_o - X_0)^2 + Y_o^2} + n_e\sqrt{(X_{oe} - X_o)^2 + (d - Y_o)^2} \\ &+ \sqrt{(X'_0 - X_{oe})^2 + (Y'_0 - d)^2} \end{aligned} \tag{13}$$

The optical path of ray eo is

$$\begin{aligned} OPL_{eo} &= n_\theta\sqrt{(X_e - X_0)^2 + Y_e^2} + n_o\sqrt{(X_{eo} - X_e)^2 + (d - Y_e)^2} \\ &+ \sqrt{(X'_0 - X_{eo})^2 + (Y'_0 - d)^2} \end{aligned} \tag{14}$$

Thus, the optical path difference between the two linearly polarized beams is

$$\Gamma = OPL_{oe} - OPL_{eo} \tag{15}$$

The phase difference is

$$\varphi = \frac{2\pi}{\lambda}\Gamma \tag{16}$$

In reflective differential interference contrast microscopy systems, as shown in Figure 2, light incident into the Nomarski prism gathers through the objective lens to the middle image plane, and then it reflects to the prism through the middle image plane and leaves the prism. The optical path difference mainly comes from the process of incidence and refraction of light in the prism.

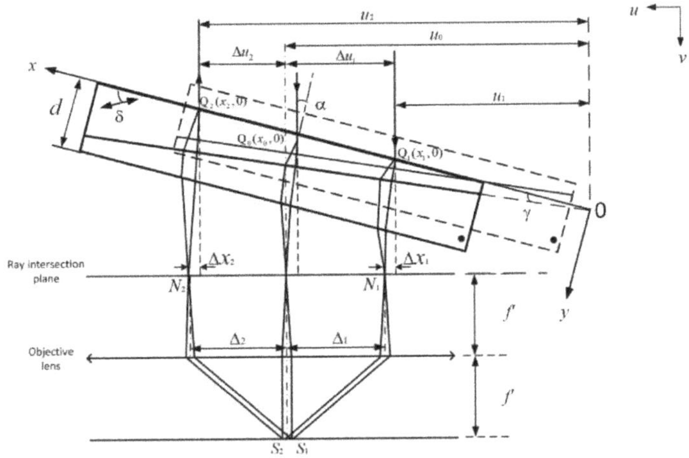

Figure 2. The schematic of the Nomarski prism, objective lens, and reflected light.

As shown in Figure 2, the light incident from point $Q_1(x_1, 0)$ into the Nomarski prism converges to the middle image plane through the objective lens and then reflects to the prism through the middle image plane and leaves the prism from point $Q_2(x_2, 0)$. In the whole process, Fermat's principle states that the actual path of a light ray between two points gives optical path length a stationary value. Since the family of rays from N_1 to N_2 are all actual paths, they all have the same optical path length. So, there is no optical path difference between the rays $N_1S_1N_2$ and $N_1S_2N_2$. The main optical path difference comes from two parts: the light incident into the prism, which refracted through the prism and reached the ray intersection plane, and the reflected light from the ray intersection plane to the prism, which emerges through the prism. Firstly, the optical path difference generated by the process of light incident from point $Q_1(x_1, 0)$ to the prism and reaching the ray intersection plane through prism refraction can be calculated using Equation (15). Because the optical path difference Γ caused by an incident on the Nomarski prism has a linear relationship with the incident coordinates x, when the incident coordinate x_1 is known, the coordinate x_2 of the reflected light on the prism can be calculated by geometric relation in Figure 2. We next calculate the optical path difference. When the prism moves horizontally along the ray intersection plane directly to the prism with the dotted line in Figure 2, the position of the incident light relative to the optical axis does not change because the optical axis of the system does not change, but the position relative to the prism surface changes, so the optical path difference also changes.

As shown in Figure 1, PI′ is the extension line of the incident light, and the incident point $P(X_0, 0)$ is set. The expression of the extension line of the incident ray can be obtained in the coordinate system based on a prism:

$$\begin{aligned} y_1 &= k_1 x + c_1 \\ k_1 &= \tan(90^\circ - \alpha) \\ c_1 &= -X_0 \tan(90^\circ - \alpha) \end{aligned} \tag{17}$$

where k_1 represents the slope and c_1 represents the intercept of the incident ray.

Because the plane of intersection of the ray of the Nomarski prism in the differential interference contrast microscope coincides with the rear focal plane of the objective lens we need to make the incident ray extend perpendicular to the plane of intersection of ray:

$$\eta - \alpha = \frac{\pi}{2} \tag{18}$$

The ray direction satisfying the above formula is the optical axis direction of the differential interference contrast optical system so that we can find the linear expression of the plane of intersection of rays:

$$\begin{aligned} y_2 &= k_2 x + c_2 \\ k_2 &= -\tfrac{1}{k_1} \\ c_2 &= Y_0' - k_2 X_0' \end{aligned} \tag{19}$$

where k_2 is the slope of the line between P′ and I′, and c_2 is the value at which the line intersects the positive y-axis.

By connecting (17) and (19), the intersection coordinate $\mathrm{I}'(x_j, y_j)$ can be obtained as

$$\begin{aligned} x_j &= -\tfrac{c_1 - c_2}{k_1 - k_2} \\ y_j &= k_1 x_j + c_1 \end{aligned} \tag{20}$$

From the coordinates of P′ and I′, the offset $\overline{\mathrm{P'I'}}$, represented by Δx, can be obtained from the geometric relationship in Figure 2.

$$\Delta x = \overline{\mathrm{P'I'}} = \frac{y_j - Y_0'}{\sin\alpha} \tag{21}$$

We take the wedge angle $\gamma = 0.4°$, the optical axis inclination angle $\delta = 100.5°$, and the prism thickness of the Nomarski prism as 2 mm; the optimized parameters are the light incidence angle $\alpha = -6.1032°$ and the plane distance between the upper surface of the prism and the ray D = 25.9643 mm. By substituting the above data into Equations (15) and (21), the offset Δx generated by different incident light positions passing through the prism can be obtained, as shown in Figure 3.

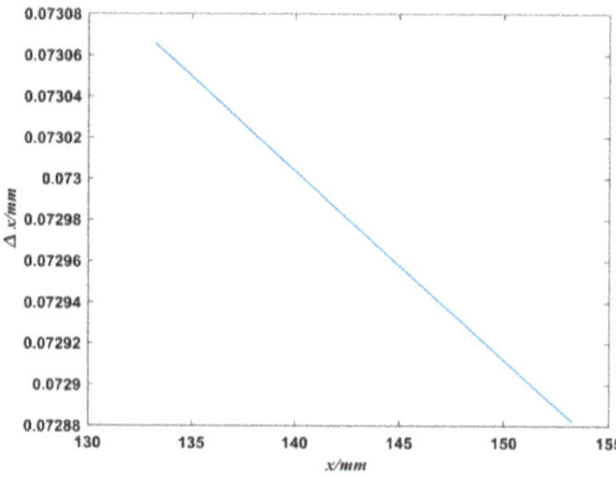

Figure 3. The relationship of the offset Δx and the ray incident position x.

As can be seen from Figure 3, Δx has a linear relationship with x. That is, the offset generated by each incident point through the prism (the difference between the incident ray extension line and its intersection point on the plane of intersection of ray) can be determined using Equation (21).

Figure 2 shows that, when the Nomarski prism is placed at an angle (for the purpose of parallel coincidence between the ray intersection plane and the objective lens), light incident into the prism passes through the objective lens to the viewing surface and then reflects from the viewing surface to the prism and out. In Figure 2, u_1 corresponds to the incident light, and u_2 corresponds to the light emitted through the prism after reflection, with the following relation expression:

$$\begin{aligned} x_0 &= \frac{u_0}{\cos\alpha} \\ x_1 &= \frac{u_1}{\cos\alpha} \\ x_2 &= \frac{u_2}{\cos\alpha} \end{aligned} \tag{22}$$

where x_0, x_1, and x_2 are the values in the x-y coordinate system; u_0, u_1, and u_2 are the values in the u-v coordinate system. According to the geometric relation in Figure 2

$$\begin{aligned} \Delta u_1 &= u_0 - u_1 = (x_0 - x_1)\cos\alpha \\ \Delta u_2 &= u_2 - u_0 = (x_2 - x_0)\cos\alpha \end{aligned} \tag{23}$$

Similarly

$$\begin{aligned} \Delta_1 &= \Delta u_1 + \Delta x_1 \\ \Delta_2 &= \Delta u_2 + \Delta x_2 \end{aligned} \tag{24}$$

Because $\Delta_2 = \Delta_1$, we have,

$$(x_0 - x_1)\cos\alpha + \Delta x_1 = (x_2 - x_0)\cos\alpha + \Delta x_2 \tag{25}$$

As can be seen from Equation (21), in the x-y coordinate system, each x value corresponds to a Δx, and the relationship between them is linear, so Δx_2 can be simplified into the following relation:

$$\Delta x_2 = kx_2 + c \tag{26}$$

From Equations (25) and (26), we can calculate the value of x in the x-y coordinate system when the ray returns.

$$x_2 = \frac{(x_0 - x_1)\cos\alpha + \Delta x_1 - c + x_0\cos\alpha}{\cos\alpha + k} \tag{27}$$

When x_1 and x_2 values are known, the corresponding optical path difference can be determined by substituting in the equation. Then, the optical path difference generated during the whole process of light incident into the Nomarski prism and exiting is

$$\Gamma_{sum} = \Gamma_{x_1} + \Gamma_{x_2} \tag{28}$$

The phase difference is

$$\chi = \frac{2\pi}{\lambda}\Gamma_{sum} \tag{29}$$

The function expression of the change in light intensity of the observed surface along with the phase is as follows [24].

$$\begin{aligned} I &= I_{\max}[Q + \tfrac{1}{2}(1 - Q)(1 - \cos\chi)] \\ Q &= \frac{I_{\min}}{I_{\max}} \end{aligned} \tag{30}$$

For an ideal optical system, we have $I_{\min} = 0$, $I_{\max} = 1$, $Q = 0$. Therefore, Equation (30) can be further simplified as

$$I = [\frac{1}{2}(1 - \cos\chi)] \tag{31}$$

Similarly, we take the Nomarski prism with wedge angle $\gamma = 0.4°$, optical axis inclination $\delta = 100.5°$, and prism thickness of 2 mm, and from Equation (28), we can obtain the relationship between the optical path difference Γ_{sum} generated during the whole process of light incident to the Nomarski prism and outgoing and the light incident coordinate x, as shown in Figure 4. Γ_{sum} is linearly related to x.

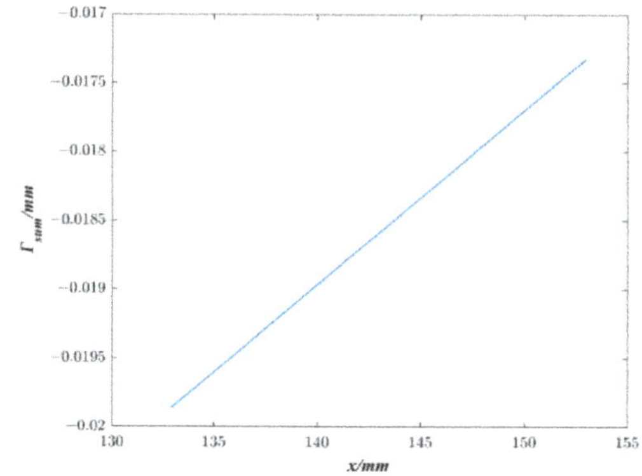

Figure 4. The relationship of optical path difference Γ_{sum} and the ray incident position x.

3. Simulation Results and Analysis

In the use of differential interference contrast (DIC) microscopy systems, it is necessary to adjust the position of the Normaski prism in the direction perpendicular to the optical axis to introduce a suitable optical path difference to obtain the best sample image. We found that the adjustment of the prism position brings about a change in the interference color. A specific analysis is presented below.

Assume that the prism length is 20 mm, and the thickness is d. As shown in Figure 5, the point where the intersecting prisms meet is considered the origin of coordinates in the absolute coordinate system. Half of the thickness is used as the center of the prism. If the prism wedge angle is γ_1, then the prism center transverse coordinate is x_1, and the prism is located at $[x_1 - 10, x_1 + 10]$, as shown in the gray area in Figure 5; if the prism wedge angle is γ_2, then the prism center transverse coordinate is x_2, and the prism is located at $[x_2 - 10, x_2 + 10]$, as shown in the green area in Figure 5 With the prism center as the initial position, it moves no further than the length of the prism itself, i.e., [−10 mm, 10 mm]. Therefore, we now use relative coordinates instead of absolute coordinates. Let the prism center be the zero point in relative coordinates, then its moving range is [−10 mm, 10 mm], as shown in Figure 6.

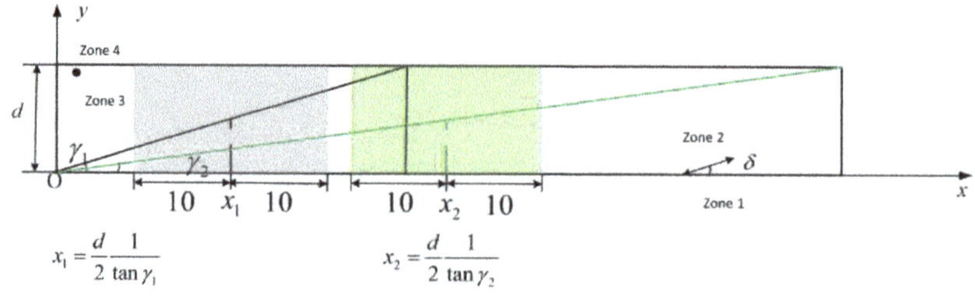

Figure 5. Schematic of the Nomarski prism.

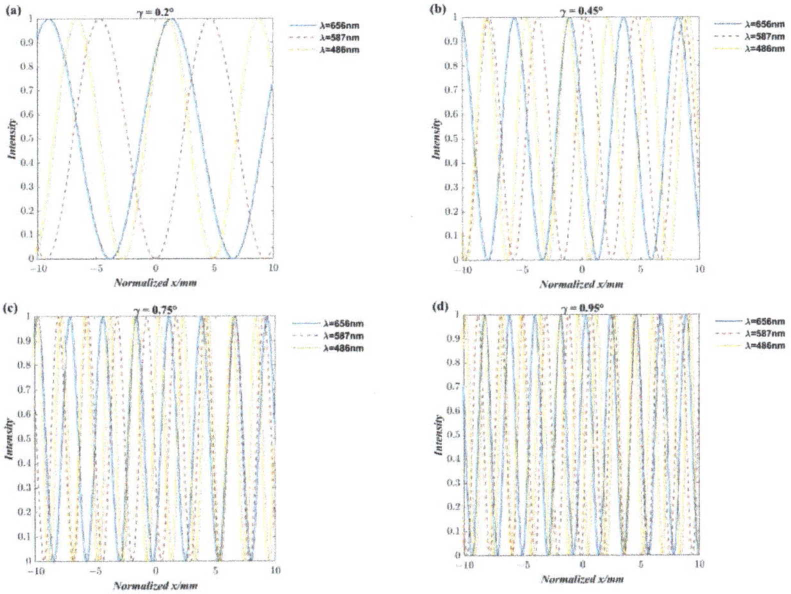

Figure 6. The relation diagram of prism position and light intensity at different wavelengths: (**a**) $\gamma = 0.2°$, (**b**) $\gamma = 0.45°$, (**c**) $\gamma = 0.75°$, and (**d**) $\gamma = 0.95°$.

Based on the above theory, three wavelengths of red light ($\lambda = 656$ nm), yellow-orange light ($\lambda = 587$ nm), and blue light ($\lambda = 486$ nm) in white light were selected. At the same time, the angle of the optical axis in the Nomarski prism was $\delta = 100.5°$, the thickness of the prism was 2 mm, and the wedge angle γ was 0.2°, 0.45°, 0.75°, and 0.95°, respectively. As shown in Figure 6, the horizontal coordinates indicate the position of the prism in the x-y coordinate system shown in Figure 2, and the difference between different coordinate values indicates the prism moves to the distance, and the vertical coordinates indicate the light intensity, where it can be seen from the comparison of (a), (b), (c) and (d) that the light intensity changes for more periods when the prism wedge angle is larger, and the prism moves the same distance.

As shown in Figure 7a, when the wedge angle $\gamma = 0.2°$, the prism moves from left to right, i.e., from -10 mm to 10 mm, and the interference color changes in the order of "red-cyan-green-black-rose-green-blue-violet." As shown in Figure 7b–d, the interference color changes faster as the wedge angle becomes larger when the prisms move the same distance. Taking the change from red to rose red as an example when the wedge angle γ is 0.2°, 0.45°, 0.75°, and 0.95°, the required moving distance is about 4.8 mm, 2.2 mm, 1.4 mm, and 1 mm, respectively. The larger the wedge angle, the shorter the moving distance required to change the same color. In other words, the larger the wedge angle is, the faster the interference color changes, provided that the same distance is moved. The results suggest that the prism wedge angle should be made as small as possible in the actual design of Nomarski. Otherwise, when the prism moves a tiny distance, it will produce multiple color changes in the observation screen, and it is not easy to find the best prism position for imaging.

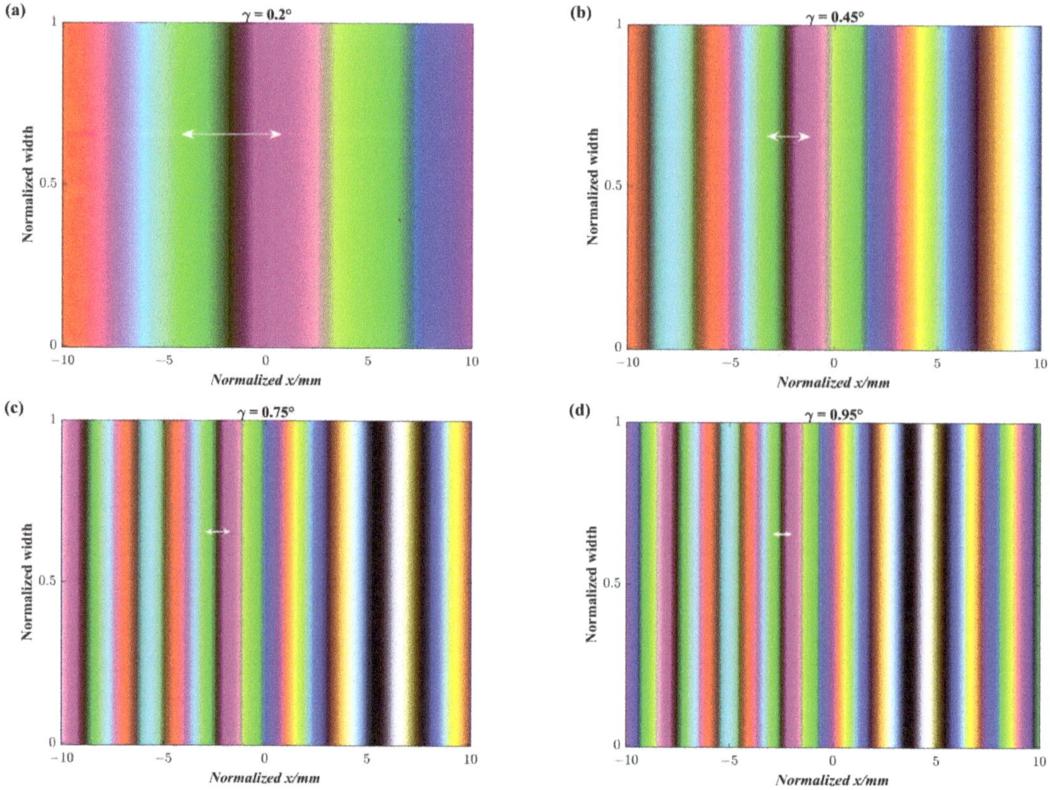

Figure 7. Schematic diagram of phase change light intensity superposition generated by prism movement: (**a**) $\gamma = 0.2°$, (**b**) $\gamma = 0.45°$, (**c**) $\gamma = 0.75°$, and (**d**) $\gamma = 0.95°$.

4. Conclusions

In differential interference contrast microscopy, the design of the Nomarski prism is a crucial technique. In practical application, it is not only necessary to make the incident light perpendicular to the plane of intersection of the light rays outside the prism but also to make the plane of intersection of the light rays coincide with the rear focal plane of the objective lens. Based on this principle, this paper deduces the optical path difference generated when the light incident on the prism is reflected from the observing surface and then comes out of the prism, and also discusses the relationship between the variation in the light intensity of the observing surface and the wedge angle of the prism when the prism moves at multiple wavelengths. The results suggest that the prism wedge angle should be made as small as possible in the actual design of Nomarski. Otherwise, when the prism moves a tiny distance, it will produce multiple color changes in the observation screen, and finding the best prism position for imaging is challenging.

Author Contributions: Conceptualization, T.Z. and F.L.; methodology, F.L.; software, F.L.; validation, F.L., and T.Z.; formal analysis, F.L.; investigation, F.L.; resources, T.Z.; data curation, F.L.; writing—original draft preparation, F.L.; writing—review and editing, F.L. and T.Z.; visualization, F.L.; supervision, T.Z.; project administration, F.L. All authors have read and agreed to the published version of the manuscript.

Funding: This research received no external funding.

Institutional Review Board Statement: Not applicable.

Informed Consent Statement: Not applicable.

Data Availability Statement: Data underlying the results presented in this paper are not publicly available at this time but may be obtained from the authors upon reasonable request.

Conflicts of Interest: The authors declare no conflict of interest.

Appendix A

When a beam of light is incident on a prism, it is divided into two beams, called o light and e light, due to the anisotropy of the light in the prism. As shown in Figure 1, linearly polarized light whose vibration direction is perpendicular to the paper surface is first discussed. This component is o light in zone 2 and has a refractive index of is n_o, which can be known from the refraction law when it passes through zone 1 and zone 2:

$$\begin{aligned} n_1 \sin \alpha &= n_o \sin \alpha_o \\ \alpha_o &= \sin^{-1}\left(\tfrac{n_1}{n_o}\sin\alpha\right) \end{aligned} \tag{A1}$$

where n_1 is the refractive index of linearly polarized light in zones 1 and 4, and since it is in air, its value is 1.

As shown in Figure 1, there is a geometric relation in ΔOPP_1

$$\alpha_o' = \alpha_o + \gamma \tag{A2}$$

In zone 3, the crystal optical axis is perpendicular to the paper surface, and the linearly polarized light component perpendicular to the paper surface is parallel to the crystal optical axis, which changes from o light to e light. Normally e light does not follow the law of refraction, so the law of refraction cannot be used. However, when the light propagation direction of e light is perpendicular to the direction of the crystal's optical axis, it follows the law of refraction. The direction of vibration of the e light in zone 3 is perpendicular to the direction of the crystal's optical axis, and the direction of the wave normal and the direction of the light coincide, so the line of polarized light follows the law of refraction, and the index of refraction at this time is n_e. Then, there is the following relationship between the interface zone 2 and zone 3:

$$\begin{aligned} n_o \sin \alpha_o' &= n_e \sin \alpha_{oe} \\ \alpha_{oe} &= \sin^{-1}\left(\tfrac{n_o}{n_e}\sin\alpha_o'\right) \end{aligned} \tag{A3}$$

In $\Delta P_1P_3O'$ by the geometric relation, there is

$$\alpha_{oe}' = \alpha_{oe} - \gamma \tag{A4}$$

The refraction law is used at the interface between zone 3 and zone 4 and can be obtained from Equations (A1)–(A4)

$$\begin{aligned} n_1 \sin \beta_{oe} &= n_e \sin \alpha_{oe}' \\ \beta_{oe} &= \sin^{-1}\left[n_e \sin\left[\sin^{-1}\left\{\tfrac{n_o}{n_e}\sin\left[\sin^{-1}\left[\tfrac{\sin\alpha}{n_o}\right]+\gamma\right]\right\}-\gamma\right]\right] \end{aligned} \tag{A5}$$

For linearly polarized light whose direction of vibration is parallel to the surface of the paper. In zone 2 of Figure 1, the vibration direction of this component is in the same plane as the optical axis, so it is ray e at this time. The ray e propagates at an angle to the optical axis of the crystal. In this case, the direction of the wave normally will be separated from the light. Let the angle between the wave normal and the incident light normal be α_k. Therefore, the relationship between the direction of the light, the direction of the optical axis, and the direction of the wave normal is shown in Figure A1. The angle between the wave normal and the optical axis is θ.

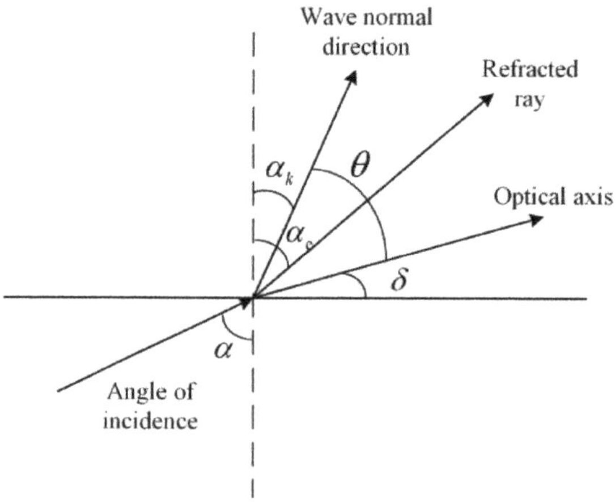

Figure A1. The relationship between the optical axis, wave normal, and the direction of light rays.

Because the direction of the wave normal obeys the law of refraction, the wave normal is used to derive the angle of *e*-light emission. The partition interface between zone 1 and zone 2 has

$$\begin{gathered} n_1 \sin\alpha = n_\theta \sin\alpha_k \\ \alpha_k = \sin^{-1}\left(\frac{n_1}{n_\theta}\sin\alpha\right) \end{gathered} \tag{A6}$$

where $n_\theta = \frac{n_o n_e}{(n_o^2 \sin^2\theta + n_e^2 \cos^2\theta)^{1/2}}$, according to the geometric relation in Figure A1, it can be obtained.

According to Figure 1, in the OPP by the geometric relationship, there is

$$\alpha_k' = \alpha_k + \gamma \tag{A7}$$

Since the vibration direction of the linearly polarized light perpendicular to the paper surface is the same as the direction of the optical axis in zone 3, the polarized light component changes from *e* light in zone 2 to *o* light in zone 3 at this time. It follows the law of refraction and has the following relationship at the interface between zone 2 and zone 3:

$$\begin{gathered} n_\theta \sin\alpha_k' = n_o \sin\alpha_{ko}, \\ \alpha_{ko} = \sin^{-1}\left(\frac{n_\theta}{n_o}\sin\alpha_k'\right). \end{gathered} \tag{A8}$$

Using the law of refraction at the partition between zones 3 and 4,

$$n_1 \sin\beta_{ko} = n_e \sin\alpha_{ko}' \tag{A9}$$

where α_{ko}' is obtained from the geometric relationship of $\Delta P_2P_4O'$ in Figure 1.

The direction of the wave normal in zone 3 is the same as the direction of the light, and the angle of emergence when the light leaves the interface between zone 3 and zone 4 can be obtained from Equations (A6)–(A9).

$$\beta_{eo} = \beta_{ko} = \sin^{-1}\left[n_o \sin\left[\sin^{-1}\left\{\frac{n_\theta}{n_o}\sin\left[\sin^{-1}\left[\frac{\sin\alpha}{n_\theta}\right] + \gamma\right]\right\} - \gamma\right]\right] \tag{A10}$$

References

1. Ni, G.; Liu, L.; Du, X.; Zhang, J.; Liu, J.; Liu, Y. Accurate AOI inspection of resistance in LCD Anisotropic Conductive Film bonding using differential interference contrast. *Optik* **2017**, *130*, 786–796. [CrossRef]
2. Zeng, Z.; Zhang, C.; Du, S.; Chen, X. Quantitative surface topography of martensitic microstructure by differential interference contrast microscopy. *J. Mech. Phys. Solids* **2018**, *124*, 102–114. [CrossRef]
3. Hamilton, S.; Regan, D.; Payne, L.; Langbein, W.; Borri, P. Sizing individual dielectric nanoparticles with quantitative differential interference contrast microscopy. *Analyst* **2022**, *147*, 1567–1580. [CrossRef] [PubMed]
4. Pluta, M. Nomarski's DIC microscopy: A review. *SPIE* **1994**, *1846*, 10–25. [CrossRef]
5. Hartman, J.S.; Gordon, R.L.; Lessor, D.L. Nomarski differential interference contrast microscopy for surface slope measurements: An examination of techniques. *Appl. Opt.* **1981**, *20*, 2665–2669. [CrossRef]
6. Lessor, D.L.; Hartman, J.S.; Gordon, R.L. Quantitative surface topography determination by Nomarski reflection microscopy I Theory. *J. Opt. Soc. Am.* **1979**, *69*, 357–366. [CrossRef]
7. Wei, Q.; Li, Y.; Vargas, J.; Wang, J.; Gong, Q.; Kong, Y.; Jiang, Z.; Xue, L.; Liu, C.; Liu, F.; et al. Principal component analysis-based quantitative differential interference contrast microscopy. *Opt. Lett.* **2018**, *44*, 45–48. [CrossRef]
8. Sochacka, M.; Staronski, L.R. Phase-stepping DIC technique for reflecting surface evaluation. Phase Contrast and Differential Interference Contrast Imaging Techniques and Applications. *SPIE* **1994**, *1846*, 222–233. [CrossRef]
9. Li, Q.; Gao, H.; Xue, S.; Li, Y. Optical profilometer based on the principle of differential interference. *Opt. Eng.* **2001**, *40*, 833–836. [CrossRef]
10. Allen, R.D.; David, G.B.; Nomarski, G. The zeiss-Nomarski differential interference equipment for transmitted-light microscopy. *Z Wiss. Mikrosk.* **1969**, *69*, 193–221. [PubMed]
11. Kagalwala, F.; Kanade, T. Simulating DIC microscope images: From physical principles to a computational model. In Proceedings of the Workshop on Photometric Modeling for Computer Vision and Graphics (Cat. No.PR00271), Fort Collins, CO, USA, 22 June 1999; pp. 48–55. [CrossRef]
12. Cogswell, C.J.; Smith, N.I.; Larkin, K.G.; Hariharan, P. Quantitative DIC microscopy using a geometric phase shifter. *Bio-Med. Opt.* **1997**, *2984*, 72–81. [CrossRef]
13. Arnison, M.R.; Larkin, K.G.; Sheppard, C.J.R.; Smith, N.I.; Cogswell, C.J. Linear phase imaging using differential interference contrast microscopy. *J. Microsc.* **2004**, *214*, 7–12. [CrossRef]
14. Kou, S.S.; Sheppard, C.J.R. Quantitative phase restoration in differential interference contrast (DIC) microscopy. *Opt. Digit. Image Process. SPIE* **2008**, *7000*, 46–53. [CrossRef]
15. Preza, C.; Snyder, D.L.; Rosenberger, F.U.; Markham, J.; Conchello, J.-A. Phase estimation from transmitted-light DIC images using rotational diversity. *SPIE—Int. Soc. Opt. Eng.* **1997**, *3170*, 97–107. [CrossRef]
16. Kagalwala, F.; Kanade, T. Reconstructing specimens using DIC microscope images. *IEEE Trans. Syst. Man Cybern. Part B* **2003**, *33*, 728–737. [CrossRef]
17. King, S.V.; Cogswell, C.J. A phase-shifting DIC technique for measuring 3D phase objects: Experimental verification. Three-Dimensional and Multidimensional Microscopy: Image Acquisition and Processing XI. *SPIE* **2004**, *5324*, 191–196. [CrossRef]
18. Ishiwata, H.; Itoh, M.; Yatagai, T. A new method of three-dimensional measurement by differential interference contrast microscope. *Opt. Commun.* **2006**, *260*, 117–126. [CrossRef]
19. Cui, X.; Lew, M.; Yang, C. Quantitative differential interference contrast microscopy based on structured-aperture interference. *Appl. Phys. Lett.* **2008**, *93*, 091113. [CrossRef]
20. Fu, D.; Oh, S.; Choi, W.; Yamauchi, T.; Dorn, A.; Yaqoob, Z.; Dasari, R.R.; Feld, M.S. Quantitative DIC microscopy using an off-axis self-interference approach. *Opt. Lett.* **2010**, *35*, 2370–2372. [CrossRef] [PubMed]
21. Holmes, T.J.; Levy, W.J. Signal-processing characteristics of differential-interference-contrast microscopy. *Appl. Opt.* **1987**, *26*, 3929–3939. [CrossRef] [PubMed]
22. Cogswell, C.J.; Sheppard, C.J.R. Confocal differential interference contrast (DIC) microscopy: Including a theoretical analysis of conventional and confocal DIC imaging. *J. Microsc.* **1992**, *165*, 81–101. [CrossRef]
23. Hartman, J.S.; Gordon, R.L.; Lessor, D.L. Development Of Nomarski Microscopy For Quantitative Determination Of Surface Topography (A). *Interferom. SPIE* **1979**, *192*, 223–230. [CrossRef]
24. Hartman, J.S.; Gordon, R.L.; Lessor, D.L. Quantitative surface topography determination by Nomarski reflection microscopy 2: Microscope modification, calibration, and planar sample experiments. *Appl. Opt.* **1980**, *19*, 2998–3009. [CrossRef] [PubMed]

photonics

MDPI

Article

Generalized Asymmetric Hermite–Gaussian and Laguerre–Gaussian Beams

Eugeny G. Abramochkin [1], Victor V. Kotlyar [2,3], Alexey A. Kovalev [2,3,*] and Sergey S. Stafeev [2,3]

[1] Lebedev Physical Institute, 221 Novo-Sadovaya St., 443034 Samara, Russia; ega@fian.smr.ru
[2] Image Processing Systems Institute of the RAS—Branch of FSRC "Crystallography & Photonics" of the RAS, 151 Molodogvardeyskaya St., 443001 Samara, Russia; kotlyar@ipsiras.ru (V.V.K.); stafeev@ipsiras.ru (S.S.S.)
[3] Samara National Research University, 34 Moskovskoe Shosse, 443086 Samara, Russia
* Correspondence: alanko@ipsiras.ru

Abstract: We derive analytical formulae for the complex amplitudes of variants of generalized Hermite–Gaussian (HG) and Laguerre–Gaussian (LG) beams. We reveal that, at particular values of parameters of the exact solution of the paraxial propagation equation, these generalized beams are converted into conventional elegant HG and LG beams. We also deduce variants of asymmetric HG and LG beams that are described by complex amplitudes in the form of Hermite and Laguerre polynomials whose argument is shifted into the complex plane. The asymmetric HG and LG beams are, respectively, shown to present the finite superposition of the generalized HG and LG beams. We also derive an explicit relationship for the complex amplitude of a generalized vortex HG beam, which is built as the finite superposition of generalized HG beams with phase shifts. Newly introduced asymmetric HG and LG beams show promise for the study of the propagation of beams carrying an orbital angular momentum through the turbulent atmosphere. One may reasonably believe that the asymmetric laser beams are more stable against turbulence when compared with the radially symmetric ones.

Keywords: asymmetric beam; Hermite–Gaussian beam; Laguerre–Gaussian beam; free space propagation

check for updates

Citation: Abramochkin, E.G.; Kotlyar, V.V.; Kovalev, A.A.; Stafeev, S.S. Generalized Asymmetric Hermite–Gaussian and Laguerre–Gaussian Beams. *Photonics* **2023**, *10*, 606. https://doi.org/10.3390/photonics10060606

Received: 24 April 2023
Revised: 17 May 2023
Accepted: 19 May 2023
Published: 23 May 2023

1. Introduction

Laser beams whose complex amplitudes can be described by explicit formulae as exact solutions of the paraxial propagation equation have invariably been the focus of attention of optical researchers. The fact is that knowledge of the complex amplitude of the beams at any distance from the source plane enables the full set of their parameters to be predicted, including the intensity distribution, intensity moments, total power of the beam, topological charge, orbital angular momentum, and beam divergence. Propagating in free space, such beams are usually either structurally stable (conserving their intensity pattern up to a scale and rotation) or change weakly, with their complex amplitude expressed by an elegant mathematical relationship. While such beams are abound, two of the most familiar are standard Hermite–Gaussian (HG) and Laguerre–Gaussian (LG) laser beams. For the first time, HG beams were discussed in Ref. [1] and elegant HG (eHG) beams were studied in Ref. [2]. Various kinds of generalized HG (gHG) beams have also been investigated [3–7]. More specifically, in Ref. [3], the generalized HG beams were defined as conventional HG beams with a parameter, whereas in Ref. [7], the generalized beams were obtained as the superposition of conventional HG beams propagating with the same phase velocity. The parameter of the generalized HG beams is in the argument of the Hermite polynomials and defines the ratio between the widths of the Gaussian envelope and of the Hermite polynomials. At a certain value of this parameter, the generalized HG beam reduces to the conventional HG beam. Combined parametric Hermite-Laguerre–Gaussian beams, introduced in Ref. [8], were shown to transform into standard HG beams or LG beams depending on the specific value of the parameter. A relationship between the HG and LG beams was established in Ref. [9], and vortex HG beams constructed as the superposition of conventional HG beams with phase shifts were studied in Ref. [10]. For the first time, the standard LG beams were analyzed as laser cavity modes in Ref. [1] before being shown to propagate in an ABCD system [11]. Issues relating to the study of LG modes include their transformation and focusing with the aid of an axicon [12,13] and their representation through Wigner distribution functions [14]. Non-paraxial variants of the HG and LG beams have also been analyzed [15], as well as studying double-frequency LG beams and their transformation into conventional LG beams [16]. Other types of beams have been studied, including asymmetric LG beams [17] and the laser-aided intra-cavity shaping of asymmetric HG beams [18], which were then transformed into asymmetric LG beams using a mode converter [9]. A technique for generating vector laser beams with the aid of LG beams has been proposed [19]. The generation of LG beams in a wide frequency band by second harmonic generation has been reported [20], whereas conventional LG beams can be realized using a standard spatial light modulator (SLM) [21].

LG and HG beams with different modifications have wide practical applications. For instance, in works [22,23], structurally stable beams are considered which are coaxial superpositions of a finite number of the conventional HG beams with complex coefficients–parameters. The authors demonstrated that changing these parameters leads to oscillations of the orbital angular momentum of such a superposition and that it can achieve large positive and negative values. Using different methods, the LG beams can be converted into HG beams. In ref. [9], using a mode converter composed of two cylindrical lenses, the HG beams were, for the first time, converted to the LG beams. Such transformation can also be performed by tilting a diffractive optical element [24]. The LG beam generated this way is elliptic. Using an elliptically squeezed axicon, a Bessel–Gaussian beam can be transformed to an HG beam, which allows us to determine the topological charge of the optical vortex [25]. Besides the asymmetric LG and HG beams, asymmetric Bessel beams [26] and Bessel–Gaussian beams [27] are known, which are also used for manipulating microparticles [28]. Ref. [20] contains a survey of works on the subject of generating the LG beams. The survey lists passive and active methods for generating the LG beams. Passive methods include cylindrical lens mode converters [29], spiral zone plates [30], fork-shaped grating [31], non-spiral phase plate [32] and q-plate [33]. Active methods for LG beams generation include addition into a laser resonator such elements as plano-convex lens [34,35], point defect spatial filter [36], special resonant cavity mirrors such as SLM [37,38] and spot-defect mirror [39–43].

In this work, we discuss variants of generalized HG and LG beams, using which we generate asymmetric gHG and gLG beams, with the former obtained in two different variants. Asymmetric Bessel–Gaussian and LG beams have been known to show high stability upon propagation in a turbulent medium [44,45], prompting the exploration of other types of asymmetric HG and LG beams as promising candidates for monitoring the atmospheric turbulences.

2. Generalized Hermite–Gaussian and Laguerre–Gaussian Beams

It is well known that the complex amplitudes of standard HG and LG beams are as follows:

$$\mathrm{HG}_{n,m}(x,y,z) = \frac{1}{|q|}\exp\left(-\frac{x^2+y^2}{w_0^2 q} - i(n+m+1)\arg q\right) H_n\left(\frac{\sqrt{2}x}{w_0|q|}\right) H_m\left(\frac{\sqrt{2}y}{w_0|q|}\right), \quad (1)$$

$$\mathrm{LG}_{n,\pm m}(r,\varphi,z) = \frac{1}{|q|}\exp\left(-\frac{r^2}{w_0^2 q} - i(2n+m+1)\arg q\right)\left(\frac{re^{\pm i\varphi}}{w_0|q|}\right)^m L_n^m\left(\frac{2r^2}{w_0^2|q|^2}\right). \quad (2)$$

Here, (x,y,z) and (r,φ,z) are Cartesian and cylindrical coordinates; $q = q(z) = 1 + iz/z_0$ is an auxiliary complex parameter; $z_0 = \pi w_0^2/\lambda$ is the Rayleigh range; λ is the wavelength; w_0 is the waist radius of the Gaussian beam, n, m are nonnegative integers; $w(z) = w_0|q|$

is the beam radius at distance z from the waist, $\gamma(z) = \arg q = \arctan(z/z_0)$ is the Gouy phase; $H_m(\cdot)$ and $L_n^m(\cdot)$ are Hermite and Laguerre polynomials.

An elegant form of the HG and LG beams can be written as

$$\mathrm{eHG}_{n,m}(x,y,z) = \frac{1}{q^{(n+m+2)/2}} \exp\left(-\frac{x^2+y^2}{w_0^2 q}\right) H_n\left(\frac{x}{w_0\sqrt{q}}\right) H_m\left(\frac{y}{w_0\sqrt{q}}\right), \quad (3)$$

$$\mathrm{eLG}_{n,\pm m}(r,\varphi,z) = \frac{1}{q^{n+1}} \exp\left(-\frac{r^2}{w_0^2 q}\right) \left(\frac{re^{\pm i\varphi}}{w_0 q}\right)^m L_n^m\left(\frac{r^2}{w_0^2 q}\right). \quad (4)$$

Generalized versions of HG and LG beams are as follows [5]:

$$\begin{aligned} \mathrm{gHG}_{n,m}(x,y,z|a,b) &= \tfrac{1}{q} \exp\left(-\tfrac{x^2+y^2}{w_0^2 q}\right) \left(\tfrac{1-aq}{q}\right)^{n/2} \left(\tfrac{1-bq}{q}\right)^{m/2} \\ &\times H_n\left(\tfrac{x}{w_0\sqrt{q(1-aq)}}\right) H_m\left(\tfrac{y}{w_0\sqrt{q(1-bq)}}\right), \end{aligned} \quad (5)$$

$$\mathrm{gLG}_{n,\pm m}(r,\varphi,z|a) = \frac{1}{q} \exp\left(-\frac{r^2}{w_0^2 q}\right) \left(\frac{1-aq}{q}\right)^n \left(\frac{re^{\pm i\varphi}}{w_0 q}\right)^m L_n^m\left(\frac{r^2}{w_0^2 q(1-aq)}\right). \quad (6)$$

In Equations (5) and (6), a and b are dimensionless parameters. Here, and in the following text, we use a vertical line to separate variables and parameters.

When $a = b = 0$, these beams are reduced to eHG and eLG beams, while for $a = b = \frac{1}{2}$, we obtain standard HG beams, $2^{(n+m)/2}\mathrm{gHG}_{n,m}\left(x,y,z\middle|\frac{1}{2},\frac{1}{2}\right) = \mathrm{HG}_{n,m}(x,y,z)$, and LG beams, $2^n \mathrm{gLG}_{n,\pm m}\left(x,y,z\middle|\frac{1}{2}\right) = \mathrm{LG}_{n,\pm m}(x,y,z)$. Other forms of generalized Gaussian beams can also be found [6,7].

3. Asymmetric Generalized Hermite–Gaussian and Laguerre–Gaussian Beams

Asymmetric LG beams obtained by shifting the argument of the complex amplitude function into the complex domain have previously been reported [17]. Laser-generated asymmetric LG beams that were then converted into asymmetric LG modes using a mode converter [9] have been realized [18]. The asymmetric LG modes have been used to steer microparticles along an arc [46], as well as for non-invasively capturing and moving biological microobjects [28]. We note that it is also possible to realize asymmetric HG beams from asymmetric LG beams using a mode converter [9]. In this work, we demonstrate that asymmetric HG and LG beams are composed of a finite set of gHG and gLG beams. Actually, arguments of Hermite and Laguerre polynomials can be shifted using the following well-known sums [47,48]:

$$\begin{aligned} &\sum_{k=0}^{n} \binom{n}{k} (2t)^{n-k} H_k(x) = H_k(x+t), \\ &\sum_{k=0}^{n} \tfrac{t^k}{k!} L_{n-k}^{m+k}(x) = L_n^m(x-t). \end{aligned} \quad (7)$$

Making use of the sums in (7), the following relationships for complex amplitudes of gLG and gHG beams can be shown to hold:

$$\begin{aligned} &\sum_{k=0}^{n} \tfrac{t^k}{k!} \mathrm{gLG}_{n-k,m+k}(r,\varphi,z|a) = \tfrac{1}{q} \exp\left(-\tfrac{r^2}{w_0^2 q}\right) \\ &\times \sum_{k=0}^{n} \tfrac{t^k}{k!} \left(\tfrac{1-aq}{q}\right)^{n-k} \left(\tfrac{re^{i\varphi}}{w_0 q}\right)^{m+k} L_{n-k}^{m+k}\left(\tfrac{r^2}{w_0^2 q(1-aq)}\right) \\ = &\tfrac{1}{q} \exp\left(-\tfrac{r^2}{w_0^2 q}\right) \left(\tfrac{1-aq}{q}\right)^n \left(\tfrac{re^{i\varphi}}{w_0 q}\right)^m L_n^m\left(\tfrac{r^2 - tqw_0 re^{i\varphi}}{w_0^2 q(1-aq)}\right) = \mathrm{gLG}_{n,m}(r,\varphi,z|a,t). \end{aligned} \quad (8)$$

$$\sum_{\nu=0}^{n}\sum_{\mu=0}^{m}\binom{n}{\nu}\binom{m}{\mu}(2\tau)^{n-\nu}(2t)^{m-\mu}\mathrm{gHG}_{\nu,\mu}(x,y,z|a,b)$$
$$=\frac{1}{q}\exp\left(-\frac{x^2+y^2}{w_0^2q}\right)\sum_{\nu=0}^{n}\binom{n}{\nu}(2\tau)^{n-\nu}\left(\frac{1-aq}{q}\right)^{\nu/2}H_\nu\left(\frac{x}{w_0\sqrt{q(1-aq)}}\right)$$
$$\times\sum_{\mu=0}^{m}\binom{m}{\mu}(2t)^{m-\mu}\left(\frac{1-bq}{q}\right)^{\mu/2}H_\mu\left(\frac{y}{w_0\sqrt{q(1-bq)}}\right) \tag{9}$$
$$=\frac{1}{q}\exp\left(-\frac{x^2+y^2}{w_0^2q}\right)\left(\frac{1-aq}{q}\right)^{n/2}\left(\frac{1-bq}{q}\right)^{m/2}H_n\left(\frac{x+\tau w_0q}{w_0\sqrt{q(1-aq)}}\right)H_m\left(\frac{y+tw_0q}{w_0\sqrt{q(1-bq)}}\right)$$
$$=\mathrm{gHG}_{n,m}(x,y,z|a,b,\tau,t).$$

In (8), the parameter a defines the ratio of the widths of the Gaussian function and Laguerre polynomial, whereas the complex parameter t defines the degree of LG beam asymmetry. Actually, it is the absolute value of t that affects the asymmetry. As seen in Equation (8), the argument of t only rotates the transverse pattern by the angle of arg t (clockwise if arg $t > 0$ and counterclockwise if arg $t < 0$). In a similar way, in (9), the parameters a and b define the relative width of the Hermite polynomial along the x- and y-axes compared to the Gaussian function width, whereas the complex parameters τ and t define the degree of HG beam asymmetry along the x- and y-axes. We note that the asymmetric LG beam in (8) is different from the similar asymmetric beam discussed in Ref. [18] because, in the latter case, the argument was shifted into the complex domain for both the Gaussian beam and the Laguerre polynomials, unlike in Equation (8), in which the argument shift into the complex plane takes place only for the Laguerre polynomial.

Using gHG beams, it is possible to create elliptical vortex HG beams different from those discussed in Ref. [10]. As a starting point, we use the summation formula of Hermite polynomials [47]:

$$\sum_{k=0}^{n}\binom{n}{k}t^kH_{n-k}(x)H_k(y)=\left(1+t^2\right)^{n/2}H_n\left(\frac{x+ty}{\sqrt{1+t^2}}\right). \tag{10}$$

Using Equations (5) and (10) for the superposition of gHG beams, the elliptic vortex HG beams can be derived in the following form:

$$\sum_{k=0}^{n}\binom{n}{k}t^k\mathrm{gHG}_{n-k,k}(x,y,z|a,b)=\frac{1}{q}\exp\left(-\frac{x^2+y^2}{w_0^2q}\right)$$
$$\times\left(\frac{(1+t^2)(1-Tq)}{q}\right)^{n/2}H_n\left(\frac{x+ty}{w_0\sqrt{(1+t^2)q(1-Tq)}}\right), \tag{11}$$

where $T = (a + bt^2)/(1 + t^2)$.

Substituting $t = i\alpha$ with a real α into (11), we obtain an elliptical optical vortex $x + i\alpha y$ on the right-hand side in the argument of the Hermite polynomial. Hence, near every real null of the Hermite polynomial, which are found on the horizontal axis (a and b are real parameters), there is an optical vortex with a topological charge of +1 ($\alpha > 0$) or −1 ($\alpha < 0$). The modulus of the total topological charge of beam (11) equals n. At $\alpha = \pm 1$, the vortex Hermite beam in (11) is converted into a single-ring LG beam.

4. Non-Orthogonality and Power of gLG Beams

With the parameter a being complex, the LG beams in (6) are not mutually orthogonal. Considering that the dot product of two solutions of a paraxial equation remains unchanged at any z, the non-orthogonality of the generalized LG beams can be proven in the source plane (at the Gaussian beam waist). Actually, it can be shown that, in the source plane, the dot product of two generalized LG beams of (6) is given by

$$\int_0^\infty \int_0^{2\pi} \mathrm{gLG}^*_{p,\ell}(r,\varphi,z|a)\mathrm{gLG}_{n,m}(r,\varphi,z|a)rdrd\varphi = \frac{\pi w_0^2}{2} \cdot \frac{(n+m)!}{2^{p+n+m}n!}\delta_{\ell,m} \times \sum_{k=0}^{\min(n,p)} \binom{n}{k}\binom{p+n+m-k}{p-k}(1-2a^*)^{p-k}(1-2a)^{n-k}\left(4\mathrm{Re}a-4|a|^2\right)^k, \tag{12}$$

where $\delta_{\ell,m}$ is the Kronecker symbol and (p, l) and (n, m) are the beams orders. In Equation (12), the generalized LG beams are seen to be orthogonal with respect to the azimuthal index, because at $m \neq \ell$, the right-hand side of (12) equals zero. Meanwhile, at different-valued radial indices, $n \neq p$, the right-hand side of (12), is non-zero. The only exception is $a = \frac{1}{2}$ when the gLG beams are reduced to LG beams, becoming orthogonal with respect to both radial and azimuthal indices. Notably, from (12) follows an expression for the power W of a gLG beam:

$$W = \int_0^\infty \int_0^{2\pi} \left|\mathrm{gLG}_{n,m}(r,\varphi,z|a)\right|^2 rdrd\varphi = \frac{\pi w_0^2}{2} \cdot \frac{(n+m)!}{2^{2n+m}n!} \times \sum_{k=0}^{n} \binom{n}{k}\binom{2n+m-k}{n-k}|1-2a|^{2n-2k}\left(4\mathrm{Re}a-4|a|^2\right)^k. \tag{13}$$

At $a = \frac{1}{2}$, Equation (13) can be seen to describe the power of standard LG beams:

$$W_{a=1/2} = 2^{2n}\int_0^\infty \int_0^{2\pi} \left|\mathrm{gLG}_{n,m}\left(r,\varphi\middle|\frac{1}{2}\right)\right|^2 rdrd\varphi = \frac{\pi w_0^2(n+m)!}{2^{m+1}n!}. \tag{14}$$

5. Topological Charge of gLG Beams and Their Anomalous Rotation upon Propagation

Following M. V. Berry [49], we define the topological charge of an arbitrary light field by the formula

$$\mathrm{TC} = \frac{1}{2\pi}\lim_{r\to\infty}\int_0^{2\pi} \arg E(r,\varphi)\, d\varphi \tag{15}$$

where $E(r, \varphi)$ is the complex amplitude in the transverse plane and (r, φ) are polar coordinates in this plane.

The superposition (8) is described by a Laguerre polynomial of the n-th degree. Thus, at $r \to \infty$, it is mostly determined by its highest-degree term, and we can suppose that the right part of Equation (8) reads as

$$\mathrm{gLG}_{n,m}(r,z|a,t << r) = \frac{1}{n!q}\exp\left(-\frac{r^2}{w_0^2 q}\right)\left(\frac{1-aq}{q}\right)^n\left(\frac{re^{i\varphi}}{w_0 q}\right)^m\left(\frac{-r^2}{w_0^2 q(1-aq)}\right)^n. \tag{16}$$

As seen in Equation (16), the TC of the whole superposition is thus equal to m, i.e., equal to the maximal TC in the superposition of the generalized LG beams. The generalized LG beams from Equation (6) have the same topological charge as of the conventional LG beams from Equation (2). In the next section, the simulation confirms this.

It can be seen in Equation (8) that if $m > 0$, then, upon propagation in free space, the transverse shape of the light beam should be rotated counterclockwise, as typically happens for most light beams.

However, the asymmetry of the intensity distribution is only affected by the Laguerre polynomial, since the absolute values of all other multipliers in Equation (8) are rotationally symmetric. For simplicity, we consider a simple case, $a = \frac{1}{2}$, since the denominator of the argument of the Laguerre polynomial becomes real-valued. The argument ξ is thus equal to

$$\xi = 2\frac{r^2 - tw(z)re^{i\varphi+i\gamma}}{w^2(z)} \tag{17}$$

where $\gamma = \arctan(z/z_0)$ is the Gouy phase and where $w(z) = w_0\,|\,q(z)\,|$. Equation (17) can be rewritten in the Cartesian coordinates:

$$\begin{aligned} \xi &= 2\frac{x^2+y^2}{w^2(z)} - 2t\frac{x+iy}{w(z)}e^{i\gamma} \\ &= 2\left(\frac{x}{w(z)} - \frac{t\cos\gamma}{2}\right)^2 + 2\left(\frac{y}{w(z)} + \frac{t\sin\gamma}{2}\right)^2 - \frac{t^2}{2} - 2it\frac{y\cos\gamma + x\sin\gamma}{w(z)}. \end{aligned} \tag{18}$$

We see that the real part of the argument is, up to the constant $-t^2/2$, the distance from the point (x,y) to the point with coordinates

$$x_c = \frac{t}{2}w(z)\cos\gamma, \quad y_c = -\frac{t}{2}w(z)\sin\gamma. \tag{19}$$

To determine the transformation of the beam transverse shape on propagation, we should choose some characteristic points whose evolution we are able to trace. Equation (18) indicates that we can choose zero-intensity points. Since all the roots of the Laguerre polynomials are real [50], Im $\xi = 0$ in these points. Actually, these points reside on the intersections of several concentric light rings (the roots of the Laguerre polynomial) with the center given by Equation (19), and of a straight line, $y\cos\gamma + x\sin\gamma = 0$. In the initial plane ($z = 0$), the Gouy phase shift is zero ($\gamma = 0$) and, thus, ξ is real at $y = 0$. The ring center is in the point $(t/2, 0)$. Thus, if $t > 0$, the considered zero-intensity points are to the right of the origin (Figure 1a). Then, after propagation over some distance z, the rings center moves to the point given by Equation (19), i.e., it is rotated clockwise by the angle $-\gamma$ with respect to the origin. The value ξ is real at $y = -x\tan\gamma = -xz/z_0$, i.e., the straight line with real values ξ is also rotated clockwise by the angle γ (Figure 1b). In the far field, $z \gg z_0$ and $\gamma = \pi/2$, the ring center is in the point $(0, -t/2)$ and ξ is real at $x = 0$ (Figure 1c).

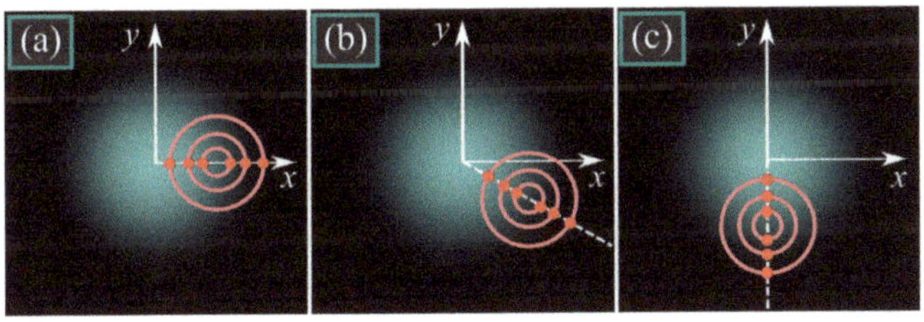

Figure 1. Schematic evolution of the asymmetric generalized Laguerre–Gaussian beam from Equation (8) shown as positions of zero-intensity points: in the initial plane (**a**), after propagation over some distance (**b**) and in the far field (**c**). Blue color denotes the rotationally symmetric part of the beam (Gaussian envelope with an optical vortex), whereas pink rings denote the rings where the Laguerre polynomial is zero (if the imaginary part of its argument is neglected). Red dots thus denote the points where the beam intensity is zero.

Thus, beam (8), consisting of the LG beams with positive topological charges $m + s$, is nevertheless rotated clockwise upon propagation. Spiral beams with this anomalous rotation have been investigated in [51], but beams (8) are different since $2(n-k) + |m+k| + \theta(m+k)$ is not a constant at $\theta = -1$ and depends on k.

6. Numerical Modeling of gLG Beams

In this section, we present the results of the computation of the asymmetric gLG beams. All the data were obtained in two ways: (1) using the numerical Fresnel transform, implemented as a convolution by adopting the fast Fourier transform, and (2) using the theoretical Equation (8). In both cases, the intensity patterns are visually indistinguishable, while the phase patterns are only different in low-intensity areas. This confirms the correctness of Equation (8) for the complex amplitudes in space propagation.

Intensity and phase distributions of the symmetric and asymmetric gLG beams for some transverse planes are shown in Figure 2.

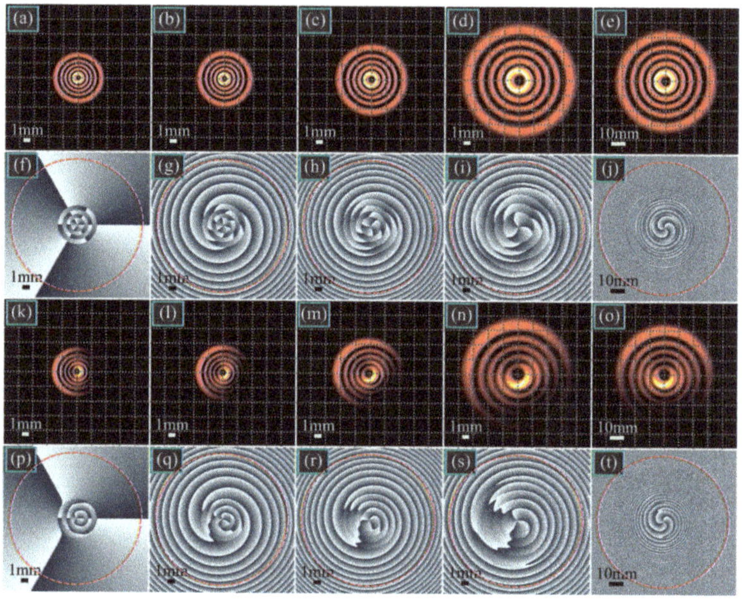

Figure 2. Intensity (**a–e**,**k–o**) and phase (**f–j**,**p–t**) distributions of the asymmetric gLG beams, regarding Equation (8), in some transverse planes for the following computation parameters: wavelength: $\lambda = 532$ nm; Gaussian beam waist radius: $w_0 = 1$ mm; radial index of the gLG beam: $n = 4$; azimuthal index (topological charge): $m = 3$; scaling factor: $a = 1/2$; asymmetry parameter: $t = 0$ (**a–j**) and $t = 0.25$ (**k–t**); propagation distances: $z = 0$ (**a**,**f**,**k**,**p**), $z = z_0/2$ (**b**,**g**,**l**,**q**), $z = z_0$ (**c**,**h**,**m**,**r**), $z = 2z_0$ (**d**,**i**,**n**,**s**) and $z = 10z_0$ (**e**,**j**,**o**,**t**). Dashed circles on the phase distributions bound the disks for the TC computation. All frames are shown in the square $[-5c, 5c] \times [-5c, 5c]$, where $c = 2$ mm (**a–d**,**f–i**,**k–n**,**p–s**) and $c = 10$ mm (**e**,**j**,**o**,**t**) are the sizes of the grid cell.

As seen in Figure 2, if $a = 1/2$ and $t = 0$, the asymmetric gLG beam reduces to the conventional shape-invariant LG beam. When $t = 0.25$, the beam becomes asymmetric and, in agreement with the theory above, rotates clockwise upon propagation in free space. For each plane, we compute the TC value. The results are 3.0000 (Figure 2p), 2.9981 (Figure 2q), 2.9971 (Figure 2r), 2.9982 (Figure 2s) and 2.9295 (Figure 2t), i.e., the theoretical value of TC = 3 is confirmed.

Now, let us consider what happens when the asymmetry parameter is further increased. Figure 3 illustrates the intensity and phase distributions of the symmetric and asymmetric gLG beams, regarding Equation (8), in some transverse planes when $t = 1$ and $t = 2$.

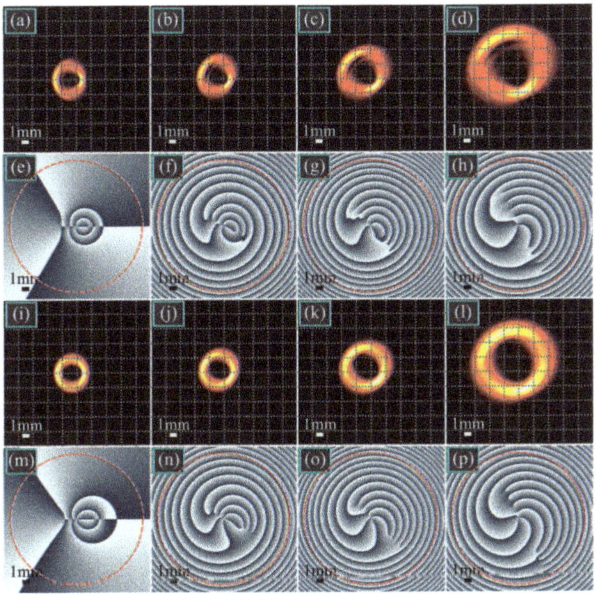

Figure 3. The same as Figure 2, but for $t = 1$ (**a–h**) and $t = 2$ (**i–p**).

According to Figure 3, when the asymmetry parameter is increased, only the central and the outermost rings become bright, while the other rings become pale. It can also be seen that the intensity trends towards a single ring shape (Figure 3i–l). This is predictable, since when t in Equation (8) grows, the most significant weight in this superposition is of the last LG beam, i.e., the beam with $k = n$. This is an exactly single-ringed LG beam (with the azimuthal index of $m + n$).

The computed TC values in each plane are as follows: 3.0000 (Figure 3e), 2.9982 (Figure 3f), 2.9972 (Figure 3g), 2.9985 (Figure 3h), 3.0000 (Figure 3m), 2.9982 (Figure 3n), 2.9972 (Figure 3o) and 2.9988 (Figure 3p), i.e., the theoretical value of TC = 3 is confirmed.

When the asymmetry parameter t becomes large enough, the single-ringed LG beam completely dominates in the superposition and the intensity distribution looks like a homogeneous ring (Figure 4). Then, looking at Equation (8), it may seem that this domination leads to the change of the TC to $m + n$. However, Figure 4 demonstrates that the TC remains equal to m, since there are additional n peripheral optical vortices of the charge -1 (insets in Figure 4i–l). For instance, when the TC is computed over a disk of radius 9 mm (circles in Figure 4e–h), the obtained values are 7.0017 (Figure 4e), 6.9980 (Figure 4f), 6.9946 (Figure 4g) and 6.9968 (Figure 4h). When the disk radius for computation has been doubled (Figure 4i–k) or tripled (Figure 4l), the obtained values became 3.0000 (Figure 4i), 2.9704 (Figure 4j), 2.9534 (Figure 4k) and 2.9625 (Figure 4l).

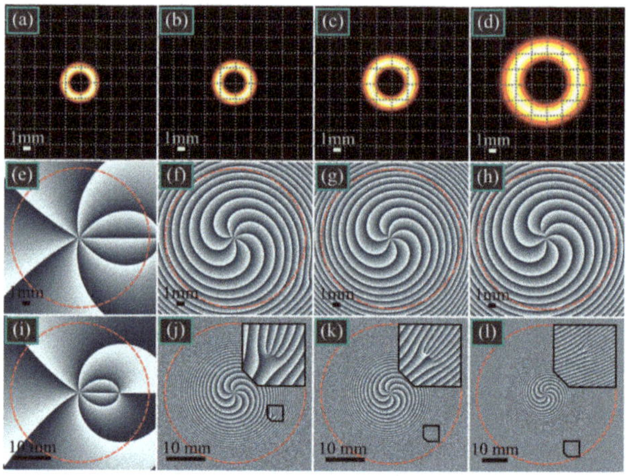

Figure 4. Intensity (**a–d**) and phase (**e–l**) distributions of the asymmetric gLG beams, regarding Equation (8), in some transverse planes for the following computation parameters: wavelength: λ = 532 nm; Gaussian beam waist radius: w_0 = 1 mm; radial index: n = 4; azimuthal index (topological charge): m = 3 of the LG beam; scaling factor: a = 1/2; asymmetric parameter: t = 10; propagation distances: z = 0 (**a**,**e**,**i**), $z = z_0/2$ (**b**,**f**,**j**), $z = z_0$ (**c**,**g**,**k**) and $z = 2z_0$ (**d**,**h**,**l**). Dashed circles on the phase distributions bound the disks of the TC computation. Phase distributions in the bottom row are the same as above but the area is enlarged. The insets (**j–l**) show the 4-fold zoomed regions with additional vortices.

7. Numerical Modeling of gHG Beams

First, we compare gHG beams with standard and elegant HG beams. Figure 5 depicts the intensity and phase distributions of HG beams $n = m = 5$, as can be seen in Equation (1), whose shape is conserved upon propagation in both transverse directions (a–h); of eHG beams $n = m = 5$, as can be seen in Equation (3), whose shape is changed (i–p); and of gHG beams $n = m = 5$, as can be seen in Equation (5), whose shape is conserved in one transverse direction and is changed in another direction (q–x).

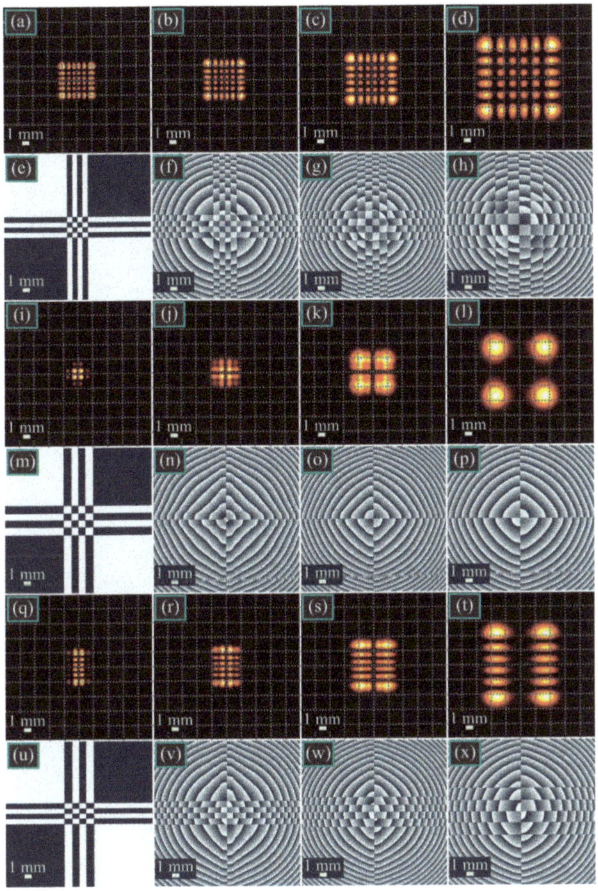

Figure 5. Intensity (**a–d**,**i–l**,**q–t**) and phase (**e–h**,**m–p**,**u–x**) distributions of gHG beams, regarding Equation (5), (i.e., without the asymmetry) in some transverse planes for the following computation parameters: wavelength: λ = 532 nm; Gaussian beam waist radius: w_0 = 1 mm; orders of the HG beams: m = 5 and n = 5; scaling factors: a = 1/2 and b = 1/2 (**a–h**), a = 0 and b = 0 (**i–p**), a = 0 and b = 1/2 (**q–x**); propagation distances: z = 0 (**a**,**e**,**i**,**m**), $z = z_0/2$ (**b**,**f**,**j**,**n**), $z = z_0$ (**c**,**g**,**k**,**o**) and $z = 2z_0$ (**d**,**h**,**l**,**p**). All frames are shown in the square $[-5c, 5c] \times [-5c, 5c]$, where c = 2 mm is the size of the grid cell.

As can be seen in Figure 5, all three beams are symmetrical with respect to the Cartesian axes.

Figure 6 illustrates the asymmetric generalized gHG beams with $n = m = 5$ that have two real-valued parameters: t and τ. Since the beam is different in different transverse directions along the axes x and y, for the changes to be more visible, we apply a different asymmetry along these axes.

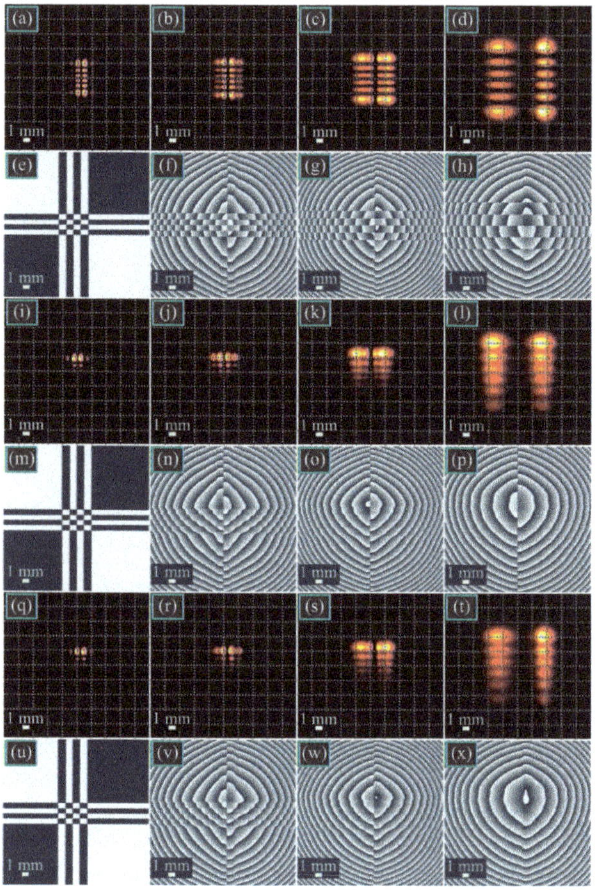

Figure 6. Intensity (**a–d**,**i–l**,**q–t**) and phase (**e–h**,**m–p**,**u–x**) distributions of asymmetric gHG beams, regarding Equation (9), in some transverse planes for the following computation parameters: wavelength: λ = 532 nm; Gaussian beam waist radius: w_0 = 1 mm; orders of the gHG beams: m = 5 and n = 5; scaling factors: a = 0 and b = 1/2; asymmetry parameters: t = 0 and τ = 1/2 (**a–h**), t = 1/4 and τ = 0 (**i–p**) and t = 1/4 and τ = 1/2 (**q–x**); propagation distances: z = 0 (**a**,**e**,**i**,**m**), $z = z_0/2$ (**b**,**f**,**j**,**n**), $z = z_0$ (**c**,**g**,**k**,**o**) and $z = 2z_0$ (**d**,**h**,**l**,**p**). All frames are shown in the square $[-5c, 5c] \times [-5c, 5c]$, where c = 2 mm is the size of the grid cell.

As expected, the nonzero parameter τ breaks the intensity asymmetry with respect to the axis y (Figure 6a–h). Similarly, the nonzero parameter t breaks the intensity asymmetry with respect to the axis x (Figure 6i–p). When both the parameters t and τ are nonzero, both asymmetries are broken (Figure 6q–x). Thus, setting the parameters t and τ allows us to change the intensity shape in the beam cross-section.

8. Experiment

Figure 7 illustrates the optical setup for the experiment. A light beam from the laser impinges onto the SLM, after which three diffraction orders occur. The diaphragm D transmits the working first diffraction order, whereas the spherical lens L_4 (f = 250 mm) focuses it. The CCD matrix registers several intensity distributions near the focus.

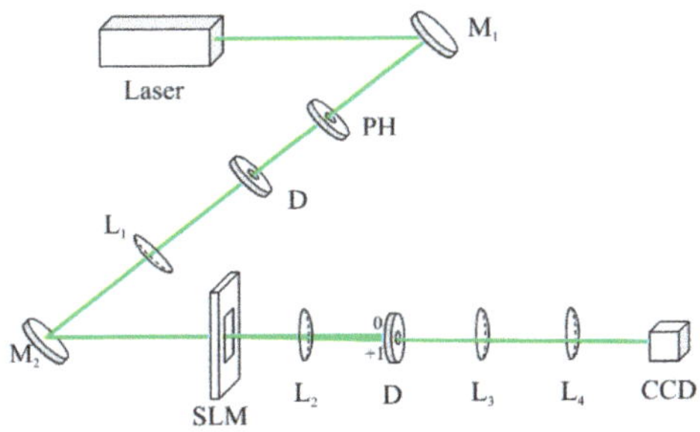

Figure 7. Experimental setup. Laser–MGL-F-532-700 (λ = 532 nm), M_1, M_2,–mirrors, PH–pinhole, L_1-L_4-spherical lenses (f_1 = 750 mm, f_2 = 150 mm, f_3 = 100mm, f_4 = 250 mm), D–diaphragm, SLM–spatial light modulator (Holoeye LC 2012), CCD–CCD camera (UCMOS10000KPA).

Figure 8a depicts the phase distribution generated in the SLM display. The phase from Figure 8a encodes the amplitude and phase of the asymmetric generalized LG beam from Equation (8) with the same parameters as in Figure 2m. The computed intensity distribution generated by the phase is shown in Figure 8b.

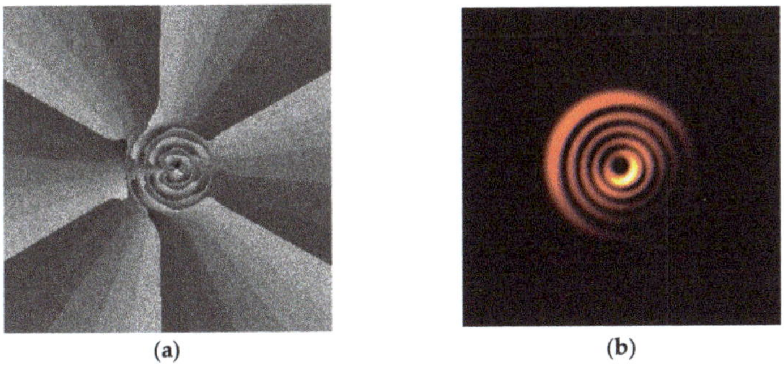

Figure 8. Encoded phase (**a**) intended for generating an asymmetric generalized LG beam from Equation (8) with the same parameters as in Figure 2m: $n = 4$, $m = 3$, $a = 1/2$, $t = 0.25$ (**b**).

Figure 9 depicts the intensity distributions at different distances before and after the focus, registered by the CCD-matrix in the optical setup from Figure 7 with using the phase from Figure 8a in the SLM.

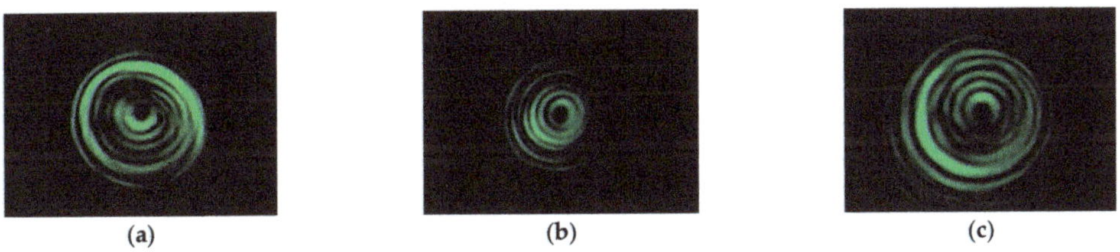

Figure 9. Intensity distributions registered by the CCD-camera in the setup from Figure 7 at a distance of 2 mm before the focus of the lens L_4 (**a**), near the focus (**b**) and at a distance of 2 mm after the focus (**c**). The frame sizes are (1 × 0.75) mm.

As can be seen in Figure 9, the experimental intensity patterns are qualitatively consistent with the computed pattern shown in Figure 8b. It can be seen that the energy goes mostly into the inner and outer light rings. Both these rings have the shape of a crescent and rotate on propagation.

9. Conclusions

Summing up, we have derived novel exact solutions of the paraxial propagation equation. Relationships have also been deduced for the complex amplitudes that describe the propagation of asymmetric gHG and gLG beams in (8) and (9). The derived relationships contain parameter sets, with the amplitude of an asymmetric gLG beam being described by two complex parameters, namely, the scale parameter and the asymmetry degree parameter. The amplitude of an asymmetric gHG beam contains four parameters, divided into pairs for each Cartesian axis. With the original generalized elegant HG (5) and LG (6) beams not retaining their intensity pattern upon free-space propagation, the derivative generalized asymmetric HG and LG beams also do not retain their intensity pattern in the course of their propagation evolution. The generalized LG beams of Equation (9) are not orthogonal with regard to the radial indices, meaning that the asymmetric LG beams of Equation (10) are also non-orthogonal with regard to the radial indices. For the generalized LG beams, we have proved that their topological charge is independent of the asymmetry and equals the upper index of the Laguerre polynomial. We have also found that the asymmetric LG beam is anomalously rotated upon propagation in free space, i.e., it is rotated clockwise when the vortex factor has a positive topological charge.

Another finding of this work is the relationship of the complex amplitude of a family of generalized vortex HG beams in Equation (11), which is characterized by three parameters, with two parameters defining the beam scale and a complex parameter defining the asymmetry degree of elliptical optical vortices of unit topological charge, with the vortices centered at the zeros of a Hermite polynomial.

The proposed generalized asymmetrical laser beams can be realized using an SLM, with the adoption of some techniques for the encoding and amplitude-phase function into a phase-only one [52,53], utilized for data transmission through atmospheric turbulence. The scale parameters and the parameter controlling the asymmetry degree offer additional degrees of freedom [54] enabling the creation of a beam least prone to distortion due to turbulence. We note that attractive beams for optical communications are non-diffractive beams such as Bessel beams [55] or Airy beams [56]. However, Laguerre and Hermite Gaussian beams are also used for these tasks [57].

The asymmetric generalized LG beams from Equation (8) can also be used for trapping and rotating microparticles along a light arc [46] or for trapping and shifting biological cell-like microobjects without damaging them [28]. Asymmetric generalized HG beams from Equation (9) constitute a four-parametric family. Changing these parameters allows us to control the intensity distribution in the beam cross-section. Such beams can be used for accelerating microparticles [58]. Asymmetric generalized LG beams from Equation (8) rotate upon propagation (Figures 2 and 3). This property can be used in optical metrology [59] for measuring the longitudinal beam shift by measuring the rotation angle of the asymmetric intensity pattern in the beam cross-section [60].

Author Contributions: Conceptualization, E.G.A.; methodology, E.G.A. and V.V.K.; software, A.A.K.; validation, V.V.K.; formal analysis, E.G.A.; investigation, E.G.A., V.V.K., A.A.K. and S.S.S.; resources, V.V.K.; data curation, V.V.K.; writing—original draft preparation, V.V.K.; writing—review and editing, E.G.A. and V.V.K.; visualization, A.A.K.; supervision, V.V.K.; project administration, V.V.K.; funding acquisition, V.V.K. and A.A.K. All authors have read and agreed to the published version of the manuscript.

Funding: The work was partly funded by the Russian Science Foundation under grant No. 22-12-00137 (Theoretical background). This work was also performed within the State assignment of Federal Scientific Research Center "Crystallography and Photonics" of Russian Academy of Sciences (in parts about numerical modeling).

Institutional Review Board Statement: Not applicable.

Informed Consent Statement: Not applicable.

Data Availability Statement: Not applicable.

Conflicts of Interest: The authors declare no conflict of interest. The funders had no role in the design of the study; in the collection, analyses, or interpretation of data; in the writing of the manuscript; or in the decision to publish the results.

References

1. Kogelnik, H.; Li, T. Laser beams and resonators. *Appl. Opt.* **1966**, *5*, 1550–1567. [CrossRef] [PubMed]
2. Siegman, A.E. Hermite-Gaussian functions of complex argument as optical beam eigenfunction. *J. Opt. Soc. Am.* **1973**, *63*, 1093–1094. [CrossRef]
3. Pratesi, R.; Ronchi, L. Generalized Gaussian beams in free space. *J. Opt. Soc. Am.* **1977**, *67*, 1274–1276. [CrossRef]
4. Zauderer, E. Complex argument Hermite-Gaussian and Laguerre-Gaussian beams. *J. Opt. Soc. Am. A* **1986**, *3*, 465–469. [CrossRef]
5. Wünsche, A. Generalized Gaussian beam solutions of paraxial optics and their connection to a hidden symmetry. *J. Opt. Soc. Am. A* **1989**, *6*, 1320–1329. [CrossRef]
6. Kotlyar, V.V.; Kovalev, A.A. Hermite-Gaussian modal laser beams with orbital angular momentum. *J. Opt. Soc. Am. A* **2014**, *31*, 274–282. [CrossRef]
7. Wang, Y.; Chen, Y.; Zhang, Y.; Chen, H.; Yu, S. Generalised Hermite–Gaussian beams and mode transformations. *J. Opt.* **2016**, *18*, 055001. [CrossRef]
8. Abramochkin, E.G.; Volostnikov, V.G. Generalized Gaussian beams. *J. Opt. A Pure Appl. Opt.* **2004**, *6*, S157–S161. [CrossRef]
9. Abramochkin, E.G.; Volostnikov, V.G. Beam transformations and nontransformed beams. *Opt. Commun.* **1991**, *83*, 123–135. [CrossRef]
10. Kotlyar, V.V.; Kovalev, A.A.; Porfirev, A.P. Vortex Hermite-Gaussian laser beams. *Opt. Lett.* **2015**, *40*, 701–704. [CrossRef]
11. Mei, Z.; Zhao, D.; Gu, J.; Mao, H. Approximate analytical expressions of Laguerre-Gaussian beams passing through a paraxial optical system with an annular aperture. *Optik* **2004**, *115*, 311–316. [CrossRef]
12. Arlt, J.; Kuhn, R.; Dholakia, K. Spatial transformation of Laguerre-Gaussian laser modes. *J. Mod. Opt.* **2001**, *48*, 783–787. [CrossRef]
13. Jarutis, V.; Paškauskas, R.; Stabinis, A. Focusing of Laguerre-Gaussian beams by axicon. *Opt. Commun.* **2000**, *184*, 105–112. [CrossRef]
14. Simon, R.; Agarwal, G.S. Wigner representation of Laguerre-Gaussian beams. *Opt. Lett.* **2000**, *25*, 1313–1315. [CrossRef] [PubMed]
15. Kim, H.C.; Lee, Y.H. Hermite-Gaussian and Laguerre-Gaussian beams beyond the paraxial approximation. *Opt. Commun.* **1999**, *169*, 9–16. [CrossRef]
16. Hasegawa, T.; Shimizu, T. Frequency-doubled Hermite-Gaussian beam and the mode conversion to the Laguerre-Gaussian beam. *Opt. Commun.* **1999**, *160*, 103–108. [CrossRef]
17. Kovalev, A.A.; Kotlyar, V.V.; Porfirev, A.P. Asymmetric Laguerre-Gaussian beams. *Phys. Rev. A* **2016**, *93*, 063858. [CrossRef]
18. Hsieh, Y.H.; Lai, Y.H.; Hsieh, M.X.; Huang, K.F.; Chen, Y.F. Generating high-power asymmetrical Laguerre-Gaussian modes and exploring topological charges distribution. *Opt. Express* **2018**, *26*, 31738–31749. [CrossRef]
19. Ghaderi Goran Abad, M.; Mahmoudi, M. Laguerre-Gaussian modes generated vector beam via nonlinear magneto-optical rotation. *Sci. Rep.* **2021**, *11*, 5972. [CrossRef]
20. Yang, Y.; Li, Y.; Wang, C. Generation and expansion of Laguerre–Gaussian beams. *J. Opt.* **2022**, *51*, 910–926. [CrossRef]
21. Matsumoto, N.; Ando, T.; Inoue, T.; Ohtake, Y.; Fukuchi, N.; Hara, T. Generation of high-quality higher-order Laguerre-Gaussian beams using liquid-crystal-on-silicon spatial light modulators. *J. Opt. Soc. Am. A* **2008**, *25*, 1642–1651. [CrossRef] [PubMed]
22. Volyar, A.; Abramochkin, E.; Akimova, Y.; Bretsko, M. Control of the orbital angular momentum via radial numbers of structured Laguerre–Gaussian beams. *Opt. Lett.* **2022**, *47*, 2402–2405. [CrossRef]
23. Volyar, A.; Abramochkin, E.; Akimova, Y.; Bretsko, M. Astigmatic-Invariant Structured Singular Beams. *Photonics* **2022**, *9*, 842. [CrossRef]
24. Kotlyar, V.V.; Khonina, S.N.; Almazov, A.A.; Soifer, V.A.; Jefimovs, K.; Turunen, J. Elliptic Laguerre-Gaussian beams. *J. Opt. Soc. Am. A* **2006**, *23*, 43–56. [CrossRef] [PubMed]
25. Dwivedi, R.; Sharma, P.; Jaiswal, V.K.; Mehrotra, R. Elliptically squeezed axicon phase for detecting topological charge of vortex beam. *Opt. Commun.* **2021**, *485*, 126710. [CrossRef]
26. Kotlyar, V.V.; Kovalev, A.A.; Soifer, V.A. Asymmetric Bessel modes. *Opt. Lett.* **2014**, *39*, 2395–2398. [CrossRef]
27. Gong, L.; Qiu, X.Z.; Ren, Y.X.; Zhu, H.Q.; Liu, W.W.; Zhou, J.H.; Zhong, M.C.; Chu, X.X.; Li, Y.M. Observation of the asymmetric Bessel beams with arbitrary orientation using a digital micromirror device. *Opt. Express* **2014**, *22*, 26763–26776. [CrossRef]
28. Rykov, M.A.; Skidanov, R.V. Modifying the laser beam intensity distribution for obtaining improved strength characteristics of an optical trap. *Appl. Opt.* **2014**, *53*, 156–164. [CrossRef]
29. Beijersbergen, M.W.; Allen, L.; van der Veen, H.E.L.O.; Woerdman, J.P. Astigmatic laser mode converters and transfer of orbital angular momentum. *Opt. Commun.* **1993**, *96*, 123–132. [CrossRef]
30. Heckenberg, N.R.; Mcduf, R.; Smith, C.P.; White, A.G. Generation of optical phase singularities by computer-generated holograms. *Opt. Lett.* **1992**, *17*, 3. [CrossRef]
31. Dennis, M.R.; O'Holleran, K.; Padgett, M.J. Chapter 5 Singular Optics: Optical Vortices and Polarization Singularities. *Prog. Opt.* **2009**, *53*, 293–363. [CrossRef]
32. Kim, G.H.; Jeon, J.H.; Ko, K.H.; Lee, J.H.; Chang, J.S. Optical vortices produced with a nonspiral phase plate. *Appl. Opt.* **1997**, *36*, 8614–8621. [CrossRef] [PubMed]

33. Marrucci, L.; Manzo, C.; Paparo, D. Optical spin-to-orbital angular momentum conversion in inhomogeneous anisotropic media. *Phys. Rev. Lett.* **2006**, *96*, 163905. [CrossRef] [PubMed]
34. Senatsky, Y.; Bisson, J.F.; Shelobolin, A.; Shirekawa, A.; Ueda, K. Circular modes selection in Yb:YAG laser using an intracavity lens with spherical aberration. *Laser Phys.* **2009**, *19*, 911–918. [CrossRef]
35. Thirugnanasambandam, M.P.; Senatsky, Y.; Ueda, K. Generation of very-high order Laguerre-Gaussian modes in Yb:YAG ceramic laser. *Laser Phys. Lett.* **2010**, *7*, 637–643. [CrossRef]
36. Tan, S.; Zhou, C.; Shirakakwa, A.; Ueda, K.; Li, J. Vortex Ti:Sapphire laser by using an intracavity spot-defect spatial filter. *Opt. Laser Technol.* **2017**, *96*, 76–80. [CrossRef]
37. Ngcobo, S.; Litvin, I.; Burger, L.; Forbes, A. A digital laser for on-demand laser modes. *Nat. Commun.* **2013**, *4*, 2289. [CrossRef]
38. Liu, S.S.; Chen, X.D.; Pu, J.X.; Lin, Z.L.; Chen, Z.Y. A V-folded digital laser for on-demand vortex beams by astigmatic transformation of Hermite-Gaussian modes. *Chinese Phys. Lett.* **2019**, *36*, 124203. [CrossRef]
39. Ito, A.; Kozawa, Y.; Sato, S. Generation of hollow scalar and vector beams using a spot-defect mirror. *J. Opt. Soc. Am. A.* **2010**, *27*, 2072–2077. [CrossRef]
40. Kano, K.; Kozawa, Y.; Sato, S. Generation of a Purely Single Transverse Mode Vortex Beam from a He-Ne Laser Cavity with a Spot-Defect Mirror. *Int. J. Opt.* **2012**, *2012*, 359141. [CrossRef]
41. Vyas, S.; Kozawa, Y.; Sato, S. Generation of a vector doughnut beam from an internal mirror He–Ne laser. *Opt. Lett.* **2014**, *39*, 2080–2082. [CrossRef] [PubMed]
42. Qiao, Z.; Xie, G.; Wu, Y.; Yuan, P.; Ma, J.; Qian, L.; Fan, D. Generating high-charge optical vortices directly from laser up to 288th order. *Laser Photon. Rev.* **2018**, *12*, 1800019. [CrossRef]
43. Uesugi, Y.; Sato, S.; Kozawa, Y. Direct generation of the lowestorder vortex beam using a spot defect mirror in the ultraviolet region. *Opt. Lett.* **2020**, *45*, 2115–2118. [CrossRef] [PubMed]
44. Yu, L.; Zhang, Y.; Wang, J. Propagation of asymmetric Bessel mode in turbulent atmosphere. *IEEE Photonics J.* **2022**, *14*, 7354106. [CrossRef]
45. Watkins, R.J.; Dai, K.; White, G.; Li, W.; Miller, J.K.; Morgan, K.S.; Johnson, E.G. Experimental probing of turbulence using a continuous spectrum of asymmetric OAM beams. *Opt. Express* **2020**, *28*, 924–935. [CrossRef] [PubMed]
46. Kovalev, A.A.; Kotlyar, V.V.; Porfirev, A.P. Optical trapping and moving of microparticles by using asymmetrical Laguerre–Gaussian beams. *Opt. Lett.* **2016**, *41*, 2426–2429. [CrossRef]
47. Prudnikov, A.P.; Brychkov, Y.A.; Marichev, O.I. *Integrals and Series. Special Functions*; Gordon and Breach: New York, NY, USA, 1986.
48. Brychkov, Y.A. *Handbook of Special Functions: Derivatives, Integrals, Series and Other Formulas*; CRC Press: Boca Raton, FL, USA, 2008.
49. Berry, M.V. Optical vortices evolving from helicoidal integer and fractional phase steps. *J. Opt. A Pure Appl. Opt.* **2004**, *6*, 259. [CrossRef]
50. Abramowitz, M.; Stegun, I.A. *Handbook of Mathematical Functions*; Dover: New York, NY, USA, 1972.
51. Abramochkin, E.G.; Volostnikov, V.G. Spiral light beams. *Phys. Usp.* **2004**, *47*, 1177–1203. [CrossRef]
52. Mendoza-Yero, O.; Mínguez-Vega, G.; Lancis, J. Encoding complex fields by using a phase-only optical element. *Opt. Lett.* **2014**, *39*, 1740–1743. [CrossRef]
53. Goorden, S.; Bertolotti, J.; Mosk, A. Superpixel-based spatial amplitude and phase modulation using a digital micromirror device. *Opt. Express* **2014**, *22*, 17999–18009. [CrossRef]
54. Singh, S.K.; Adachi, Y.; Kinashi, K.; Tsutsumi, N.; Sakai, W.; Jackin, B.J. Tailoring Large Asymmetric Laguerre–Gaussian Beam Array Using Computer-Generated Holography. *Photonics* **2023**, *10*, 247. [CrossRef]
55. Durnin, J. Exact solutions for nondiffracting beams. I. The scalar theory. *J. Opt. Soc. Am. A* **1987**, *4*, 651–654. [CrossRef]
56. Siviloglou, G.A.; Broky, J.; Dogariu, A.; Christodoulides, D.N. Observation of Accelerating Airy Beams. *Phys. Rev. Lett.* **2007**, *99*, 213901. [CrossRef] [PubMed]
57. Cox, M.A.; Maqondo, L.; Kara, R.; Milione, G.; Cheng, L.; Forbes, A. The Resilience of Hermite– and Laguerre–Gaussian Modes in Turbulence. *J. Lightwave Technol.* **2019**, *37*, 3911–3917. [CrossRef]
58. Bochove, E.J.; Moore, G.T.; Scully, M.O. Acceleration of particles by an asymmetric Hermite-Gaussian laser beam. *Phys. Rev. A* **1992**, *46*, 6640–6653. [CrossRef]
59. Guo, M.; Le, W.; Wang, C.; Rui, G.; Zhu, Z.; He, J.; Gu, B. Generation, Topological Charge, and Orbital Angular Momentum of Off-Axis Double Vortex Beams. *Photonics* **2023**, *10*, 368. [CrossRef]
60. Pavani, S.R.P.; Piestun, R. High-efficiency rotating point spread functions. *Opt. Express* **2008**, *16*, 3484. [CrossRef]

photonics

Communication

Radially Phased-Locked Hermite–Gaussian Correlated Beam Array and Its Properties in Oceanic Turbulence

Peiying Zhu, Guiqiu Wang *, Yan Yin, Haiyang Zhong, Yaochuan Wang and Dajun Liu *

Department of Physics, College of Science, Dalian Maritime University, Dalian 116026, China
* Correspondence: gqwang@dlmu.edu.cn (G.W.); liudajun@dlmu.edu.cn (D.L.)

Abstract: The descriptions of a radially phased-locked Hermite–Gaussian correlated beam array are introduced, the equation of this beam array in oceanic turbulence is derived, and the intensity profiles of this beam array are shown and analyzed. The results imply that the evolutions of the sub-beam of this beam array in free space are the same as the Hermite–Gaussian correlated beam, while the intensity of this beam array can be adjusted by controlling the initial beam radius R and the coherence length. The intensity profiles of this beam array in free space have multiple spots during propagation, while the same beam array in oceanic turbulence can become a beam spot due to the influences of R and oceanic turbulence. The beam array with smaller coherence length in oceanic turbulence retains the splitting properties better during propagation.

Keywords: beam array; correlation function; intensity; oceanic turbulence

check for updates

Citation: Zhu, P.; Wang, G.; Yin, Y.; Zhong, H.; Wang, Y.; Liu, D. Radially Phased-Locked Hermite–Gaussian Correlated Beam Array and Its Properties in Oceanic Turbulence. *Photonics* **2023**, *10*, 551. https://doi.org/10.3390/photonics10050551

Received: 10 April 2023
Revised: 29 April 2023
Accepted: 7 May 2023
Published: 9 May 2023

1. Introduction

The characteristics of laser transmitting in underwater turbulence affect the performances of underwater optical communication [1], and the propagation of beams in underwater turbulence has recently been widely studied. Partially coherent beams (PCBs) reduce the degrading effects of turbulence [2], and the structured light can remain invariant during propagation [3]. Previously, the scintillation properties of waves in underwater turbulence have been analyzed [4–6]. To examine the performances of underwater optical systems, the BER properties of laser in oceanic turbulence were investigated [7,8]. PCBs with unique correlation functions have been introduced into the underwater turbulence environments. The propagation of a multi-sinc Schell-model beam in anisotropic oceanic turbulence has been investigated [9]. The self-splitting structured beam can lower the negative influences of oceanic turbulence, which is achieved by controlling the parameters of the beams [10]. The influences of oceanic turbulence on non-uniformly correlated beams have been studied by numerical simulation [11,12]. The propagation of vortex beams in oceanic turbulence has been investigated, and neural networks have also been used in studies of vortex beams in underwater turbulence [13–17]. The beam array has a lattice-like intensity, and the properties of the beam array have been widely analyzed [18,19].

The PCBs' shaping can be modulated by coherence structure engineering [20], and PCBs can be experimentally generated [21]. The cross-spectral density (CSD) of PCBs with different coherence function has been widely discussed. The multi-Gaussian Schell-model beam (multi-GSM) has a flat-topped intensity profile in the far field [22]. The Gaussian intensity array can be obtained by anisotropic GSM beam array [23]. The Laguerre-Gaussian correlated beam can obtain the ring-shaped intensity distribution for propagation in free space [24]. The Hermite–Gaussian correlated Schell-model (HGCSM) beam can split from one beam spot to multiple spots [25]. Non-uniform PCB modulated by a non-uniform HGCSM function has been described, with the position of maximum intensity adjusted using the beam parameters [26]. The twist HGCSM beam can be obtained by introducing the twist phase [27]. In practice, the beam array can achieve high power, and can be composed of off-axis beams [28,29]. The radial intensity distribution of the beam array can be obtained by radially phase-locked PCBs [30–32]. The splitting properties of HGCSM are very interesting. However, the beam array composed of HGCSM sub-beams has not yet been reported. In this work, the HGSM beam was extended into beam array. The CSD of a radially phased-locked HGCSM (RPLHGSM) beam array is introduced based on the HGCSM beam, and the intensity distributions of this beam array transmitting in oceanic turbulence are numerically analyzed. The paper is organized as follows. In Section 2, the CSD of an RPLHGSM beam array on the source plane is introduced. In Section 3, the propagation equation of an RPLHGSM beam array in oceanic turbulence is derived. In Section 4, the intensity profiles of this beam array in oceanic turbulence are discussed.

2. Model of an RPLHGSM Beam Array

The CSD of the HGCSM beam on the source plane $z = 0$ in the space-frequency domain is written as [24]:

$$W(\mathbf{r}_1,\mathbf{r}_2,0) = C_0 \exp\left(-\frac{\mathbf{r}_1^2+\mathbf{r}_2^2}{4w_0^2}\right) \exp\left[-\frac{(x_2-x_1)^2}{2\delta_{0x}^2}\right] \frac{H_{2m}\left[\frac{x_2-x_1}{\sqrt{2}\delta_{0x}}\right]}{H_{2m}(0)} \exp\left[-\frac{(y_2-y_1)^2}{2\delta_{0y}^2}\right] \frac{H_{2n}\left[\frac{y_2-y_1}{\sqrt{2}\delta_{0y}}\right]}{H_{2n}(0)} \tag{1}$$

In Equation (1), $\mathbf{r} = (x,y)$ is the position vector, C_0 is a constant, w_0 represents beam width of Gauss part, δ_{0x} and δ_{0y} denote coherence length along x and y directions, respectively. H_m is the Hermite polynomial of order m and can be given as [33]:

$$H_m(x) = \sum_{l=0}^{[m/2]} \frac{(-1)^l m!}{l!(m-2l)!}(2x)^{m-2l} \tag{2}$$

Recalling the descriptions of beam array [27,34–36], the center of sub-beam of a beam array is set as (r_x, r_y), the CSD of an RPLHGSM beam array composed of N HGCSM sub-beams with radial distributions at source plane $z = 0$ can be given as:

$$\begin{aligned} W(\mathbf{r}_1,\mathbf{r}_2,0) = {} & C_0 \sum_{h_1=1}^{N} \exp\left[-\frac{\left(x_1-r_{h_1x}\right)^2+\left(y_1-r_{h_1y}\right)^2}{4w_0^2}\right] \\ & \times \sum_{h_2=1}^{N} \exp\left[-\frac{\left(x_2-r_{h_2x}\right)^2+\left(y_2-r_{h_2y}\right)^2}{4w_0^2}\right] \\ & \times \exp\left[-\frac{\left[\left(x_1-r_{h_1x}\right)-\left(x_2-r_{h_2x}\right)\right]^2}{2\delta_{0x}^2}\right] \exp\left[-\frac{\left[\left(y_1-r_{h_1y}\right)-\left(y_2-r_{h_2y}\right)\right]^2}{2\delta_{0y}^2}\right] \\ & \times \frac{H_{2m}\left[\frac{\left(x_2-r_{h_2x}\right)-\left(x_1-r_{h_1x}\right)}{\sqrt{2}\delta_{0x}}\right]}{H_{2m}(0)} \frac{H_{2n}\left[\frac{\left(y_2-r_{h_2y}\right)-\left(y_1-r_{h_1y}\right)}{\sqrt{2}\delta_{0y}}\right]}{H_{2n}(0)} \\ & \times \exp\left[i\left(\varphi_{h_1}-\varphi_{h_2}\right)\right] \end{aligned} \tag{3}$$

where N is the number of sub-beams, R is the radius, $r_{hx} = R\cos\varphi_h$, $r_{hy} = R\sin\varphi_h$, $\varphi_h = h2\pi/N, (h = 1,2,\ldots N)$ is the phase of the h-th sub-beam.

3. Propagation of an RPLHGSM Beam Array in Oceanic Turbulence

The CSD of PCBs propagating in oceanic turbulence is described by the extended Huygens–Fresnel principle [9–19]:

$$\begin{aligned} W(\boldsymbol{\rho}_1,\boldsymbol{\rho}_2,z) = \frac{k^2}{4\pi^2 z^2}\int_{-\infty}^{+\infty}\int_{-\infty}^{+\infty}\int_{-\infty}^{+\infty}\int_{-\infty}^{+\infty} W(\mathbf{r}_1,\mathbf{r}_2,0) \\ \times \exp\left[-\frac{ik}{2z}(\boldsymbol{\rho}_1-\mathbf{r}_1)^2+\frac{ik}{2z}(\boldsymbol{\rho}_2-\mathbf{r}_2)^2\right] \\ \times \langle \exp[\psi(\mathbf{r}_1,\boldsymbol{\rho}_1)+\psi^*(\mathbf{r}_2,\boldsymbol{\rho}_2)]\rangle d\mathbf{r}_1 d\mathbf{r}_2 \end{aligned} \tag{4}$$

with

$$\langle \exp[\psi(\mathbf{r}_1,\boldsymbol{\rho}_1)+\psi^*(\mathbf{r}_2,\boldsymbol{\rho}_2)]\rangle = \exp\left[-\frac{(\mathbf{r}_1-\mathbf{r}_2)^2+(\mathbf{r}_1-\mathbf{r}_2)\cdot(\boldsymbol{\rho}_1-\boldsymbol{\rho}_2)+(\boldsymbol{\rho}_1-\boldsymbol{\rho}_2)^2}{\Lambda_0^2}\right] \tag{5}$$

where $1/\Lambda_0^2 = \pi^2 k^2 z T_0/3$, and $T_0 = \int \kappa^3 \Phi(\kappa) d\kappa$. $\Phi(\kappa)$ is the power spectrum of oceanic turbulence and is given as [37]:

$$\begin{aligned} \Phi(\kappa) = 0.388\times 10^{-8}\varepsilon^{-1/3}\kappa^{-11/3}\chi_T\left[1+2.35(\kappa\eta)^{2/3}\right] \\ \times\left[\exp(-A_T\Theta)+\varsigma^{-2}\exp(-A_S\Theta)-2\varsigma^{-1}\exp(-A_{TS}\Theta)\right] \end{aligned} \tag{6}$$

The κ is the spatial frequency of power spectrum of oceanic turbulence. η is the Kolmogorov micro scale. χ_T, ε and ς are parameters of oceanic turbulence [36]. The parameters in Equation (6) are $A_T = 1.863\times 10^{-2}$, $A_S = 1.9\times 10^{-4}$, $A_{TS} = 9.41\times 10^{-3}$, $\Theta = 8.284(\kappa\eta)^{4/3}+12.978(\kappa\eta)^2$. The T_0 can be given as [12]:

$$T_0 = 0.388\times 10^{-8}(\varepsilon\eta)^{-1/3}\left(47.5708\varsigma^{-2}-17.6701\varsigma^{-1}+6.78335\right) \tag{7}$$

Substituting Equation (3) into Equation (4), the CSD of an RPLHGSM beam array in oceanic turbulence can be derived as:

$$\begin{aligned} W(\boldsymbol{\rho}_1,\boldsymbol{\rho}_2,z) = \frac{k^2}{4\pi^2 z^2}\exp\left[-\frac{ik}{2z}\left(\boldsymbol{\rho}_1^2-\boldsymbol{\rho}_2^2\right)\right]\exp\left[-\frac{(\rho_{1x}-\rho_{2x})^2+(\rho_{1y}-\rho_{2y})^2}{\Lambda_0^2}\right] \\ G_0\sum_{h_1=1}^{N}\sum_{h_2=1}^{N}\frac{1}{H_{2m}(0)}\frac{1}{H_{2n}(0)}\exp\left[i\left(\varphi_{h_1}-\varphi_{h_2}\right)\right] \\ \sum_{l_x=0}^{m}\frac{(-1)^{l_x}(2m)!}{l_x!(2m-2l_x)!}\left(\frac{2}{\sqrt{2}\delta_{0x}}\right)^{2m-2l_x}\sum_{l_y=0}^{n}\frac{(-1)^{l_y}(2n)!}{l_y!(2n-2l_y)!}\left(\frac{2}{\sqrt{2}\delta_{0y}}\right)^{2n-2l_y} \\ W_x(\rho_x,z)W_y(\rho_y,z) \end{aligned} \tag{8}$$

where

$$\begin{aligned} W_x(\rho_x,z) = \exp\left[-\frac{\left(r_{h_1x}-r_{h_2x}\right)^2}{\Lambda_0^2}\right]\exp\left[-\frac{ik}{2z}\left(r_{h_1x}^2-r_{h_2x}^2\right)\right]\exp\left[\frac{ik}{z}\left(\rho_{1x}r_{h_1x}-\rho_{2x}r_{h_2x}\right)\right] \\ \exp\left[-\frac{(\rho_{1x}-\rho_{2x})\left(r_{h_1x}-r_{h_2x}\right)}{\Lambda_0^2}\right]\sum_{s_x=0}^{2m-2l_x}\frac{(2m-2l_x)!(-1)^{s_x}}{s_x!(2m-2l_x-s_x)!} \\ \sqrt{\frac{\pi}{a_x}}s_x!\left(\frac{1}{a_x}\right)^{s_x}\exp\left[\frac{1}{a_x}\left(-\frac{r_{h_1x}-r_{h_2x}}{\Lambda_0^2}-\frac{\rho_{1x}-\rho_{2x}}{2\Lambda_0^2}-\frac{ik}{2z}r_{h_1x}+\frac{ik}{2z}\rho_{1x}\right)^2\right] \\ \sum_{t_x=0}^{[\frac{s_x}{2}]}\frac{1}{t_x!(s_x-2t_x)!}\left(\frac{a_x}{4}\right)^{t_x}\sum_{d_x=0}^{s_x-2t_x}\frac{(s_x-2t_x)!}{d_x!(s_x-2t_x-d_x)!} \\ \left(-\frac{r_{h_1x}-r_{h_2x}}{\Lambda_0^2}-\frac{\rho_{1x}-\rho_{2x}}{2\Lambda_0^2}-\frac{ik}{2z}r_{h_1x}+\frac{ik}{2z}\rho_{1x}\right)^{s_x-2t_x-d_x} \\ \left(\frac{1}{2\delta_{0x}^2}+\frac{1}{\Lambda_0^2}\right)^{d_x}\sqrt{\frac{\pi}{b_x}}\left(\frac{i}{2\sqrt{b_x}}\right)^{2m-2l_x-s_x+d_x}\exp\left(\frac{c_x^2}{b_x}\right)H_{2m-2l_x-s_x+d_x}\left(-\frac{ic_x}{\sqrt{b_x}}\right) \end{aligned} \tag{9}$$

$$W_y(\rho_y, z) = \exp\left[-\frac{\left(r_{h_1y}-r_{h_2y}\right)^2}{\Lambda_0^2}\right] \exp\left[-\frac{ik}{2z}\left(r_{h_1y}^2 - r_{h_2y}^2\right)\right] \exp\left[\frac{ik}{z}\left(\rho_{1y}r_{h_1y} - \rho_{2y}r_{h_2y}\right)\right]$$
$$\exp\left[-\frac{\left(\rho_{1y}-\rho_{2y}\right)\left(r_{h_1y}-r_{h_2y}\right)}{\Lambda_0^2}\right] \sum_{s_y=0}^{2n-2l_y} \frac{(2n-2l_y)!(-1)^{s_y}}{s_y!(2n-2l_y-s_y)!}$$
$$\sqrt{\frac{\pi}{a_y}} s_y! \left(\frac{1}{a_y}\right)^{s_y} \exp\left[\frac{1}{a_y}\left(-\frac{r_{h_1y}-r_{h_2y}}{\Lambda_0^2} - \frac{\rho_{1y}-\rho_{2y}}{2\Lambda_0^2} - \frac{ik}{2z}r_{h_1y} + \frac{ik}{2z}\rho_{1y}\right)^2\right]$$
$$\sum_{t_y=0}^{[\frac{s_y}{2}]} \frac{1}{t_y!(s_y-2t_y)!}\left(\frac{a_y}{4}\right)^{t_y} \sum_{d_y=0}^{s_y-2t_y} \frac{(s_y-2t_y)!}{d_y!(s_y-2t_y-d_y)!} \tag{10}$$
$$\left(-\frac{r_{h_1y}-r_{h_2y}}{\Lambda_0^2} - \frac{\rho_{1y}-\rho_{2y}}{2\Lambda_0^2} - \frac{ik}{2z}r_{h_1y} + \frac{ik}{2z}\rho_{1y}\right)^{s_y-2t_y-d_y}$$
$$\left(\frac{1}{2\delta_{0y}^2} + \frac{1}{\Lambda_0^2}\right)^{d_y} \sqrt{\frac{\pi}{b_y}} \left(\frac{i}{2\sqrt{b_y}}\right)^{2n-2l_y-s_y+d_y} \exp\left(\frac{c_y^2}{b_y}\right) H_{2n-2l_y-s_y+d_y}\left(-\frac{ic_y}{\sqrt{b_y}}\right)$$

with

$$a_\beta = \frac{1}{4w_0^2} + \frac{1}{2\delta_{0\beta}^2} + \frac{1}{\Lambda_0^2} + \frac{ik}{2z}, (\beta = x, y) \tag{11}$$

$$b_\beta = \frac{1}{4w_0^2} + \frac{1}{2\delta_{0\beta}^2} + \frac{1}{\Lambda_0^2} - \frac{ik}{2z} - \frac{1}{a_\beta}\left(\frac{1}{2\delta_{0\beta}^2} + \frac{1}{\Lambda_0^2}\right)^2 \tag{12}$$

$$c_\beta = \frac{r_{h_1\beta}-r_{h_2\beta}}{\Lambda_0^2} + \frac{\rho_{1\beta}-\rho_{2\beta}}{2\Lambda_0^2} + \frac{ik}{2z}r_{h_2\beta} - \frac{ik}{2z}\rho_{2\beta}$$
$$+ \frac{1}{a_\beta}\left(\frac{1}{2\delta_{0\beta}^2} + \frac{1}{\Lambda_0^2}\right)\left(-\frac{r_{h_1\beta}-r_{h_2\beta}}{\Lambda_0^2} - \frac{\rho_{1\beta}-\rho_{2\beta}}{2\Lambda_0^2} - \frac{ik}{2z}r_{h_1\beta} + \frac{ik}{2z}\rho_{1\beta}\right) \tag{13}$$

Based on the derived equations, the average intensity of an RPLHGSM beam array in oceanic turbulence is written as [9–19]:

$$I(\mathbf{r}, z) = W(\mathbf{r}, \mathbf{r}, z) \tag{14}$$

4. Numerical Results and Analysis

In this section, the average intensity characteristics of an RPLHGSM beam array transmitting in oceanic turbulence are analyzed. In the simulations, the parameters were chosen as $w_0 = 4$ mm, $\lambda = 532$ nm, $R = 6$ cm, $N = 4$, $m = n = 2$, $\varsigma = -2.5$, $\chi_T = 8 \times 10^{-8}$ K^2/s, $\varepsilon = 10^{-7}$ m^2 s^{-1} and $\eta = 1$ mm without other descriptions in the figures.

The intensity of an RPLHGSM beam array with $N = 4$ and $\delta_{0x} = \delta_{0y} = 3$ mm transmitting in free space is given in Figure 1. When the propagation distance is short, the intensity profile of an RPLHGSM beam array retains the four beam spots (Figure 1a). As the distance increases further, the sub-beam of an RPLHGSM beam array exhibits splitting properties, and the intensity distribution of an RPLHGSM beam array in free space becomes multiple beam spots, due to the influences of coherence $\delta_{0x} = \delta_{0y}$. The self-splitting phenomenon of the sub-beam is the same as the HGCSM beam [24]. Finally, the intensity profile of an RPLHGSM beam array shows multiple spots at $z = 200$ m (Figure 1d). Therefore, the intensity profile of an RPLHGSM beam array in free space is controlled by the initial beam number N and Hermite correlation function.

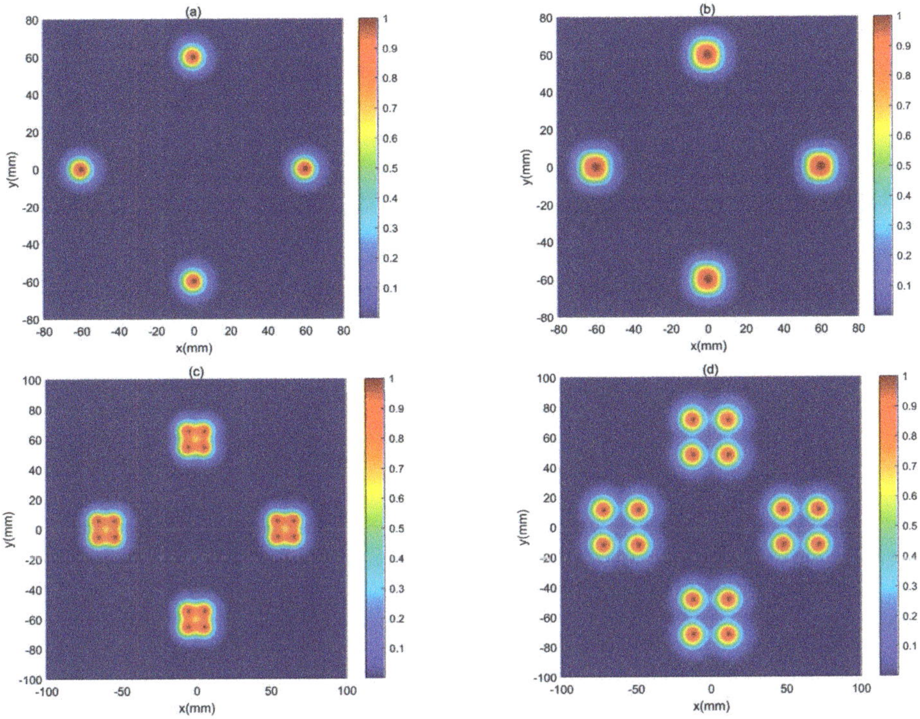

Figure 1. The intensity of an RPLHGSM beam array with $\delta_{0x} = \delta_{0y} = 3$ mm in free space. (**a**) $z = 30$ m, (**b**) $z = 60$ m, (**c**) $z = 100$ m, (**d**) $z = 200$ m.

The characteristics of an RPLHGSM beam array with $\delta_{0x} = \delta_{0y} = 3$ mm for the different initial beam parameters N, R, m and n transmitting in free space at $z = 200$ m are illustrated in Figure 2. From Figures 1d and 2a, it is found that the intensity profile of an RPLHGSM beam array has multiple spots, N affects the distribution of sub-beams, and the intensity pattern can be modulated by controlling the sub-beam number N. When the initial parameters m, n of sub-beams change, the intensity pattern of an RPLHGSM beam array has different intensity distributions. The intensity profile of the sub-beam is the same as the HGCSM beam with the same parameters in free space, however, the whole intensity pattern of an RPLHGSM beam array in free space is composed of different HGCSM beamlets (Figure 2a–c). When the radius R decreases, the multiple spots of the sub-beam of an RPLHGSM beam array overlap with the near sub-beams (Figure 2d), and the separate spot distributions are destroyed. Hence, we can adjust the intensity profiles of an RPLHGSM beam array in free space by modifying the initial beam parameters.

We considered the effects of an increase in the coherence length $\delta_{0x} = \delta_{0y}$. The intensity of an RPLHGSM beam array with larger $\delta_{0x} = \delta_{0y} = 5$ mm in free space is shown in Figure 3. It shows that the beamlet of an RPLHGSM beam array with larger $\delta_{0x} = \delta_{0y}$ in free space retains the single spot better at $z = 100$ m (Figure 3a), while the beam array with $\delta_{0x} = \delta_{0y} = 3$ mm splits into multiple spots (Figure 1c), an RPLHGSM beam array with smaller coherence length split fasters. Comparing Figure 3b with Figure 1d, an RPLHGSM beam array with larger $\delta_{0x} = \delta_{0y} = 5$ mm also split at $z = 200$ m, but the splitting phenomenon is not more apparent than for the beam array with $\delta_{0x} = \delta_{0y} = 3$ mm. Thus, the intensity distributions of an RPLHGSM beam array with smaller $\delta_{0x} = \delta_{0y}$ can retain the array profile better during propagation.

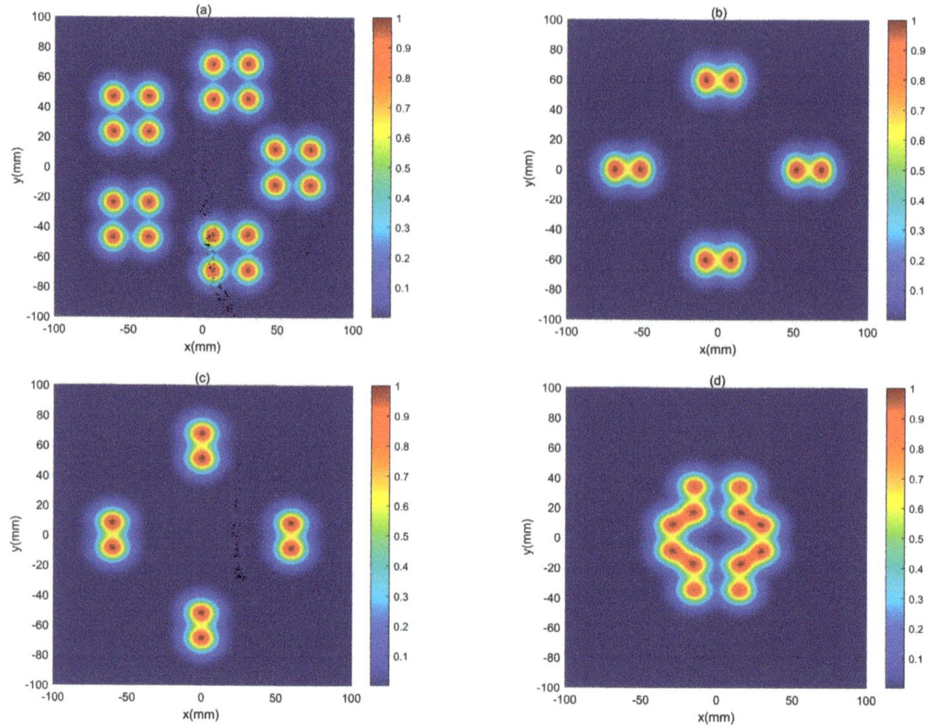

Figure 2. The intensity of an RPLHGSM beam array in free space at $z = 200$ m. (**a**) N = 5, R = 6 cm, $m = n = 2$, (**b**) $N = 4$, $R = 6$ cm, $m = 1$, $n = 0$, (**c**) N = 4, R = 6 cm, m = 0, n = 1, (**d**) $N = 4$, $R = 3$ cm, $m = 0$, $n = 1$.

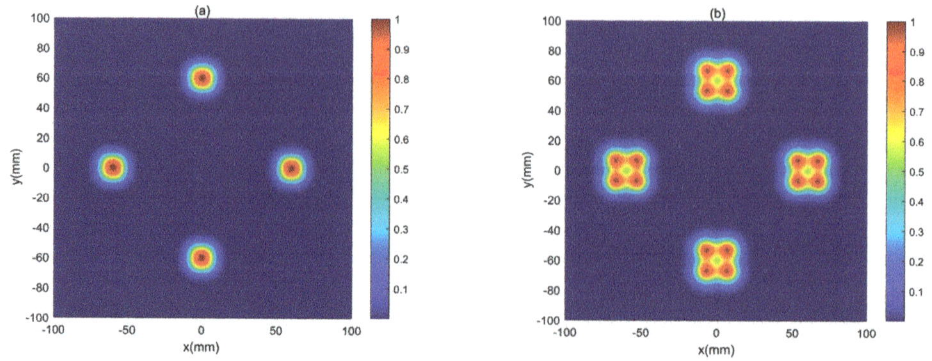

Figure 3. The intensity of an RPLHGSM beam array with $\delta_{0x} = \delta_{0y} = 5$ mm in free space. (**a**) z = 100 m , (**b**) $z = 200$ m.

Figures 4 and 5 show an RPLHGSM beam array with different $\delta_{0x} = \delta_{0y}$ propagating in oceanic turbulence. In Figure 4, the sub-beam of an RPLHGSM beam array with $\delta_{0x} = \delta_{0y} = 3$ mm in oceanic turbulence has combining properties (Figure 4b), while the same beam array in free space has separate spot profiles (Figure 1d). when $\delta_{0x} = \delta_{0y}$ increases, the sub-beam of an RPLHGSM beam array with $\delta_{0x} = \delta_{0y} = 5$ mm in oceanic turbulence becomes a spot, and this beam array in free space can retain multiple spot distribution (Figure 3b). Comparing Figures 3–5, the beam array with smaller $\delta_{0x} = \delta_{0y}$ better retained its splitting properties during propagation on oceanic turbulence (Figures 4b and 5b). The combining properties of an RPLHGSM beam array can be seen in this example of beam array propagation in oceanic turbulence. (Figures 3b and 5b). Thus, the combining phenomenon of an RPLHGSM beam array in oceanic turbulence is affected by oceanic turbulence and coherence length, and a beam array with low coherence length retains its splitting properties better during propagation in oceanic turbulence.

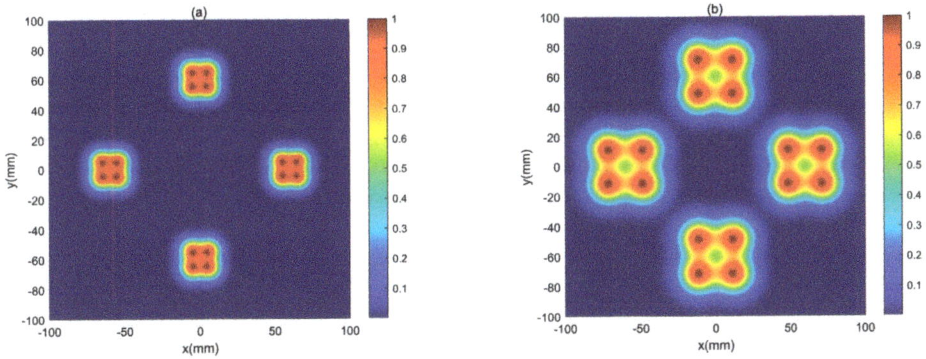

Figure 4. The intensity of an RPLHGSM beam array with $\delta_{0x} = \delta_{0y} = 3$ mm in oceanic turbulence. (**a**) $z = 100$ m, (**b**) $z = 200$ m.

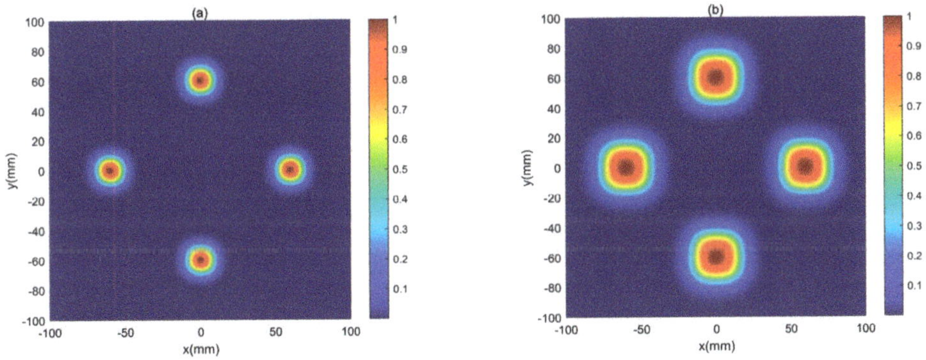

Figure 5. The intensity of an RPLHGSM beam array with $\delta_{0x} = \delta_{0y} = 5$ mm in oceanic turbulence. (**a**) $z = 100$ m, (**b**) $z = 200$ m.

When the radius R decreases, the intensity profiles of an RPLHGSM beam array with $N = 4$ and $R = 2$ cm transmitting through free space and oceanic turbulence are given in Figures 6 and 7. At $z = 100$ m, the intensity profile of sub-beam of an RPLHGSM beam array remains a spot profile. The sub-beam of this beam array splits at $z = 200$ m (Figure 6b), but the sub-beams of an RPLHGSM beam array in free space overlap with the surrounding sub-beams (Figure 6b) due to the reduction of R. While the same beam array in oceanic turbulence can evolve into one beam spot at $z = 200$ m, the combining properties of this beam array in oceanic turbulence are affected by radius R (Figures 5b and 7b). Therefore, the intensity profile of an RPLHGSM beam array in free space and oceanic turbulence can be modulated by controlling the parameter R and oceanic turbulence.

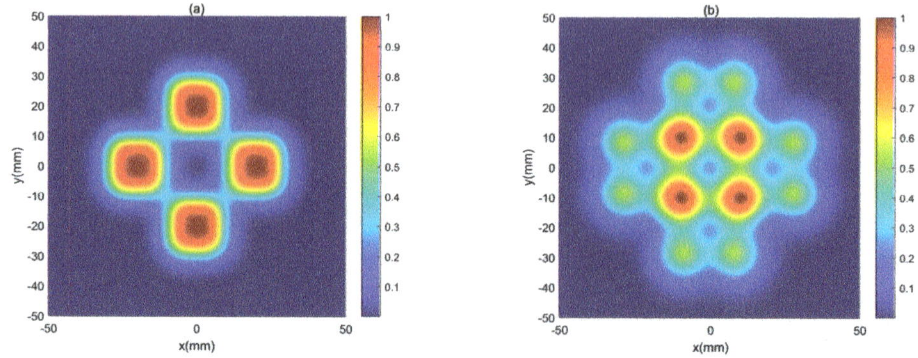

Figure 6. The intensity of an RPLHGSM beam array in free space. (**a**) $z = 100$ m, (**b**) $z = 200$ m.

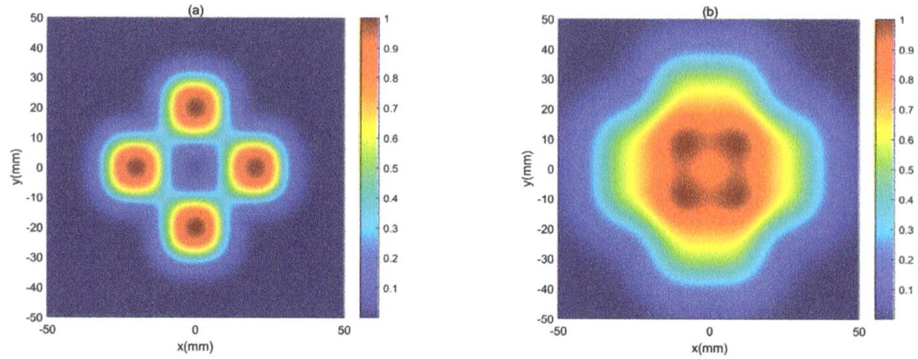

Figure 7. The intensity of an RPLHGSM beam array in oceanic turbulence. (**a**) $z = 100$ m, (**b**) $z = 200$ m.

5. Conclusions

A new type of beam array, named RPLHGSM beam array, composed of N HGCSM sub-beams with radial distribution is proposed. The CSD of this beam array in oceanic turbulence was derived, and the influences of initial beam parameters and oceanic turbulence on evolution characteristics of this beam array are numerically discussed. The beamlets of an RPLHGSM beam array retain self-splitting properties in free space, the RPLHGSM beam array in free space can attain multiple spot distribution during propagation, and an RPLHGSM beam array with low coherence retains splitting properties better. The sub-beam of the same beam array transmitting in oceanic turbulence will overlap, and as R decreases, the profile of this beam array in oceanic turbulence becomes a beam spot. An RPLHGSM beam array with smaller coherence length in oceanic turbulence will retain its splitting properties better. The results confirm that an RPLHGSM beam array can have applications in underwater laser sensing and laser radar.

Author Contributions: Conceptualization, D.L.; writing—original draft preparation, P.Z. and Y.Y.; writing—review and editing, G.W. and D.L.; software, Y.W.; data curation, H.Z.; project administration, G.W. and D.L. All authors have read and agreed to the published version of the manuscript.

Funding: This work was supported by National Natural Science Foundation of China (11604038, 11875096, 11404048).

Institutional Review Board Statement: Not applicable.

Informed Consent Statement: Not applicable.

Data Availability Statement: Data is contained within the article.

Conflicts of Interest: The authors declare no conflict of interest.

References

1. Baykal, Y.; Ata, Y.; Gökçe, M.C. Underwater turbulence, its effects on optical wireless communication and imaging: A review. *Opt. Laser Technol.* **2022**, *156*, 108624. [CrossRef]
2. Wang, F.; Liu, X.L.; Cai, Y.J. Propagation of Partially Coherent Beam in Turbulent Atmosphere: A Review. *Prog. Electromagn. Res.* **2015**, *150*, 123–143. [CrossRef]
3. Klug, A.; Peters, C.; Forbes, A. Robust structured light in atmospheric turbulence. *Adv. Photonics* **2023**, *5*, 016006. [CrossRef]
4. Korotkova, O.; Farwell, N.; Shchepakina, E. Light scintillation in oceanic turbulence. *Waves Random Complex Media* **2012**, *22*, 260–266. [CrossRef]
5. Baykal, Y. Scintillation index in strong oceanic turbulence. *Opt. Commun.* **2016**, *375*, 15–18. [CrossRef]
6. Xu, G.; Lai, J. Scintillation index and BER performance for optical wave propagation in anisotropic underwater turbulence under the effect of eddy diffusivity ratio. *Appl. Opt.* **2020**, *59*, 2551–2558. [CrossRef]
7. Gercekcioglu, H. Bit error rate of focused Gaussian beams in weak oceanic turbulence. *J. Opt. Soc. Am. A* **2014**, *31*, 1963–1968. [CrossRef]
8. Zheng, Y.; Yang, D.Y.; Qin, S.Q.; Zhang, Y.X. Received Probability of Orbital-Angular-Momentum Modes Carried by Diffraction- and Attenuation- Resistant Beams in Weak Turbulent Oceans. *J. Mar. Sci. Eng.* **2020**, *8*, 701. [CrossRef]
9. Liu, X.; Zhou, G.; Shen, Y. Effect of oceanic turbulence with anisotropy on the propagation of multi-sinc Schell-model beams. *Results Phys.* **2022**, *36*, 105447. [CrossRef]
10. Liu, L.; Liu, Y.; Chang, H.; Huang, J.; Zhu, X.; Cai, Y.; Yu, J. Second-Order Statistics of Self-Splitting Structured Beams in Oceanic Turbulence. *Photonics* **2023**, *10*, 339. [CrossRef]
11. Xu, J.; Tang, M.M.; Zhao, D.M. Propagation of electromagnetic non-uniformly correlated beams in the oceanic turbulence. *Opt. Commun.* **2014**, *331*, 1–5. [CrossRef]
12. Wei, D.; Wang, K.; Xu, Y.; Du, Q.; Liu, F.; Liu, J.; Dong, Y.; Zhang, L.; Yu, J.; Cai, Y.; et al. Propagation of a Lorentz Non-Uniformly Correlated Beam in a Turbulent Ocean. *Photonics* **2023**, *10*, 49. [CrossRef]
13. Wang, X.; Wang, L.; Zhao, S. Research on Hypergeometric-Gaussian Vortex Beam Propagating under Oceanic Turbulence by Theoretical Derivation and Numerical Simulation. *J. Mar. Sci. Eng.* **2021**, *9*, 442. [CrossRef]
14. Liu, D.; Wang, G.; Yin, H.; Zhong, H.; Wang, Y. Propagation properties of a partially coherent anomalous hollow vortex beam in underwater oceanic turbulence. *Opt. Commun.* **2019**, *437*, 346–354. [CrossRef]
15. Ye, F.; Zhang, J.B.; Xie, J.T.; Deng, D.M. Propagation properties of the rotating elliptical chirped Gaussian vortex beam in the oceanic turbulence. *Opt. Commun.* **2018**, *426*, 456–462. [CrossRef]
16. Hu, Y.; Zhang, M.; Dou, J.; Zhao, J.; Li, B. Influences of salinity and temperature on propagation of radially polarized rotationally-symmetric power-exponent-phase vortex beams in oceanic turbulence. *Opt. Express* **2022**, *30*, 42772–42783. [CrossRef]
17. Zhan, H.; Peng, Y.; Chen, B.; Wang, L.; Wang, W.; Zhao, S. Diffractive deep neural network based adaptive optics scheme for vortex beam in oceanic turbulence. *Opt. Express* **2022**, *30*, 23305–23317. [CrossRef]
18. Liu, D.; Zhong, H.; Wang, G.; Yin, H.; Wang, Y. Radial phased-locked multi-Gaussian Schell-model beam array and its properties in oceanic turbulence. *Opt. Laser Technol.* **2020**, *124*, 106003. [CrossRef]
19. Tang, M.; Li, H. Statistical properties of twisted Gaussian Schell-model array beams in anisotropic ocean. *Optik* **2020**, *211*, 164612. [CrossRef]
20. Chen, Y.H.; Wang, F.; Cai, Y.J. Partially coherent light beam shaping via complex spatial coherence structure engineering. *Adv. Phys. X* **2022**, *7*, 2009742. [CrossRef]
21. Wang, H.; Peng, X.; Zhang, H.; Liu, L.; Chen, Y.; Wang, F.; Cai, Y. Experimental synthesis of partially coherent beam with controllable twist phase and measuring its orbital angular momentum. *Nanophotonics* **2022**, *11*, 689–696. [CrossRef]
22. Korotkova, O.; Sahin, S.; Shchepakina, E. Multi-Gaussian Schell-model beams. *J. Opt. Soc. Am. A* **2012**, *29*, 2159–2164. [CrossRef] [PubMed]
23. Zheng, S.; Huang, J.; Ji, X.; Cheng, K.; Wang, T. Rotating anisotropic Gaussian Schell-model array beams. *Opt. Commun.* **2021**, *484*, 126684. [CrossRef]
24. Chen, Y.H.; Liu, L.; Wang, F.; Zhao, C.L.; Cai, Y.J. Elliptical Laguerre-Gaussian correlated Schell-model beam. *Opt. Express* **2014**, *22*, 13975–13987. [CrossRef] [PubMed]
25. Chen, Y.H.; Gu, J.X.; Wang, F.; Cai, Y.J. Self-splitting properties of a Hermite-Gaussian correlated Schell-model beam. *Phys. Rev. A* **2015**, *91*, 013823. [CrossRef]
26. Song, Z.; Liu, Z.; Zhou, K.; Sun, Q.; Liu, S. Propagation characteristics of a non-uniformly Hermite–Gaussian correlated beam. *J. Opt.* **2016**, *18*, 015606. [CrossRef]
27. Liu, L.; Wang, H.; Liu, L.; Dong, Y.; Wang, F.; Hoenders, B.J.; Chen, Y.; Cai, Y.; Peng, X. Propagation Properties of a Twisted Hermite-Gaussian Correlated Schell-Model Beam in Free Space. *Front. Phys.* **2022**, *10*, 847649. [CrossRef]
28. Cai, Y.J.; Lin, Q.; Baykal, Y.; Eyyuboglu, H.T. Off-axis Gaussian Schell-model beam and partially coherent laser array beam in a turbulent atmosphere. *Opt. Commun.* **2007**, *278*, 157–167. [CrossRef]

29. Zhou, P.; Wang, X.L.; Ma, Y.X.; Ma, H.T.; Xu, X.J.; Liu, Z.J. Propagation of partially coherent partially phase-locked laser array in turbulent atmosphere. *Opt. Commun.* **2010**, *283*, 1071–1074. [CrossRef]
30. Yan, Y.; Wang, G.; Yin, Y.; Yin, H.; Wang, Y.; Liu, D. Radial phased-locked Laguerre-Gaussian correlated schell-model beam array *Heliyon* **2022**, *8*, e11295. [CrossRef]
31. Lu, L.; Wang, Z.; Cai, Y. Propagation properties of phase-locked radially-polarized vector fields array in turbulent atmosphere *Opt. Express* **2021**, *29*, 16833–16844. [CrossRef] [PubMed]
32. Ma, B.B.; Sun, C.; Lv, X.; Zhang, J.B.; Yang, X.B.; Wang, G.H.; Hong, W.Y.; Deng, D.M. Effect of turbulent atmosphere on the propagation of a radial phased-locked rotating elliptical Gaussian beam array. *J. Opt. Soc. Am. A* **2019**, *36*, 1690–1698. [CrossRef] [PubMed]
33. Jeffrey, A.; Dai, H.H. *Handbook of Mathematical Formulas and Integrals*, 4th ed.; Academic Press Inc.: Cambridge, MA, USA, 2008.
34. Zhang, Y.; Yang, K.; Li, P.; Wen, F.; Gu, Y.; Wu, Z. Generation of off-axis phased Gaussian optical array along arbitrary curvilinear arrangement. *Opt. Commun.* **2023**, *527*, 128967. [CrossRef]
35. Zhang, Y.; Hou, T.; Chang, Q.; Chang, H.; Long, J.; Ma, P.; Zhou, P. Propagation Properties of Gaussian Schell-Model Beam Array in the Jet Engine Exhaust Induced Turbulence. *IEEE Photonics J.* **2020**, *12*, 1–13. [CrossRef]
36. Zhao, J.; Wang, G.; Ma, X.; Zhong, H.; Yin, H.; Wang, Y.; Liu, D. Intensity and Coherence Characteristics of a Radial Phase-Locked Multi-Gaussian Schell-Model Vortex Beam Array in Atmospheric Turbulence. *Photonics* **2020**, *8*, 5. [CrossRef]
37. Nikishov, V.V.; Nikishov, V.I. Spectrum of Turbulent Fluctuations of the Sea-Water Refraction Index. *Int. J. Fluid Mech. Res.* **2000**, *27*, 82–98. [CrossRef]

photonics

MDPI

Communication

Statistical Properties of Partially Coherent Higher-Order Laguerre-Gaussian Power-Exponent Phase Vortex Beams

Zhiyuan Ma [1,2], Yuqi Pan [2], Jiantai Dou [2], Jiang Zhao [3], Bo Li [1,*] and Youyou Hu [2,*]

1 School of Optical and Electronic Information & Wuhan National Laboratory for Optoelectronics, Optics Valley Laboratory National Engineering Research Center for Laser Processing, Huazhong University of Science and Technology, Wuhan 430074, China; 202050007@stu.just.edu.cn
2 Department of Optoelectronic Information of Science and Engineering, School of Science, Jiangsu University of Science and Technology, Zhenjiang 212100, China; 211110501102@stu.just.edu.cn (Y.P.); jiantai.dou@just.edu.cn (J.D.)
3 Key Laboratory of Ferro and Piezoelectric Materials and Devices of Hubei Province, Faculty of Physics and Electronic Science, Hubei University, Wuhan 430062, China; zhaojiang@hubu.edu.cn
* Correspondence: libohust@hust.edu.cn (B.L.); yyhu@just.edu.cn (Y.H.)

Abstract: In this paper, partially coherent radially polarized (RP) Laguerre-Gaussian (LG) rotationally symmetrical power-exponent phase vortex (RSPEPV) beams with the LG-correlated Schell-model (LGSM) were introduced. The statistical properties of the tightly focused beams, including intensity distribution, degrees of polarization and coherence, and Stokes vector, were studied based on vectorial Richards-Wolf diffraction integral theory. Moreover, when the distance between focal plane and the observation plane $z = 0$, the relationships between the tight-focusing properties of RP-LG-RSPEPV beams with LGSM and the order of LGSM p', topological charges l, power exponent n, spatial correlation δ, and radial index p were investigated. The results show that by changing the order of LGSM, topological charge, power exponent, spatial correlation length, and radial index, the focal spot distribution of various shapes can be obtained. This work provides ideas for the application of partially coherent beams in particle capture and optical tweezers.

Keywords: radially polarized; partially coherent; stokes parameter; tight-focusing

Citation: Ma, Z.; Pan, Y.; Dou, J.; Zhao, J.; Li, B.; Hu, Y. Statistical Properties of Partially Coherent Higher-Order Laguerre-Gaussian Power-Exponent Phase Vortex Beams. *Photonics* **2023**, *10*, 461. https://doi.org/10.3390/photonics10040461

Received: 10 March 2023
Revised: 13 April 2023
Accepted: 14 April 2023
Published: 17 April 2023

1. Introduction

Power-exponent phase vortex (PEPV) beams are special optical vortexes with different phase velocities within the transverse dimension [1–6]. According to the intensity distribution, PEPV beams can be categorized as asymmetric and rotationally symmetric. In the phase distribution of an asymmetric power-exponential phase vortex (APEPV) beam, the angle of each sector is different, so that the phase velocity of each sector is different, which leads to its optical intensity distribution resembling a "C" shape [1,6]. These unique properties give it some advantages in optical manipulation, and a study has shown that particles can be moved along specified curved paths by tuning the topological charge and power exponent of the APEPV beams [7]. Further, for rotationally symmetric power-exponent phase vortex (RSPEPV) beams, the angle of each sector in its phase distribution is the same. However, the phase velocity within the same sector is different, thus forming a rotationally symmetric and blade-like intensity distribution [8,9]. The number of the "blade" is equal to the topological charges of the RSPEPV beams, which makes it possible to manipulate multiple particles of the same type simultaneously [10]. At present, the research on RSPEPV beams is mainly in the field of fully coherent beams, such as self-focusing Airy [11], radially polarized or scalar RSPEPV beams [6,9], as mentioned in our previous works. However, there is little research into partially coherent higher-order LG-RSPEPV beams in this field.

In addition, apart from controlling the light intensity and phase to achieve specific practical requirements, polarization and coherence are also two important controllable parameters [12–14]. For example, a very strong longitudinal component can be generated by using radially polarized (RP) beams to pass through a high-numerical aperture (NA) focusing system [15,16], which has potential applications in particle trapping [17]. Further, the size of the tightly focused beam is generally in the subwavelength size, which also promotes the development of an optical needle [18], optical field manipulation [19], and quantum optics [20]. Therefore, the tight-focusing characteristics of radially polarized beams have always been the focus of research [15,16]. Partially coherent vector beams have advantages over fully coherent beams in the fields of light trapping [21,22], free-space optical (FSO) communication [23] and high-resolution imaging [24]. Specifically, by tuning the spatial coherence of radially polarization beams, it is possible to trap types of particles with different sizes and refractive indices [21,22]. In previous works, we have investigated the propagation properties of partially coherent [3] Laguerre-Gaussian (LG) APEPV beams [1], tightly focused properties of RSPEPV beams [2], and 0-order LG-RSPEPV [25] beams, and pointed out some of their application advantages in optical trapping [10]. However, they may not provide the opportunity to manipulate different types of quantitative particles simultaneously. The high-order LG-RSPEPV beam has a controllable number of outer rings and complex structures. By adjusting the topological charge, the radial index of LG beams, the order of the LGSM and coherence length, the radiation force of the beams can be modulated, which provides a way to solve this problem.

In this paper, partially coherent radially polarized RP-LG RSPEPV beams with an LG-correlated Schell-model (LGSM) were introduced. Based on vectorial Richards-Wolf diffraction integral theory, the intensity distribution, degree of polarization and coherence, and Stokes vector of the tightly focused beams are investigated. Moreover, by adjusting the parameters including power exponent n, radial index p, topological charges l, spatial correlation δ, radial index p, and the order of LGSM p', the statistical characteristics of the beams in the focal plane were investigated. Combing the characteristics of RP-LG-RSPEPV and LGSM, the beam shape can be controlled from multiple parameters. This work presents the opportunity to manipulate different types of quantitative particles simultaneously.

2. Theory

2.1. Theory for RP-LG-RSPEPV Beams with LGSM

Firstly, in the polar coordinates' system, the RSPEPV beams at the source plane can be characterized as follows [2]:

$$E(r,\varphi) = A_0 E(r) \exp\left[2\pi\left(\frac{rem(l\varphi, 2\pi)}{2\pi}\right)^n\right], \quad (1)$$

where r and φ are polar coordinates; l denotes topological charges; n is the power exponent; A_0 is the amplitude of the incident beam, which is equal to 1 for convenience; $rem(x,y)$ represent remainder function; $E(r)$ is the expression of different beam modes, and $E(r)$ could be equal to the Airy function [11], the Lorentz-Gaussian function [26], or the Bessel-Gaussian [27] function to obtain different forms of RSPEPV beams. However, the Laguerre-Gaussian function is introduced in this paper,

$$E_{\mathrm{LG}}(r) = \frac{\sqrt{2p!/\pi(|l|+p)!}}{w}\left(\frac{\sqrt{2}r}{w}\right)^{|l|} L_p^{|l|}\left(\frac{2r^2}{w^2}\right)\exp\left(\frac{-r^2}{w^2}\right), \quad (2)$$

where $L_p^{|l|} = (\cdot)$ is the Laguerre polynomials, and w is the beam width.

$$L_p^{|l|}(x) = \sum_{m=0}^{p}(-1)^m \frac{(|l|+p)!}{(p-m)!(|l|+m)!m!}x^m, \quad (3)$$

Then, substituting Equation (2) into Equation (1), the LG-PEPV beams at the source plane can be written as

$$E_{\mathrm{LG-PEPV}}(r,\varphi) = \frac{\sqrt{2p!/\pi(|l|+p)!}}{w}\left(\frac{\sqrt{2}r}{w}\right)^{|l|} L_p^{|l|}\left(\frac{2r^2}{w^2}\right)\exp\left(\frac{-r^2}{w^2}\right)\exp\left[2\pi\left(\frac{rem(l\varphi,2\pi)}{2\pi}\right)^n\right], \tag{4}$$

Secondly, the coherent RP beams in the Cartesian coordinate system can be characterized [16]:

$$\mathbf{E}(r,\varphi) = E_0(r,\varphi)\left[r\left(\cos m\varphi\cdot\mathbf{e}_x + \sin m\varphi\cdot\mathbf{e}_y\right)\right], \tag{5}$$

where $\mathbf{e}_x$ and $\mathbf{e}_y$ are the unit vectors in the x and y directions, respectively; $E_0(r, \varphi)$ is the electric field expression of the beams; m is the order of the cylindrical vector beams. In this paper, we only consider the RP beam, i.e., the case of $m = 1$. Substituting Equation (4) into Equation (5), we can obtain the mathematical model of the RP-LG-RSPEPV beams at the source plane.

Further, according to the unified theory of coherent and polarization, the second-order statistical properties of the vector partially coherent beams can be characterized by a 2 × 2 cross-spectral density (CSD) matrix, which can be expressed as follows [16]:

$$W(\mathbf{r}_1,\mathbf{r}_2) = \begin{bmatrix} W_{xx}(\mathbf{r}_1,\mathbf{r}_2) & W_{xy}(\mathbf{r}_1,\mathbf{r}_2) \\ W_{yx}(\mathbf{r}_1,\mathbf{r}_2) & W_{yy}(\mathbf{r}_1,\mathbf{r}_2)\end{bmatrix}. \tag{6}$$

where $\mathbf{r}_1$, $\mathbf{r}_2$ are any two position vectors on the source plane, and the expression of CSD with LGSM is [28]

$$\begin{aligned} W_{\alpha\beta}(\mathbf{r}_1,\mathbf{r}_2) &= \left\langle E_\alpha(\mathbf{r}_1)E_\beta^*(\mathbf{r}_2)\right\rangle, (\alpha = x,y; \beta = x,y), \\ &= A_\alpha A_\beta B_{\alpha\beta}\exp\left[-\frac{\mathbf{r}_1^2+\mathbf{r}_2^2}{w_0{}^2} - \frac{(\mathbf{r}_1-\mathbf{r}_2)^2}{2\delta_{\alpha\beta}{}^2}\right] L_{p'}^0\left[\frac{(\mathbf{r}_1-\mathbf{r}_2)^2}{2\delta_{\alpha\beta}{}^2}\right]. \end{aligned} \tag{7}$$

In Equation (7), $E_x(\mathbf{r})$ and $E_y(\mathbf{r})$ denote two orthogonal random electric field components along the x and y directions, respectively. '*' denotes a complex conjugate; '< >' represents the ensemble average; A_x and A_y are the amplitudes in the x and y directions of electric field, respectively; and B_{xy} represents the correlation coefficient of the single point. It is worth mentioning that to achieve radial polarization, the conditions of $B_{xy(yx)} = B_{xx(yy)} = 1$, $\delta_{xy(yx)} = B_{xx(yy)} = \delta$. δ is the spatial correlation length of the partially coherent beam [16].

Combining Equations (4) and (7), the theoretical model of RP-LG-RSPEPV beams on the source plane can be given as

$$\begin{aligned} W_{\alpha\beta}(\mathbf{r}_1,\mathbf{r}_2,0) &= \frac{2p!/\pi(|l|+p)!}{w^2}\left(\frac{2r_1r_2}{w}\right)^{|l|} L_p^{|l|}\left(\frac{2r_1{}^2}{w^2}\right) L_p^{|l|}\left(\frac{2r_2{}^2}{w^2}\right) \\ &\exp\left(-\frac{\mathbf{r}_1{}^2+\mathbf{r}_2{}^2}{w_0{}^2} - \frac{(\mathbf{r}_1-\mathbf{r}_2)^2}{2\delta}\right) L_{p'}^0\left(\frac{(\mathbf{r}_1-\mathbf{r}_2)^2}{2\delta^2}\right) \\ &\exp\left[i2\pi\left(\frac{rem(l\varphi_1,2\pi)}{2\pi}\right)^n - i2\pi\left(\frac{rem(l\varphi_2,2\pi)}{2\pi}\right)^n\right], \end{aligned} \tag{8}$$

where the p and p' are the radial index of the RP-LG-RSPEPV beams and the order of LGSM, respectively. When the $p' = 0$, the LGSM reduces to the conventional Gaussian-correlated Schell-model (GSM).

2.2. Theory for the Tight-Focusing of RP-LG-RSPEPV Beams with LGSM

In this part, the theory for the tight-focusing of RP-RSPEPV beams with LGSM is introduced. Firstly, the schematic diagram of RP beams through a lens with high numerical aperture (NA) is shown in Figure 1.

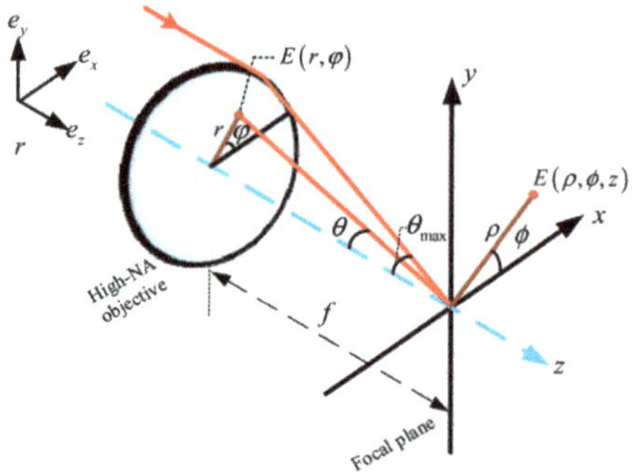

Figure 1. Schematic diagram of cylindrical vector beams through a lens with high numerical aperture.

According to the Richards-Wolf vectorial diffraction integral theory and the sine condition of the aplanatic lens, that is, $\rho = f \cdot \sin\theta$, the vectorial electric field of the RP beam at the focal plane can be characterized [16]:

$$E_f(\rho,\phi,z) = \begin{bmatrix} E_{fx} \\ E_{fy} \\ E_{fz} \end{bmatrix} = -\frac{ikn_1 f}{2\pi}\int_0^{\theta_{\max}}\int_0^{2\pi}\begin{bmatrix} A_x(\theta,\varphi) \\ A_y(\theta,\varphi) \\ A_z(\theta,\varphi) \end{bmatrix}\sqrt{\cos\theta}\sin\theta \times \exp[ikn_1(z\cos\theta + \rho\sin\theta\cos(\varphi-\phi))]d\varphi d\theta, \tag{9}$$

With

$$\begin{bmatrix} A_x(\theta,\varphi) \\ A_y(\theta,\varphi) \\ A_z(\theta,\varphi) \end{bmatrix} = \begin{bmatrix} l_x(\theta,\varphi)(\cos\theta + \sin^2\varphi(1-\cos\theta)) + l_y(\theta,\varphi)\cos\varphi\sin\varphi(\cos\theta - 1) \\ l_x(\theta,\varphi)\cos\varphi\sin\varphi(\cos\theta-1) + l_y(\theta,\phi)(\cos\theta + \cos^2\varphi(1-\cos\theta)) \\ -l_x(\theta,\phi)\sin\theta\cos\varphi - l_y(\theta,\phi)\sin\theta\sin\varphi \end{bmatrix} \tag{10}$$

where ρ, ϕ, z are the cylindrical coordinates of the focal plane; f is the focal length of the lens; $k = 2\pi/\lambda$ is the wave number; n_1 is the refractive index of the surrounding medium; θ is the NA angle; the maximal NA angle is defined as $\theta_{\max} = \arcsin(\mathrm{NA}/n)$; $l_x(\theta$, and $\varphi)$ and $l_y(\theta, \varphi)$ are the apodization functions of the aperture surface, which can be obtained by taking $x = \rho\cos\varphi$ and $y = \rho\sin\varphi$ into the $E_\alpha(x,y)$.

Further, when the RP-LG-RSPEPV beam with LGSM is focused by a high-NA objective, the statistical properties of the beam near the focused region can be given by a 3×3 CSD matrix:

$$W(r_1,r_2,\phi_1,\phi_2,z) = \begin{bmatrix} W_{xx}(r_1,r_2,\phi_1,\phi_2,z) & W_{xy}(r_1,r_2,\phi_1,\phi_2,z) & W_{xz}(r_1,r_2,\phi_1,\phi_2,z) \\ W_{yx}(r_1,r_2,\phi_1,\phi_2,z) & W_{yy}(r_1,r_2,\phi_1,\phi_2,z) & W_{yz}(r_1,r_2,\phi_1,\phi_2,z) \\ W_{zx}(r_1,r_2,\phi_1,\phi_2,z) & W_{zy}(r_1,r_2,\phi_1,\phi_2,z) & W_{zz}(r_1,r_2,\phi_1,\phi_2,z) \end{bmatrix}. \tag{11}$$

where

$$W_{f\alpha\beta}(r_1,\phi_1,r_2,\phi_2,z) = \left\langle E_{f\alpha}(r_1,\phi_1,z)E_{f\beta}{}^*(r_2,\phi_2,z)\right\rangle. \tag{12}$$

In the focal plane, the expression for the diagonal elements of the CSD matrix can represent the average intensity distribution of the tightly focused RP-LG-RSPEPV beam with LGSM near the focal region. Specifically, it can be written as

$$\begin{aligned} W_{xx}(r_1,\phi_1,r_2,\phi_2,z) &= \frac{f^2}{\lambda^2}\int_0^{\alpha}\int_0^{\alpha}\int_0^{2\pi}\int_0^{2\pi} W_0(\theta_1,\theta_2,\varphi_1,\varphi_2) \times (\cos\theta_1\cos\theta_2)^{3/2}(\sin\theta_1\sin\theta_2)^2 \\ &\times \exp[-ikz\cos\theta_1 - ikr_1\sin\theta_1\cos(\varphi_1-\phi_1)] \\ &\times \exp[ikz\cos\theta_2 + ikr_2\sin\theta_2\cos(\varphi_2-\phi_2)] \\ &\times \cos\varphi_1\cos\varphi_2 d\theta_1 d\theta_2 d\varphi_1 d\varphi_2, \end{aligned} \tag{13}$$

$$W_{yy}(r_1,\phi_1,r_2,\phi_2,z) = \frac{f^2}{\lambda^2}\int_0^{\alpha}\int_0^{\alpha}\int_0^{2\pi}\int_0^{2\pi} W_0(\theta_1,\theta_2,\varphi_1,\varphi_2) \times (\cos\theta_1\cos\theta_2)^{3/2}(\sin\theta_1\sin\theta_2)^2 \times \exp[-ikz\cos\theta_1 - ikr_1\sin\theta_1\cos(\varphi_1-\phi_1)] \times \exp[ikz\cos\theta_2 + ikr_2\sin\theta_2\cos(\varphi_2-\phi_2)] \times \sin\varphi_1\sin\varphi_2 d\theta_1 d\theta_2 d\varphi_1 d\varphi_2, \tag{14}$$

$$W_{zz}(r_1,\phi_1,r_2,\phi_2,z) = \frac{f^2}{\lambda^2}\int_0^{\alpha}\int_0^{\alpha}\int_0^{2\pi}\int_0^{2\pi} W_0(\theta_1,\theta_2,\varphi_1,\varphi_2) \times (\cos\theta_1\cos\theta_2)^{1/2} \times \exp[-ikz\cos\theta_1 - ikr_1\sin\theta_1\cos(\varphi_1-\phi_1)] \times \exp[ikz\cos\theta_2 + ikr_2\sin\theta_2\cos(\varphi_2-\phi_2)] \times (\sin\theta_1\sin\theta_2)^3 d\theta_1 d\theta_2 d\varphi_1 d\varphi_2. \tag{15}$$

where

$$W_0(\theta_1,\theta_2,\varphi_1,\varphi_2) = \frac{2p!/\pi(|l|+p)!}{w^2}\left(\frac{2f^2\sin\theta_1\sin\theta_2\cos(\varphi_1-\varphi_2)}{w^2}\right)^l \times \exp\left[-\frac{f^2(\sin^2\theta_1+\sin^2\theta_2)}{w^2}\right] L_p^{|l|}\left(\frac{2f(\sin\theta_1)^2}{w^2}\right) L_p^{|l|}\left(\frac{2f(\sin\theta_2)^2}{w^2}\right) \times \exp\left[i2\pi\left(\frac{rem(l\varphi_1,2\pi)}{2\pi}\right)^n - i2\pi\left(\frac{rem(l\varphi_2,2\pi)}{2\pi}\right)^n\right] \times \exp\left[-\frac{f^2(\sin^2\theta_1+\sin^2\theta_2-\sin\theta_1\sin\theta_2\cos(\varphi_1-\varphi_2))}{2\delta^2}\right] \times L_{p'}^0\left(\frac{f^2(\sin^2\theta_1+\sin^2\theta_2-\sin\theta_1\sin\theta_2\cos(\varphi_1-\varphi_2))}{2\delta^2}\right). \tag{16}$$

Setting $r_1 = r_2 = r$, $\phi_1 = \phi_2 = \phi$, the intensity of transverse, longitudinal and total component at the focal plane can be expressed as

$$I_{tra}(\rho,\phi,z) = E_{fx}(\rho,\phi,z)\cdot E_{fx}^*(\rho,\phi,z) + E_{fy}(\rho,\phi,z)\cdot E_{fy}^*(\rho,\phi,z) \tag{17}$$

$$I_z(\rho,\phi,z) = E_z(\rho,\phi,z)\cdot E_z^*(\rho,\phi,z) \tag{18}$$

$$I_{total}(\rho,\phi,z) = I_{tra}(\rho,\phi,z) + I_z(\rho,\phi,z) \tag{19}$$

3. Results and Discussions

3.1. The Relationships between Tight-Focusing Properties and Topological Charges, the Power Exponent and Radial Index

In this section, we will investigate how the topological charge l, the power exponent n and the radial index p of the RP-LG-RSPEPV beams affect the intensity distribution of the beams passing through a high-NA lens by numerical simulation. In the following simulation calculation, the wavelength of the beams is set as 632.8 nm, the NA of the lens is set as 0.98, the beams' width is w = 2 mm, the spatial correlation length of the beams is set as 5 mm, the refractive index of the surrounding medium n_1 = 1, and the focal length of the lens is f = 3 mm.

Figure 2 shows the normalized intensity distribution of each component of the different topological charges l of the tight focusing RP-LG-RSPEPV beams at the focal plane when the power exponent n = 4, the spatial correlation length δ = 0.005 m, the radial order p = 1, and the order of LGSM p' = 1. It can be seen clearly that when the topological charge increases, the intensity of the transverse component decreases, while the intensity of the longitudinal component increases, which may mean that the beams with large waist width are more likely to generate longitudinal components with powerful intensity when passing through a high-NA lens. Additionally, the size of the beams becomes larger, and the outer ring of the total intensity distribution gradually becomes clear as the topological charges increases. This is because in the process of tight-focusing, when the beam size increases, the transverse component of light intensity weakens while the longitudinal component gradually increases. From Figure 2, the intensity of the outer ring is mainly concentrated in the longitudinal component of light intensity. In addition, Figure 2a1–a3 is an APEPV, because when the topological charge is 1, the expression of APEPV and RSPEPV is the same. In optical capture, particles can be quantitatively captured by adjusting the topological charge number [10].

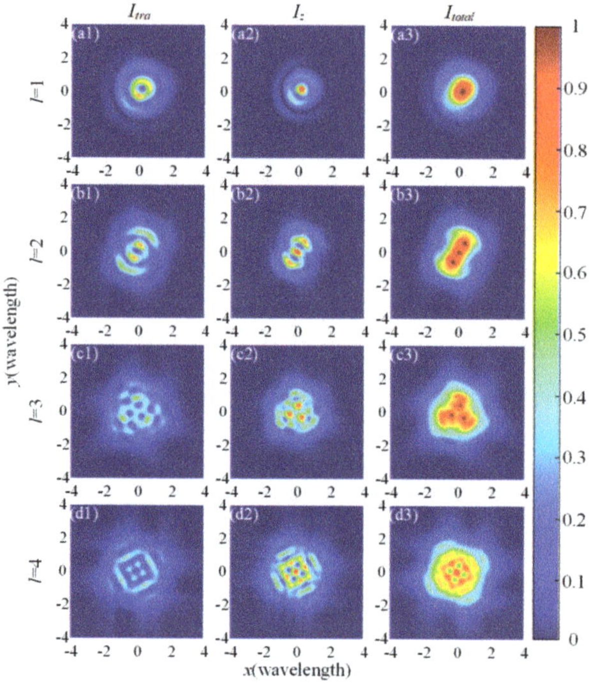

Figure 2. The normalized intensity distribution of each component (n = 4, p = 1, and p' = 1). (**a1–d1**) The transverse intensity distribution, (**a2–d2**) the longitudinal intensity distribution, and (**a3–d3**) the total intensity distribution of the RP-LG-RSPEPV beams with topological charges of 1–4.

To further reveal the influence of power exponent n and radial index p on the intensity distribution of RP-LG-RSPEPV beams at the focal plane, the intensity distribution of RP-LG-RSPEPV beams with a different power exponent n and radial index p, when the topological charges l = 3, the spatial correlation length δ = 0.005 m, and the order of LGSM p' = 1, are shown in Figure 3. The results showed that the number of outer rings is equal to the radial index p. However, when the topological charge of the beam is small, it cannot be found in the total field distribution. As the topological charge increases, the outer ring becomes clearer. The extra outer ring is caused by the Bessel function in the tight-focusing mathematical model, and the larger the beam size, the more obvious the influence of the Bessel function. With the increase in the power index n, the light intensity is fan-like, and the larger the n, the clearer the contour of the fan-like light intensity. Observing Figure 3a2,b2, it can be found that the total field distribution of beam intensity was more concentrated with the increase in radial order p. It is worth noting that the hollow focal field of the rotationally symmetric focal spot may help to trap irregular particles or manipulate particles.

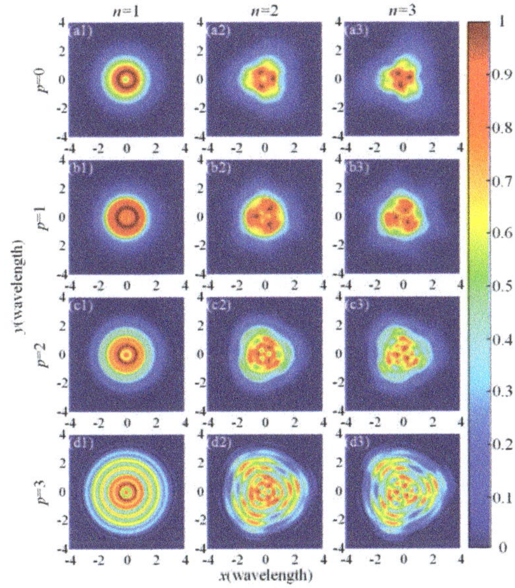

Figure 3. The normalized intensity distribution of different power exponents n and radial index p ($l = 3$, $\delta = 0.005$ m, and $p' = 1$). When the radial index p is 0 to 3, the total intensity distribution of **(a1–d1)** the power exponents $n = 1$, **(a2–d2)** the power exponents $n = 2$, and **(a3–d3)** the power exponents $n = 3$.

3.2. The Relationships between Tight-Focusing Properties and the Spatial Correlation Length, and the Order of LGSM

To learn more about the influence of the spatial correlation length δ and the order of LGSM p' on the intensity distribution of RP-LG-RSPEPV after it is tightly focused, we calculated the normalized focal intensity distribution of topological charge $l = 3$, spatial correlation length $\delta = 0.005$ m, and $p = 1$, with the change in the order of LGSM p' and power exponent n in Figure 4. One can see from Figure 4 that the total intensity of the beam decreases when the order of LGSM p' increases, which is more obvious when the power exponent n increases. In fact, the weakening of light intensity may reflect the weakening of coherence, because the lasers always have strong correlation and light intensity. However, this alone cannot explain why the RS-LG-RSPEPV beams have different degrees of coherence under different orders of LGSM. This can be described by the degree of coherence at the plane of $z = 0$, which itself can be given by

$$\mu_{\alpha\beta} = \frac{W_{\alpha\beta}(\mathbf{r}_1, \mathbf{r}_2, 0)}{\sqrt{W_{\alpha\alpha}(\mathbf{r}_1, \mathbf{r}_2, 0) W_{\beta\beta}(\mathbf{r}_1, \mathbf{r}_2, 0)}} (\alpha, \beta = x, y, z). \tag{20}$$

Figure 5 shows the normalized focal intensity distribution of topological charge $l = 3$, power exponent $n = 4$, and $p = 1$, with the change in the order of LGSM p' and the spatial correlation length. It is interesting to observe in Figure 5 that when the orders of LGSM p' are 0 and 1, with the increase in the spatial coherence length δ, the total light intensity distribution profile at the focal plane gradually becomes clear, and the normalized light intensity is gradually reduced. On the contrary, the beam energy is more concentrated. When p' is 2 and 3, the normalized light intensity gradually increases. Both the orders of LGSM and the spatial coherence length will affect the total intensity distribution of the focal spot in the RP-LG-RSPEPV with LGSM.

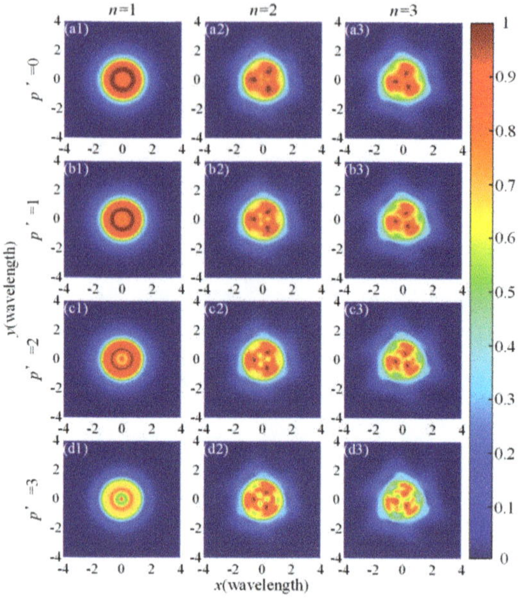

Figure 4. The normalized intensity distribution of different power exponents n and the order of LGSM p' ($l = 3$, $\delta = 0.005$ m, and $p = 1$). When the order of LGSM p' is 0 to 3, the total intensity distribution of (**a1–d1**) the power exponents $n = 1$, (**a2–d2**) the power exponents $n = 2$, and (**a3–d3**) the power exponents $n = 3$.

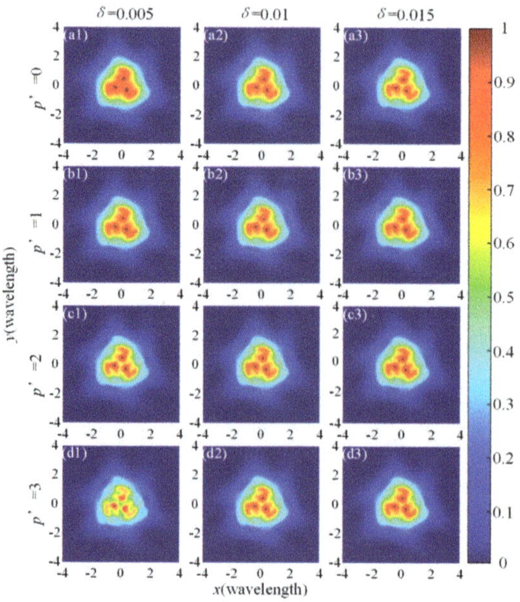

Figure 5. The normalized intensity distribution of different the spatial correlation lengths δ and the order of LGSM p' ($l = 3$, $n = 4$, and $p = 1$). When the order of LGSM p' is 0 to 3, the total intensity distribution of (**a1–d1**) the spatial correlation lengths $\delta = 0.005$ m, (**a2–d2**) the spatial correlation lengths $\delta = 0.01$ m, and (**a3–d3**) the spatial correlation lengths $\delta = 0.015$ m.

3.3. Degree of Polarization and Degree of Coherence of Tightly Focused RP-LG-RSPEPV Beams in the Transverse Direction

The degree of polarization and degree of coherence in the transverse mode of the beam in the focal plane will directly affect the total field distribution of the beam. Therefore, in this section, we discuss the Stokes parameters of the beam in the transverse section and the degree of coherence between the components. The Stokes parameters in the transverse plane of the RP-LG-RSPEPV are given as follows [29]:

$$\begin{cases} S_0(\mathbf{r}_1,\mathbf{r}_2,0) = W_{xx}(\mathbf{r}_1,\mathbf{r}_2,0) + W_{yy}(\mathbf{r}_1,\mathbf{r}_2,0) \\ S_1(\mathbf{r}_1,\mathbf{r}_2,0) = W_{xx}(\mathbf{r}_1,\mathbf{r}_2,0) - W_{yy}(\mathbf{r}_1,\mathbf{r}_2,0) \\ S_2(\mathbf{r}_1,\mathbf{r}_2,0) = W_{xy}(\mathbf{r}_1,\mathbf{r}_2,0) + W_{yx}(\mathbf{r}_1,\mathbf{r}_2,0) \\ S_3(\mathbf{r}_1,\mathbf{r}_2,0) = i(W_{yx}(\mathbf{r}_1,\mathbf{r}_2,0) + W_{xy}(\mathbf{r}_1,\mathbf{r}_2,0)) \end{cases} \quad (21)$$

Firstly, the Stokes parameters of different power exponents n in the transverse plane when the $l = 3$, $p' = 1$, and $\delta = 0.005$ m are given in Figure 6. It can be found that the increase in power exponent n makes the linear polarization component of the tightly focused RP-LG-RSPEPV beam irregular. However, there is still a spatially inhomogeneous polarization distribution, that is, it still has vector characteristics. In addition, subsequently, the tightly focused RP-LG-RSPEPV beam has a circularly polarized component, as shown in Figure 6d1–d3. This may be related to the conservation of total angular momentum, which produces spin angular momentum in the process of tight-focusing [30].

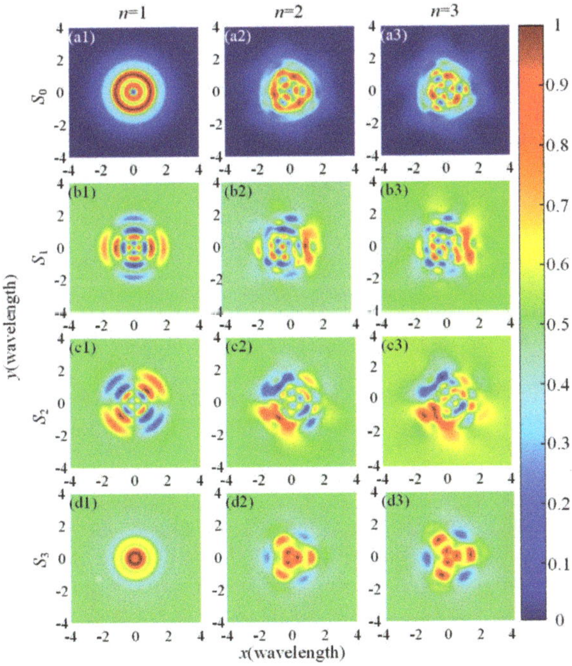

Figure 6. The Stokes parameters of different power exponents n in the transverse plane. ($l = 3$, $p' = 1$, and $\delta = 0.005$ m). The Stokes parameters of (**a1–d1**) the power exponents $n = 1$, (**a2–d2**) the power exponents $n = 2$, and (**a3–d3**) the power exponents $n = 3$.

Further, we can learn more about the polarization of the RP-LG-RSPEPV beams with different spatial coherence lengths δ and $l = 3$, $p' = 1$, and $n = 4$ after tight-focusing by looking to Figure 7. It can be found that the values of the S_1, S_2, S_3 in the majority of the tightly focused RP-LG-RSPEPV beams are less than 1, which indicates that the tight-focusing will depolarize the RP-LG-RSPEPV beams. This phenomenon is more obvious for the S_3 component of the RP-LG-RSPEPV beams in the transverse plane, when the spatial coherence is small. However, with the change in the spatial coherence length δ, depolarization has little effect on S_1 and S_2.

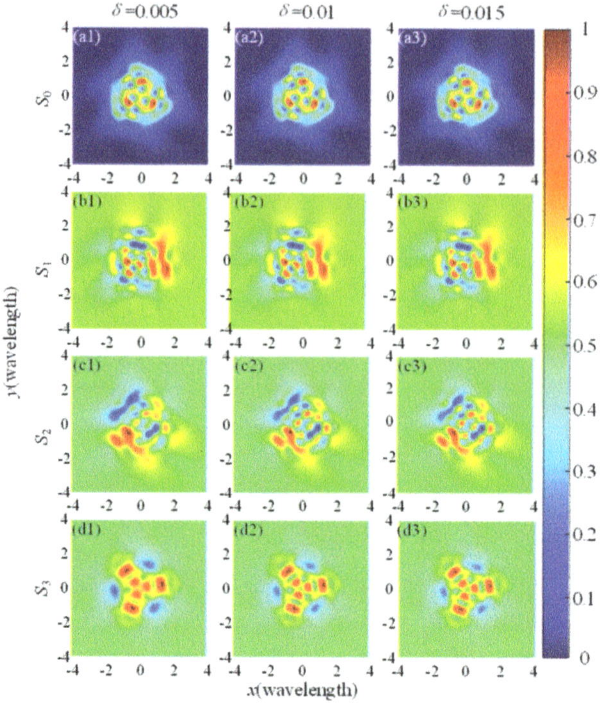

Figure 7. The Stokes parameters of different the spatial coherence lengths δ in the transverse plane. ($l = 3$, $p' = 1$, and $n = 4$). The Stokes parameters of (**a1–d1**) the spatial correlation lengths $\delta = 0.005$ m, (**a2–d2**) the spatial correlation lengths $\delta = 0.01$ m, and (**a3–d3**) the spatial correlation lengths $\delta = 0.015$ m.

Then, the degree of coherence between each component is given in Figure 8, when the spatial coherence length δ changes with topological charge $l = 3$, radial index $p = 1$, the order of LGSM $p' = 1$, and the power exponent $n = 4$. It can be found that the spatial coherence length will affect the coherence between the components of the LG beam. The larger the spatial coherence length, the higher the degree of the coherence in the RP-LG-RSPEPV beams. Besides, any two components of a tightly focused beam are partially coherent, which indicates that tight-focusing will reduce the correlation between any two components of the three orthogonal electric field components. The influence of the coherence width on the coherence degree may be affected by the special correlation function of the incident beam. Controlling the structure of the correlation function can determine the transverse field distribution and the longitudinal field distribution of the closely focused radially polarized partially coherent beam. In practical applications, we can adjust multiple parameters of the RP-LG-RSPEPV beam to adjust the light field distribution of the beam.

In addition, the degree of coherence between each component is shown in Figure 9, when the order of LGSM p' changes with topological charge $l = 3$, radial index $p = 1$, the spatial coherence length $\delta = 0.015$ m, and the power exponent $n = 4$. The results indicated that as the order of LGSM increases, the coherence of the beam decreases. The increase in the order of LGSM had a greater impact on the degree of coherence U_{xz} and U_{yz}, but had a smaller impact on the degree of coherence U_{xy}.

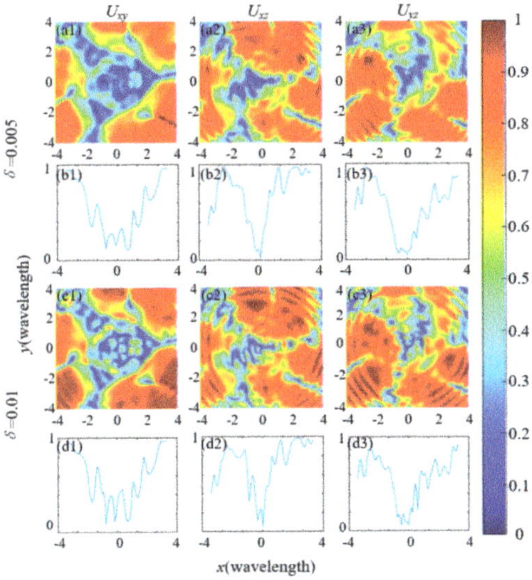

Figure 8. The degree of coherence between every component with different spatial coherence lengths δ. ($l = 3$, $p = 1$, $p' = 1$, and $n = 4$). Where, (a1–d1) the component U_{xy}, (a2–d2) the component U_{xz}, (a3–d3) the component U_{yz}.

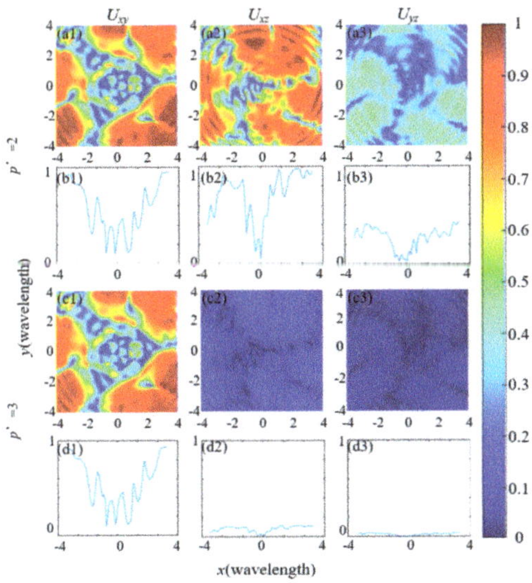

Figure 9. The degree of coherence between every component with different the order of LGSM p'. ($l = 3$, $p = 1$, $\delta = 0.015$ m, and $n = 4$). Where, (a1–d1) the component U_{xy}, (a2–d2) the component U_{xz}, (a3–d3) the component U_{yz}.

4. Conclusions

In this paper, RP-LG-RSPEPV beams with LGSM were introduced. The tight-focusing properties of the beams, including intensity distribution, the degree of polarization and coherence, and Stokes parameters, are studied based on Richards-Wolf diffraction vectorial integral theory. Moreover, the relationship between the tight-focusing properties of RP-LG-RSPEPV beams with LGSM and its power exponent n, radial index p, and topological charges l was explored. The results show that tight-focusing will cause the depolarization of the RP-LG beam, especially the component S_3 in the transverse plane. The decrease in spatial coherence length δ and the increase in the order of LGSM p' will reduce the coherence of RP-LG-PSPEPV beams, especially the components U_{xz} and U_{yz}. The RP-LG-RSPEPV beams' intensity can be adjusted by changing the parameters of the beam. This work may be of great significance for optical tweezers and particle manipulation.

Author Contributions: Conceptualization, Z.M. and Y.H.; Formal analysis, Z.M. and J.Z.; Funding acquisition, B.L. and Y.H.; Methodology, Z.M.; Software, Y.P.; Supervision, B.L. and Y.H.; Validation, J.D.; Writing—original draft, Z.M. and Y.H.; Writing—review and editing, B.L. All authors have read and agreed to the published version of the manuscript.

Funding: This research was funded by Postgraduate Research & Practice Innovation Program of Jiangsu Province (Grant No. KYCX22_3812); the National Natural Science Foundation of China (Grant No. 62205133); the Jiangsu Provincial Key Research and Development Program (Grant No. BE2022143); and the Natural Science Foundation of Jiangsu Province (Grant No. BK20190953).

Institutional Review Board Statement: Not applicable.

Informed Consent Statement: Not applicable.

Data Availability Statement: The data presented in this study are available on request from the corresponding author.

Conflicts of Interest: The authors declare no conflict of interest.

References

1. Ma, Z.Y.; Chen, K.; Zhang, M.M.; Dou, J.T.; Hu, Y.Y. Propagation characteristics of Laguerre-Gaussian power-exponentphase-vortex beams. *Acta. Opt. Sin.* **2022**, *42*, 0526001. [CrossRef]
2. Chen, K.; Ma, Z.Y.; Zhang, M.M.; Dou, J.T.; Hu, Y.Y. The tight-focusing properties of radially polarized symmetrical power-exponent-phase vortex beam. *J. Opt.* **2022**, *24*, 055602. [CrossRef]
3. Chen, K.; Ma, Z.Y.; Zhang, M.M.; Dou, J.T.; Hu, Y.Y. Propagation properties of partially coherent power-exponent-phase vortex beam. *Acta. Phys. Sin.* **2022**, *71*, 014203. [CrossRef]
4. Chen, K.; Ma, Z.Y.; Hu, Y.Y. Tightly focused properties of a partially coherent radially polarized power-exponent-phase vortex beam. *Chin. Phys. B* **2023**, *32*, 024208. [CrossRef]
5. Lao, G.M.; Zhang, Z.H.; Zhao, D.M. Propagation of the power-exponent-phase vortex beam in paraxial ABCD system. *Opt. Express* **2016**, *24*, 18082–18094. [CrossRef] [PubMed]
6. Pan, Y.Q.; Zhao, M.L.; Zhang, M.M.; Dou, J.T.; Zhao, J.; Li, B.; Hu, Y.Y. Propagation properties of rotationally-symmetric power-exponent-phase vortex beams through oceanic turbulence. *Opt. Laser Technol.* **2023**, *159*, 109024. [CrossRef]
7. Pei, Z.H.; Huang, S.J.; Chen, Y.; Yan, C. Comparison of microparticle manipulating characteristics of canonical vortex beam and power-exponent-phase vortex beam. *J. Mod. Opt.* **2021**, *68*, 224–232. [CrossRef]
8. Shen, D.H.; Wang, K.; Zhao, D.M. Generation and propagation of a new kind of power-exponent–phase vortex beam. *Opt. Express* **2019**, *27*, 24642–24653. [CrossRef]
9. Hu, Y.Y.; Zhang, M.; Dou, J.T.; Zhao, J.; Li, B. Influences of salinity and temperature on propagation of radially polarized rotationally-symmetric power-exponent-phase vortex beams in oceanic turbulence. *Opt. Express* **2022**, *30*, 42772–42783. [CrossRef]
10. Wu, Z.H.; Zhao, J.; Dou, J.T.; Liu, J.; Jing, Q.L.; Li, B.; Hu, Y.Y. Optical trapping of multiple particles based on rotationally-symmetric power-exponent-phase vortex beams. *Opt. Express* **2022**, *30*, 42892–42901. [CrossRef]
11. Yan, X.; Guo, L.X.; Cheng, M.J.; Li, J.; Huang, Q.; Sun, R. Probability density of orbital angular momentum mode of autofocusing Airy beam carrying power-exponent-phase vortex through weak anisotropic atmosphere turbulence. *Opt. Express* **2017**, *25*, 15286–15298. [CrossRef] [PubMed]
12. Ma, Z.Y.; Zhao, W.Q.; Zhao, J.; Liu, J.; Jing, Q.; Dou, J.T.; Li, B.; Hu, Y.Y. Generation of arbitrary higher-order Poincaré sphere beams from a ring fiber laser with cascaded Q-plates. *Opt. Laser Technol.* **2022**, *156*, 108552. [CrossRef]
13. Hu, Y.Y.; Ma, Z.Y.; Zhao, W.Q.; Zhao, J.; Liu, J.; Jing, Q.L.; Dou, J.T.; Li, B. Controlled generation of mode-switchable nanosecond pulsed vector vortex beams from a Q-switched fiber laser. *Opt. Express* **2022**, *30*, 33195–33207. [CrossRef] [PubMed]
14. Miao, D.; Zhao, C.L.; Cai, Y.J.; Yang, Y. Partially coherent vortex beams: Fundamentals and applications. *Sci. China Phys. Mech. Astron.* **2021**, *64*, 224201. [CrossRef]
15. Ping, C.C.; Liang, C.H.; Wang, F.; Cai, Y.J. Radially polarized multi-Gaussian Schell-model beam and its tight focusing properties. *Opt. Express* **2017**, *25*, 32475–32490. [CrossRef]
16. Xu, H.F.; Zhang, R.; Sheng, Z.Q.; Qu, J. Focus shaping of partially coherent radially polarized vortex beam with tunable topological charge. *Opt. Express* **2019**, *27*, 23959–23969. [CrossRef]
17. Fan, C.J.; Liu, Y.X.; Wang, X.Y.; Chen, Z.; Pu, J. Trapping two types of particles by using a tightly focused radially polarized power-exponent-phase vortex beam. *J. Opt. Soc. Am. A* **2018**, *35*, 903–907. [CrossRef]

18. Man, Z.S.; Min, C.J.; Du, L.P.; Zhang, Y.; Zhu, S.; Yuan, X. Sub-wavelength sized transversely polarized optical needle with exceptionally suppressed side-lobes. *Opt. Express* **2016**, *24*, 874–882. [CrossRef] [PubMed]
19. Chen, J.; Wan, C.H.; Kong, L.J.; Zhan, Q. Experimental generation of complex optical fields for diffraction limited optical focus with purely transverse spin angular momentum. *Opt. Express* **2017**, *25*, 8966–8974. [CrossRef]
20. Chang, D.; Vuletić, V.; Lukin, M. Quantum nonlinear optics-photon by photon. *Nat. Photon.* **2014**, *8*, 685–694. [CrossRef]
21. Miao, D.; Jiang, D.G.; Luo, N.H.; Yang, Y. Trapping two types of Rayleigh particles using a focused partially coherent anomalous vortex beam. *Appl. Phys. B* **2019**, *125*, 55. [CrossRef]
22. Zhao, C.L.; Cai, Y.J.; Lu, X.H.; Eyyuboğlu, H.T. Radiation force of coherent and partially coherent flat-topped beams on a Rayleigh particle. *Opt. Express* **2009**, *17*, 1753–1765. [CrossRef] [PubMed]
23. Cheng, W.; Haus, J.W.; Zhan, Q.W. Propagation of vector vortex beams through a turbulent atmosphere. *Opt. Express* **2009**, *17*, 17829–17836. [CrossRef] [PubMed]
24. Kozawa, Y.; Matsunaga, D.; Sato, S. Superresolution imaging via superoscillation focusing of a radially polarized beam. *Optica* **2018**, *5*, 86–92. [CrossRef]
25. Chen, K.; Li, S.Z.; Pan, Y.Q.; Zhang, M.; Yang, Y.Q.; Hu, Y.Y. Tight Focusing Properties of Partially Coherent Radially Polarized Rotationally-Symmetric Power-Exponent-Phase Vortex Beam. *Acta. Opt. Sin.* **2022**, *42*, 2226002. [CrossRef]
26. Eyyuboğlu, H.T. Partially coherent Lorentz Gaussian beam and its scintillations. *Appl. Phys. B* **2011**, *103*, 755–762. [CrossRef]
27. Zhu, K.C.; Zhou, G.Q.; Li, X.G.; Zheng, X.; Tang, H. Propagation of Bessel-Gaussian beams with optical vortices in turbulent atmosphere. *Opt. Express* **2008**, *16*, 21315–21320. [CrossRef]
28. Chen, Y.H.; Wang, F.; Zhao, C.L.; Cai, Y.J. Experimental demonstration of a Laguerre-Gaussian correlated Schell-model vortex beam. *Opt. Express* **2014**, *22*, 5826–5838. [CrossRef]
29. Khan, S.N.; Stuti, J.S.; Kanseri, B.; Senthilkumaran, P. Detection of partially coherent polarization singular vector beams using Stokes polarimetry. *Appl. Phys. Lett.* **2021**, *118*, 051104. [CrossRef]
30. Bomzon, Z. Angular momentum and geometrical phases in tight-focused circularly polarized plane waves. *Appl. Phys. Lett.* **2006**, *118*, 051104. [CrossRef]

photonics

Communication

Second-Order Statistics of Self-Splitting Structured Beams in Oceanic Turbulence

Liming Liu [1,2], Yulu Liu [1], Hao Chang [1,2], Jifei Huang [1,2], Xinlei Zhu [1,2], Yangjian Cai [1,2,*] and Jiayi Yu [1,2,*]

1 Shandong Provincial Key Laboratory of Optics and Photonic Devices, Shandong Provincial Engineering and Technical Center of Light Manipulation, School of Physics and Electronics, Shandong Normal University, Jinan 250358, China
2 Collaborative Innovation Center of Light Manipulation and Applications, Shandong Normal University, Jinan 250358, China
* Correspondence: yangjiancai@sdnu.edu.cn (Y.C.); jiayiyu0528@sdnu.edu.cn (J.Y.)

Abstract: Free-space optical communication is restricted by random media-stimulated beam degradation. However, partially coherent structured beams modulated by the coherence structure can potentially mitigate this negative effect. By employing the extended Huygens–Fresnel integral, we provide an examination of the second-order statistical features of a common type of partly coherent structured beams, self-splitting structured beams, in a turbulent ocean. The implications of turbulence parameters relating to the ocean and beginning beam parameters corresponding to the progression of such beam propagation attributes are fully investigated. Our numerical outcomes show that, for turbulence with a low-dissipation kinetic energy rate per unit mass of fluid, small Kolmogorov inner scale, large relative strength of temperature to salinity undulations, and large dissipation rate of mean-square temperature has a greater negative effect on the structured beams. In addition, we suggest an effective approach, enhancing the order of the beam and reducing the coherence length of the beams, to lower the oceanic turbulence-induced negative effects, and thus have future extensive possibilities in free-space optical communication.

Keywords: partially coherent structured beam; oceanic turbulence; propagation

Citation: Liu, L.; Liu, Y.; Chang, H.; Huang, J.; Zhu, X.; Cai, Y.; Yu, J. Second-Order Statistics of Self-Splitting Structured Beams in Oceanic Turbulence. *Photonics* **2023**, *10*, 339. https://doi.org/10.3390/photonics10030339

Received: 27 February 2023
Revised: 14 March 2023
Accepted: 19 March 2023
Published: 22 March 2023

1. Introduction

One of the most important laser uses is free-space optical (FSO) communication, which utilizes the wireless transmission of data to receivers via light beam propagations. The environment where the light beam is located is not a perfect vacuum; it will be affected by random media, causing beam quality degradation, and limiting the development of FSO communication [1]. As a result, in recent decades, there has been a great deal of interest in studying beams of light and their propagation characteristics in random media and developing techniques to mitigate the deleterious impacts of random media on light beams [2–4].

A generally understood fact is that altering beam coherence is a useful method for improving light beam performance in random media [3,5]. Numerous studies have shown that partially coherent structured beams, whose spatial coherence has been reduced [6], have smaller turbulence-stimulated negative effects compared to their completely coherent counterparts [7–13]. Shirai et al. presented a scientific interpretation for this resistance using the coherent mode representation, implying that a light beam is delivered across many non-interfering channels at the same time [14]. With the development of modulation technology of the coherence structure of light beams [15], more light beams with specific coherence structures are tailored [16]. Such beams display unusual propagation features caused by prescribed coherence structure, examples include self-focusing and lateral shifting of the intensity maximum, far-field beam profile shaping, self-splitting, furthermore, such beams also possess better resistance to random media [8–10,17–20], and it is also a strategy for encoding information into the coherence structure to transmission [21,22].

Coherence manipulation of light beams brings vitality to studies on the propagation of electromagnetic radiation in random media. However, scholars have previously preferred to study the propagation attributes of electromagnetic beams within a turbulent atmosphere and thereby often overlook the turbulent ocean that encompasses 71% of the total area of the planet. This pattern could be explained by the fact that the spectra of refractive index (relating to power) changes in both the ocean and atmosphere to a similar extent and involve a comparable system [1,23]. The authors are aware that refraction is the primary cause of light beam disturbance in the water, other than absorption and scattering [4,23], and the primary drivers of turbulence in the ocean are fluctuations in humidity, salinity, temperature, and irregular movement of water. Light beam propagation in a turbulent ocean is more complicated than that in a turbulent atmosphere. Therefore, as underwater technologies advance, it is important and vital to examine how oceanic turbulence has an effect on the characteristics of light beams as they propagate and to look for a solution to lessen the impact that oceanic turbulence has on beams of light. In 2019, it was addressed how a rotating elliptical Gaussian optical coherence lattice propagated in oceanic turbulence. Such findings demonstrated that every sub-beam in such a lattice retains a regulated rotation within a specific distance; this is crucial for alleviating turbulence effects [24]. Recently, the effect of the parameters for the oceanic turbulence on multi-sinc Schell-model beams was explored explicitly [25]. The best we can tell, though, is that there are many reports on the oceanic turbulence effects on partially coherent structured beams, but few of them discuss in detail the impact of the coherence structure parameters of beams on turbulence resistance.

In this paper, we choose a self-splitting structured (SSS) beam as the research object and focus on discussing its propagation properties in the presence of oceanic turbulence, and the consequences of the parameters of this type of turbulence on the second-order statistical characteristics of such beams. Additionally, we offer a method for modifying the coherence length, the beam order, and the starting beam parameters in order to preserve the self-splitting propagation features and lessen the detrimental impacts brought on by ocean turbulence. In Section 2, the analytical expressions of the cross-spectral density of SSS beams in a turbulent ocean are derived based on the extended Huygens–Fresnel principle. In Section 3, the analytical expressions of propagation factor and relative radius of curvature of SSS beams in turbulence are derived based on the Wigner distribution function. In Section 4, the numerical calculation and results are presented and discussed. In Section 5, the conclusion and observation are presented.

2. Cross-Spectral Density of SSS Beams in Oceanic Turbulence

The cross-spectral density of the SSS beam at the source plane ($z = 0$) is written as [26]

$$W_0(\mathbf{r}_1, \mathbf{r}_2) = \tau^*(\mathbf{r}_1)\tau(\mathbf{r}_2)\mu(\mathbf{r}_1, \mathbf{r}_2), \tag{1}$$

in which $\mathbf{r}_1 = (x_1, y_1)$ and $\mathbf{r}_2 = (x_2, y_2)$ represent the two transverse position vectors in the source plane. Here, the asterisk signifies a complex conjugate and the amplitude function $\tau(\mathbf{r}) = \exp(-\mathbf{r}^2/w^2)$ with w as the beam width. Moreover, $\mu(\mathbf{r}_1, \mathbf{r}_2)$ represents the spectral degree of coherence (SDOC), which is explicitly written as

$$\mu(\mathbf{r}_1 - \mathbf{r}_2) = \prod_{\xi = x,y} \left[G_0 H_{2m}\left(\frac{\xi_1 - \xi_2}{\sqrt{2}\delta_\xi}\right) \exp\left(-\frac{(\xi_1 - \xi_2)^2}{2\delta_\xi^2}\right)\right], \tag{2}$$

where $\prod$ represents a product over a set of terms denoted by $\xi = x, y$, a constant $G_0 = 1/H_{2m}(0)$, which incorporates H_{2m} and signifies the Hermite polynomial of order $2m$, and δ_ξ denotes the spatial coherence length along the ξ-direction.

With respect to the paraxial approximation, a partly coherent structured beam propagating along the positive half plane (i.e., $z > 0$) in a random medium requires the employment of the generalized Huygens–Fresnel integral [1]

$$W(\boldsymbol{\rho}_1, \boldsymbol{\rho}_2, z) = \frac{1}{\lambda^2 z^2} \iint_{-\infty}^{+\infty} W_0(\mathbf{r}_1, \mathbf{r}_2) \exp\left\{ -\frac{ik}{2z} \left[(\boldsymbol{\rho}_1 - \mathbf{r}_1)^2 - (\boldsymbol{\rho}_2 - \mathbf{r}_2)^2 \right] \right\} \times \langle \exp[\Psi^*(\mathbf{r}_1, \boldsymbol{\rho}_1, z) + \Psi(\mathbf{r}_2, \boldsymbol{\rho}_2, z)] \rangle d^2\mathbf{r}_1 d^2\mathbf{r}_2, \tag{3}$$

where $k = 2\pi/\lambda$ is the wave number, in which λ is the wavelength, and $\boldsymbol{\rho}_1 = (\rho_{1x}, \rho_{1y})$ and $\boldsymbol{\rho}_2 = (\rho_{2x}, \rho_{2y})$ are two spatial positions in the target plane. The random transmission medium's fluctuating refractive index between the source and destination planes results in a complex perturbation phase called $\Psi(\mathbf{r}, \boldsymbol{\rho}, z)$. $\langle \quad \rangle$ denotes ensemble average, and the following formula can be used to express the ensemble average term [1]

$$\begin{aligned} &\langle \exp[\Psi^*(\mathbf{r}_1, \boldsymbol{\rho}_1, z) + \Psi(\mathbf{r}_2, \boldsymbol{\rho}_2, z)] \rangle \\ &= \exp\left\{ -4\pi^2 k^2 z \int_0^1 d\psi \int_0^\infty \boldsymbol{\kappa}\Phi(\boldsymbol{\kappa}) d^2\boldsymbol{\kappa} [1 - J_0[\boldsymbol{\kappa} | \psi(\mathbf{r}_1 - \mathbf{r}_2) + (1 - \psi)(\boldsymbol{\rho}_1 - \boldsymbol{\rho}_2) |]] \right\}, \end{aligned} \tag{4}$$

with J_0 is the zero-order Bessel function, which can be approximated as

$$J_0[\boldsymbol{\kappa} | \psi(\mathbf{r}_1 - \mathbf{r}_2) + (1 - \psi)(\boldsymbol{\rho}_1 - \boldsymbol{\rho}_2) |] \sim 1 - \frac{1}{4} [(1 - \psi)(\boldsymbol{\rho}_1 - \boldsymbol{\rho}_2) + \psi(\mathbf{r}_1 - \mathbf{r}_2)]^2 \boldsymbol{\kappa}^2. \tag{5}$$

Substituting the relation Equation (5) into Equation (4), one acquires the ensemble average term in Equation (3) as

$$\begin{aligned} &\langle \exp[\Psi^*(\mathbf{r}_1, \boldsymbol{\rho}_1, z) + \Psi(\mathbf{r}_2, \boldsymbol{\rho}_2, z)] \rangle \\ &= \exp\left\{ -k\Omega T z^{-1} \left[(\boldsymbol{\rho}_1 - \boldsymbol{\rho}_2)^2 + (\boldsymbol{\rho}_1 - \boldsymbol{\rho}_2) \cdot (\mathbf{r}_1 - \mathbf{r}_2) + (\mathbf{r}_1 - \mathbf{r}_2)^2 \right] \right\}, \end{aligned} \tag{6}$$

where $\Omega = k\pi^2 z^2/3$, $T = \int_0^\infty \boldsymbol{\kappa}^3 \Phi(\boldsymbol{\kappa}) d^2\boldsymbol{\kappa}$, in which $\Phi(\boldsymbol{\kappa})$ signifies the spectrum (relating to power) of the refractive fluctuations due to oceanic turbulent. The latter is defined as [8,27]

$$\begin{aligned} \Phi(\boldsymbol{\kappa}) &= C_0 \varepsilon^{-1/3} \kappa^{-11/3} \chi_T \left[1 + 2.35(\kappa\eta)^{2/3} \right] \\ &\times \left[\exp(-A_T \sigma) + \frac{\exp(-A_S \sigma)}{\omega^2} - \frac{2\exp(-A_{TS}\sigma)}{\omega} \right], \end{aligned} \tag{7}$$

where $C_0 = 0.388 \times 10^{-8}$. Moreover, ε is the dissipation kinetic energy rate per unit mass of fluid, which has a value between $10^{-1}\mathrm{m}^2/\mathrm{s}^3$ and $10^{-10}\mathrm{m}^2/\mathrm{s}^3$. And χ_T is the dissipation rate of mean-square temperature (between $10^{-4}\mathrm{K}^2/\mathrm{s}$ and $10^{-10}\mathrm{K}^2/\mathrm{s}$). In addition, ω is the relative strength of temperature to salinity fluctuations, (from -5 to 0) and η is the Kolmogorov inner scale. The quantities used are $A_T = 1.863 \times 10^{-2}$, $A_S = 1.9 \times 10^{-4}$, $A_{TS} = 9.41 \times 10^{-3}$ and $\sigma = 8.284(\kappa\eta)^{4/3} + 12.978(\kappa\eta)^2$. Thus, it is possible to express T as [23]

$$T = C_0 \varepsilon^{-1/3} \chi_T \eta^{-1/3} \left(45.5708\omega^{-2} - 17.6701\omega^{-1} + 6.78335 \right). \tag{8}$$

By inserting Equation (1) into Equation (3), which is followed by a rather tedious integration, the cross-spectral density of the SSS beams in the target plane is obtained

$$\begin{aligned} W(\boldsymbol{\rho}_1, \boldsymbol{\rho}_2, z) = \prod_{\xi = x,y} &\left\{ \frac{G_0 P \delta_\xi}{a_\xi^{m+1/2}} (a_\xi - 1)^m H_{2m} \left[-\frac{b_\xi}{2\sqrt{a_\xi^2 - a_\xi}} \right] \right. \\ &\left. \times \exp\left[\frac{b_\xi^2}{4a_\xi} - \frac{iP}{w} \Re_{s\xi} \Re_{d\xi} - \left(\frac{2P}{w}\Omega T + \frac{P^2}{2} \right) \Re_{d\xi}^2 \right] \right\}, \end{aligned} \tag{9}$$

with

$$a_\xi = 2\delta_\xi^2\left(\frac{1}{8w^2} + \frac{P\Omega T}{w} + \frac{1}{2\delta_\xi^2} + \frac{P^2}{2}\right);$$
$$b_\xi = \sqrt{2}\delta_\xi\left(\frac{iP}{w}\Re_{s\xi} + P^2\Re_{d\xi}\right); \quad P = \frac{kw}{z}. \tag{10}$$

To simplify the operation in the above integral, "sum" and "difference" coordinates are introduced, which are defined as

$$\begin{aligned}\mathbf{r}_s &= (\mathbf{r}_1 + \mathbf{r}_2)/2; \quad \mathbf{r}_d = \mathbf{r}_1 - \mathbf{r}_2;\\ \Re_s &= (\Re_1 + \Re_2)/2; \quad \Re_d = \Re_1 - \Re_2,\end{aligned} \tag{11}$$

with

$$\Re_1 = \rho_1 + i\Omega T(\rho_1 - \rho_2); \quad \Re_2 = \rho_2 + i\Omega T(\rho_1 - \rho_2). \tag{12}$$

As a result, we can use Equation (9) to obtain the cross-spectral density of the SSS beams within the target plane and use the definition to examine the spectral strength of such beams, that is, [6]

$$S(\boldsymbol{\rho}, z) = W(\boldsymbol{\rho}, \boldsymbol{\rho}, z). \tag{13}$$

The SDOC of such beams within the target plane is obtained from [6]

$$\mu(\boldsymbol{\rho}_1, \boldsymbol{\rho}_2, z) = \frac{W(\boldsymbol{\rho}_1, \boldsymbol{\rho}_2, z)}{\sqrt{W(\boldsymbol{\rho}_1, \boldsymbol{\rho}_1, z)W(\boldsymbol{\rho}_2, \boldsymbol{\rho}_2, z)}}. \tag{14}$$

3. Propagation Factors and Relative Radius of Curvature of SSS Beams in Oceanic Turbulence

Because the physics of partially coherent structured beams propagating in oceanic turbulence is complex, a number of different variables can be examined to determine how the turbulence affects partially coherent structured beams. Studying the beam's global attributes, such as its relative radius of curvature and propagation factor, is the simplest strategy. These can all be estimated using the Wigner distribution function (WDF) and represented with respect to the second-order moments of the beams. Expressions for the relative radius of curvature and the propagation factor of such beams, under oceanic turbulence conditions, are then derived in this section. This occurs after the analytical formulations are first derived for the second-order moments of the WDF of the SSS beams under the effects of oceanic turbulence. The impact of turbulence on the propagation of the beam is well described by these global factors, which we shall explore in more detail in the next section. By using the following formula, the WDF of a partially coherent structured beam is expressible in terms of the cross-spectral density [28]

$$h(\boldsymbol{\rho}_s, \boldsymbol{\theta}, z) = \frac{1}{\lambda^2}\iint_{-\infty}^{+\infty} W(\boldsymbol{\rho}_s, \boldsymbol{\rho}_d, z)\exp(-ik\boldsymbol{\theta}\cdot\boldsymbol{\rho}_d)d^2\boldsymbol{\rho}_d, \tag{15}$$

where $\boldsymbol{\theta} = (\theta_x, \theta_y)$ represents the angle of the relevant vector with respect to the z-direction, while $k\theta_x$ and $k\theta_y$ are the wave vector components aligning with the x-axis and y-axis, respectively. Using Equation (3), the term $W(\boldsymbol{\rho}_s, \boldsymbol{\rho}_d, z)$ is expressible in the alternative form that follows

$$\begin{aligned}W(\boldsymbol{\rho}_s, \boldsymbol{\rho}_d, z) = \frac{1}{4\pi^2}&\iiiint_{-\infty}^{+\infty} W_0\left(\mathbf{r}_s, \boldsymbol{\rho}_d + \frac{w}{P}\boldsymbol{\kappa}_d\right)d^2\mathbf{r}_s d^2\boldsymbol{\kappa}_d\\ &\times\exp\left[-i\boldsymbol{\rho}_s\cdot\boldsymbol{\kappa}_d + i\mathbf{r}_s\cdot\boldsymbol{\kappa}_d - \Theta\left(\boldsymbol{\rho}_d, \boldsymbol{\rho}_d + \frac{w}{P}\boldsymbol{\kappa}_d\right)\right],\end{aligned} \tag{16}$$

with

$$\Theta\left(\boldsymbol{\rho}_d, \boldsymbol{\rho}_d + \frac{w}{P}\boldsymbol{\kappa}_d\right) = 3\Omega T\left(\frac{P}{w}\boldsymbol{\rho}_d^2 + \boldsymbol{\rho}_d\cdot\boldsymbol{\kappa} + \frac{w}{3P}\boldsymbol{\kappa}^2\right), \tag{17}$$

where $\boldsymbol{\rho}_s$ and $\boldsymbol{\rho}_d$ also satisfy the "sum" and "difference" coordinate forms as shown is Equation (11).

Applying Equations (15)–(17), the following is how we arrive at the equation for the WDF of the SSS beams traversing into oceanic turbulence

$$h(\boldsymbol{\rho}_s, \boldsymbol{\theta}, z) = \prod_{\xi=x,y} h(\rho_{s\xi}, \theta_\xi, z), \tag{18}$$

where

$$\begin{aligned} h(\rho_{s\xi}, \theta_\xi, z) = \frac{G_0 w}{\lambda\sqrt{2\pi}} \iint_{-\infty}^{+\infty} H_{2m}\left[-\frac{1}{\sqrt{2}\delta_\xi}\left(\rho_{d\xi} + \frac{w}{P}\kappa_{d\xi}\right)\right] \\ \times \exp\left(-c_\xi\rho_{d\xi}^2 - d_\xi\kappa_{d\xi}^2 - e_\xi\rho_{d\xi}\kappa_{d\xi} - ik\theta_\xi\rho_{d\xi}\right)d\kappa_{d\xi}d\rho_{d\xi}, \end{aligned} \tag{19}$$

with

$$\begin{aligned} c_\xi = \frac{1}{2\delta_\xi^2} + \frac{1}{8w^2} + \frac{3P\Omega T}{w}; \quad d_\xi = \frac{w^2}{2\delta_\xi^2 P^2} + \frac{1}{8P^2} + \frac{w^2}{2} + \frac{3w\Omega T}{P}; \\ e_\xi = \frac{w}{\delta_\xi^2 P} + \frac{1}{4wP} + 3\Omega T. \end{aligned} \tag{20}$$

The orderly moments $n_1 + n_2 + m_1 + m_2$ of the WDF of a beam consists of [28]

$$\langle x^{n_1} y^{n_1} \theta_x^{m_1} \theta_y^{m_2} \rangle = \frac{1}{\Lambda} \iiiint_{-\infty}^{+\infty} x^{n_1} y^{n_1} \theta_x^{m_1} \theta_y^{m_2} h(\boldsymbol{\rho}_s, \boldsymbol{\theta}, z) d^2\boldsymbol{\rho}_s d^2\boldsymbol{\theta}, \tag{21}$$

where

$$\Lambda = \iiiint_{-\infty}^{+\infty} h(\boldsymbol{\rho}_s, \boldsymbol{\theta}, z) d^2\boldsymbol{\rho}_s d^2\boldsymbol{\theta}. \tag{22}$$

Then, substituting Equation (19) into Equation (21), one obtains the second-order moments of WDF of the SSS beams traversing through oceanic turbulence, that is,

$$\langle \rho_\xi^2 \rangle = -\frac{2mz^2}{k^2\delta_\xi^2} - 2b_\xi; \quad \langle \theta_\xi^2 \rangle = -\frac{2m}{k^2\delta_\xi^2} - \frac{2a_\xi}{k^2}; \quad \langle \rho_\xi\theta_\xi \rangle = -\frac{2mz}{k^2\delta_\xi^2} - \frac{c_\xi}{k}. \tag{23}$$

The aforementioned second-order moments enable an assessment of the beam quality metric, which is expressed as [29]

$$Q = \langle \boldsymbol{\rho}^2 \rangle \langle \boldsymbol{\theta}^2 \rangle - \langle \boldsymbol{\rho} \cdot \boldsymbol{\theta} \rangle^2. \tag{24}$$

This variable correlates with the beam propagation factor $M^2 = \sqrt{Q}$. Taking a certain beam width at the source plane, lower beam divergence, and the correspondingly reduced M^2 are related to improved beam quality. Next, we consider the propagation factors that align with the x- and y-directions, due to the beam being rectangularly symmetric. Substituting Equations (23) into Equation (24), we obtain an explicit expression for the propagation factor of the SSS beams in the oceanic turbulence.

We can also study the curvature radius that align with the x- and y-directions of such beams in a turbulent ocean by considering the second-order moments of WDF, so that [30]

$$R_\xi = \frac{\langle \rho_\xi^2 \rangle}{\langle \rho_\xi\theta_\xi \rangle}. \tag{25}$$

In order to better conduct comparative studies, we define the relative radius of curvature

$$\Delta R_\xi = \frac{|R_{tur} - R_{free}|}{R_{free}}, \tag{26}$$

where R_{tur} and R_{free} represent the beam's curvature radius in oceanic turbulence and in free space, respectively.

4. Numerical Calculation and Analysis of Second-Order Statistics of SSS Beams

The second-order statistics of SSS beams traversing through oceanic turbulence are examined using the expressions that were derived in earlier sections. In the examples that follow, unless stated otherwise, the initial settings for the turbulence and the beam parameters are given as follows: λ = 632.8 nm, w= 1 cm, $\delta_x = \delta_y = \delta$ = 5 mm, $\varepsilon = 10^{-3}$ m^2/s^3, $\chi_T = 10^{-8}$ K^2/s, η = 1 mm, and ω = −3.0. For convenience, the wavelength is taken as that of a common HeNe-type laser diode; the results presented here are qualitatively similar for optical communication wavelengths.

Figure 1 depicts the evolution of an SSS beam's SDOC modulus, at various propagation lengths, in oceanic turbulence and for free space. The diagrams within this figure demonstrate that the SDOC of the SSS beam is spread in an array at the source plane and, while the beam traverses across free space, the array distribution gradually vanishes before changing into a diamond distribution. However, in oceanic turbulence, the SDOC evolves into an array distribution in the short propagation distance, and then eventually develops in the far field into a Gaussian distribution. From which we can conclude that the SDOC of the SSS beam is degraded due to the negative effects caused by oceanic turbulence.

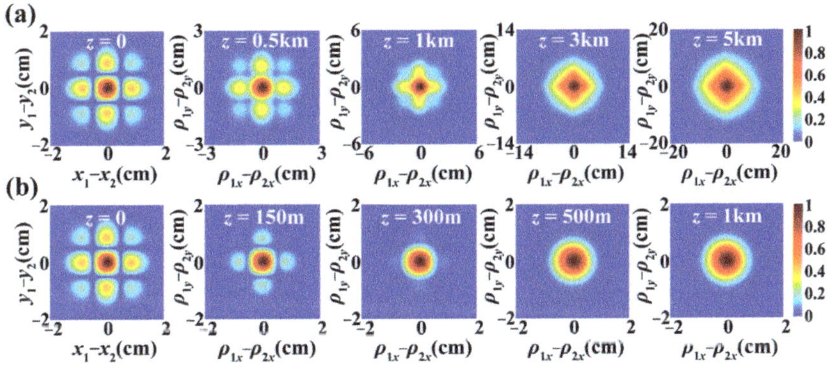

Figure 1. Evolution of the SDOC modulus of an SSS beam when traversing through (**a**) free space and (**b**) oceanic turbulence.

The evolution of the SSS beam propagating in free space, in terms of intensity, is shown in Figure 2a. It displays normalized density plots of the transverse spectral intensity for various propagation distances. As the propagation distance grows, it is clearly observable that four beam spots eventually develop from the original single beam spot while propagating, i.e., the SSS beam displays splitting characteristics as it travels. Figure 2b depicts how the spectral intensity changes over time in oceanic turbulence. The fact that such a beam still seems to exhibit splitting properties is shown when propagating over a short distance, while the intensity distribution of the spectra steadily develops into a Gaussian form over longer distances. This is due to the fact that, when the propagation distance is low, the beam's free-space diffraction dominates, whereas the isotropic effect of oceanic turbulence progressively builds up and dominates as the propagation distance grows.

Following that, we will talk about how turbulence parameters affect the SDOC and spectrum intensity of the suggested beams and how the initial beam parameters help the beams resist the negative impacts of oceanic turbulence.

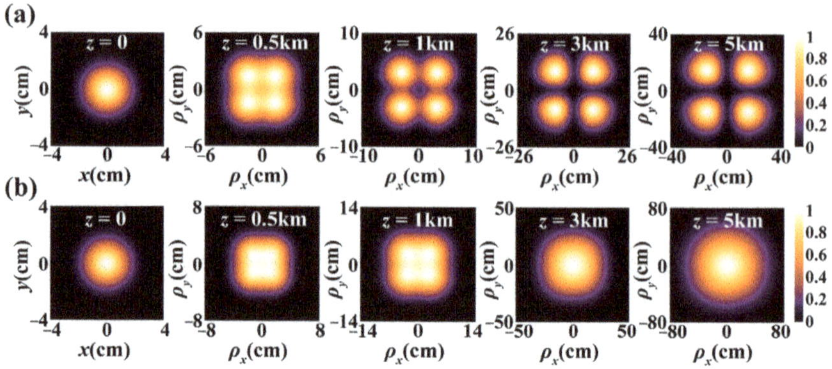

Figure 2. Development of the normalized spectral intensity of an SSS beam traversing through (**a**) free space and (**b**) oceanic turbulence.

Figure 3 illustrates the SDOC modulus of the SSS beam for a propagation distance of z = 300 m, while assuming various oceanic turbulence parameters. On the SDOC degradation in oceanic turbulence, the impacts of different dissipation rate values for the kinetic energy per unit mass of the fluid are compared in Figure 3a. The SDOC distribution in oceanic turbulence exhibits side robes when the kinetic energy dissipation rate per unit mass of the fluid takes a high value, and it degenerates more quickly to a Gaussian form as the dissipation rate drops. Thus, it may be concluded that the development of the spectral degree of coherence of the SSS beam is significantly harmed by oceanic turbulence, which has a low kinetic energy dissipation rate per unit mass of the fluid. The slower degradation of SDOC can be attributed to reduced oceanic turbulence, which is caused by the faster dissipation rate of kinetic energy per unit mass of fluid. With a similar analysis, the following conclusions are drawn from Figure 3b–d. The oceanic turbulence containing a small Kolmogorov inner scale, alternatively with an extensive dissipation rate for the mean-square temperature or with a substantial relative strength for the temperature to salinity undulations, has a sizeable negative effect on the SSS beam from the perspective of the degeneration of the SDOC. It makes sense that the significant light scattering will be caused by the smaller inner scale of turbulence. As the light scattering increases, the degeneration of the SDOC caused by oceanic turbulence increases. The more energetic turbulence is often determined by the mean-squared temperature dissipation rate, which leads to enhanced SDOC degeneration. Regarding the case, the temperature and salinity undulations' relative strength $\omega \to -5$, the temperature field's volatility dominates. Whereas when $\omega \to 0$, the fluctuation of the salinity field is dominant. This suggests that the deterioration of the SDOC of the SSS beams is greatly influenced by the salinity fluctuations in comparison to the temperature fluctuations. Therefore, the SSS beams propagating in deep-sea regions can maintain the beam performance better in shallow-sea regions.

Our understanding of the impact of various turbulence characteristics on the propagation properties of the SSS beams in oceanic turbulence is based on the analysis provided earlier. Realistically, modifying the characteristics of oceanic turbulence is extremely difficult. Hence, we require an investigation into methods that preserve the beam's extensive propagation properties when traversing through oceanic turbulence. Namely, we determine procedures by which oceanic turbulence resistance can be improved by altering the initial beam parameters. We proceed to discuss the SDOC distribution of the proposed beams traversing through turbulence.

Figure 4 displays the SDOC modulus of the SSS beams at propagation distance of z = 300 m in oceanic turbulence for various beam order values and coherence lengths. It is possible to confirm that the SDOC distribution of the SSS beams is altered by both the beam order and coherence length. The Gaussian to non-Gaussian distribution changes when the beam order and coherence change, suggesting that SSS beams containing low coherence and higher beam order are little impacted by oceanic turbulence.

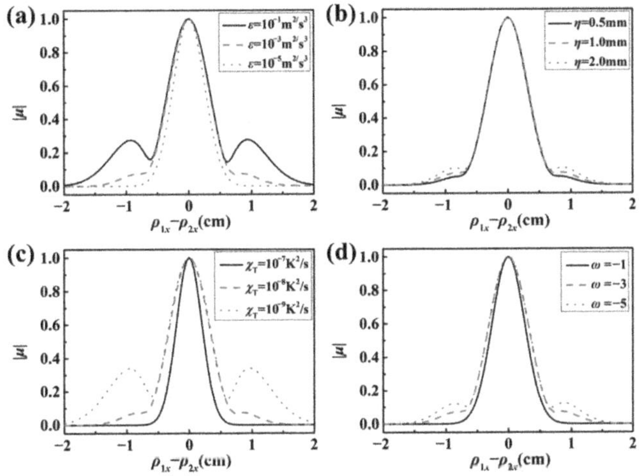

Figure 3. Cross line ($\rho_{1y} - \rho_{2y} = 0$) of the SDOC modulus at a propagation distance of z = 300 m for various oceanic turbulence parameters: (**a**) the kinetic energy dissipation rate per unit mass of the fluid; (**b**) the Kolmogorov inner scale; (**c**) the dissipation rate of the mean-square temperature; (**d**) the relative strength of the temperature to salinity undulations.

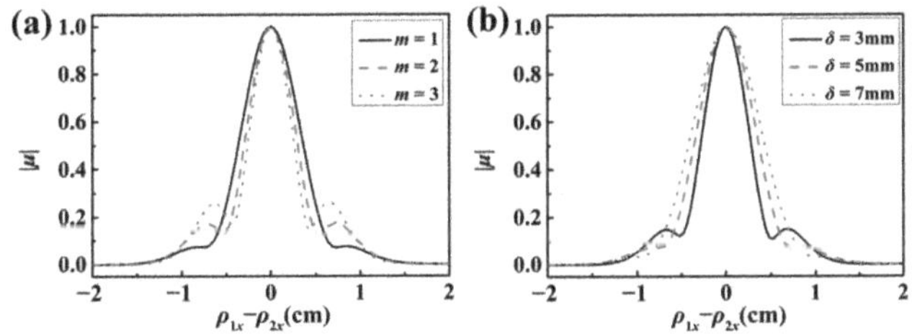

Figure 4. Cross line ($\rho_{1y} - \rho_{2y} = 0$) of the SDOC modulus at a propagation distance of z = 300 m for various (**a**) beam orders and (**b**) coherence lengths.

We then concentrate on the development of the spectral intensity of SSS beams traveling through ocean turbulence. Figure 5 depicts the relationship between the maximum spectral intensity within the transverse plane and the optical axis ratio for the SSS beams at propagation distance z = 1 km for various beginning beam characteristics. According to Figure 5a, the ratio is equal to one when the dissipation rate is very low, and it gradually declines as the dissipation rate increases. This means that when the dissipation rate is high, the beam profile remains split but, as it falls, the spots of the four beams combine to create one spot that represents a single beam. We provide new evidence that oceanic turbulence with low fluid kinetic energy dissipation rates per unit mass significantly hinders the propagation of SSS beams from the standpoint of spectral intensity deterioration. Furthermore, under the same oceanic turbulence, such ratio of the large-beam order and low-coherence SSS beams is smaller than that of the small-beam order and high-coherence SSS beams. This suggests that the SSS beams that contain a high-beam order and low coherence are more resistant to turbulence in the ocean. Likewise, we can get the same conclusion from Figure 5b–d.

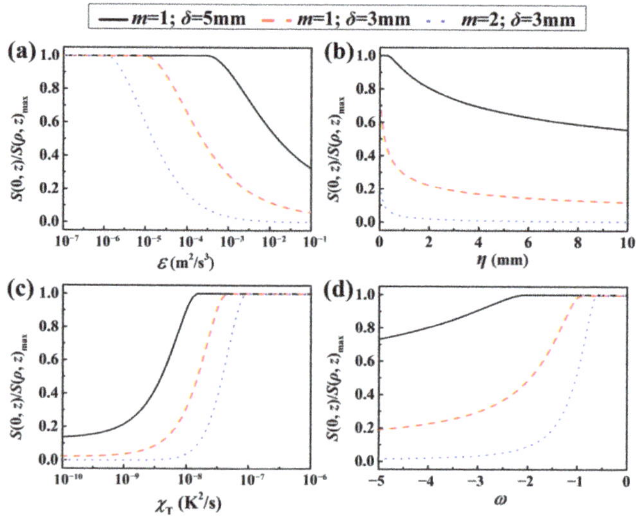

Figure 5. Ratio of the spectral intensity in the optical axis to the maximum intensity in the transverse plane of SSS beams for different beam orders and coherence lengths versus various oceanic turbulence parameters: (**a**) the kinetic energy dissipation rate per unit mass of the fluid; (**b**) the Kolmogorov inner scale; (**c**) the dissipation rate of the mean-square temperature; (**d**) the relative strength of the temperature to salinity undulations.

To more fully demonstrate the results of the coherence length or beam order on maintaining the self-splitting propagation characteristics during structured light propagation in oceanic turbulent, for various choices of beam order and coherence duration, we plot the ratio of the spectral intensity for the optical axis ($\rho = 0$) against the maximum spectral intensity in the transverse plane for the SSS beams with changing propagation distance (Figure 6). It is evident that higher-order beams with low coherence can travel further when the ratio is less than 1, that is, it exhibits improved self-splitting propagation properties, which means possessing better resistance of oceanic turbulence.

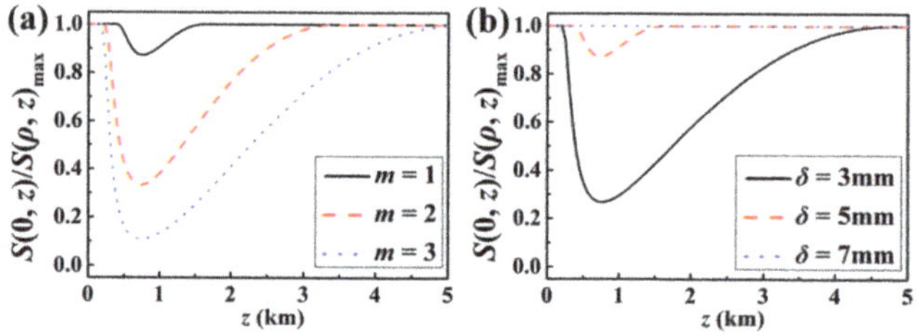

Figure 6. Ratio of the spectral intensity for the optical axis (ρ = 0) against the maximum intensity in the transverse plane for the SSS beams traversing through oceanic turbulence for various (**a**) beam orders and (**b**) coherence lengths.

To fully persuade readers that the SSS beam with large beam order and low coherence propose greater turbulence resistance, we search for more angles to examine the effects of beam order and coherence length on turbulence resistance. The normalized propagation factor for SSS beams traveling through oceanic turbulence for various beam orders and coherence lengths is shown in Figure 7 as it changes over time. As a control group, we also show the normalized propagation factor for Gaussian Schell-model (GSM) beam (w = 1 cm, δ = 5 mm) under the same turbulence parameters. GSM beam is well-known and simple partially coherent beam, and there are many studies on its propagation properties. In Ref. [31], it is confirmed that such beam can be evaluated oceanic turbulent optical wireless communication performance from the perspective of the beam wander. From Figure 7a, it can be shown that the beam order has a substantial impact on the variance of the propagation factor in turbulence. This variation grows more quickly with lower beam order, suggesting that a higher-order SSS beam is less altered by turbulence. Additionally, from Figure 7b it is determined that the coherence length has a substantial impact on the propagation factor, i.e., SSS beams containing a low coherence have reduced propagation factors. As a result, we draw the conclusion from Figure 7 that higher-order SSS beams with low coherence are less susceptible to turbulence and have superior resistance to oceanic turbulence when accounting for the propagation factor. Furthermore, we noticed that the normalized propagation factor of the GSM beam increases faster than that of the SSS beam on propagation, which means that the SSS beam modulated by the coherence structure can effectively mitigate the negative effects of turbulence.

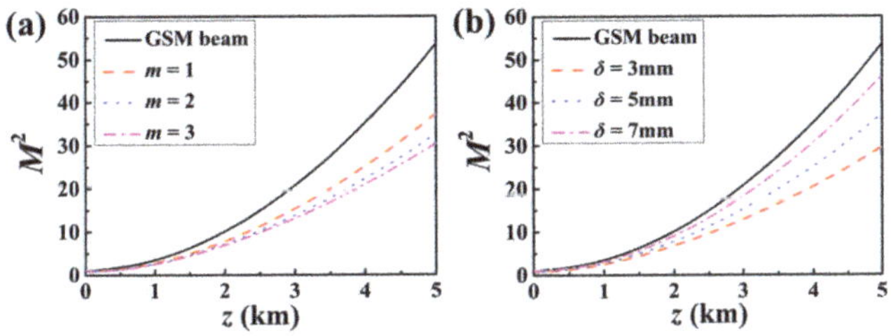

Figure 7. Normalized propagation factor for SSS beams traversing through oceanic turbulence for various (**a**) beam orders and (**b**) coherence lengths. The dark line on the graphs denotes the corresponding result for a GSM beam.

Figure 8 illustrates the relative radius of curvature for various beam orders and coherence lengths of SSS beams and GSM beam propagating in oceanic turbulence. We discover that at small propagation distances, the relative radius of curvature increases quickly, whereas at large propagation distances, it increases gradually. With the same initial beam parameters, the SSS beam has a smaller relative radius of curvature than GSM beam, furthermore, the SSS beams with large beam order and low coherence experience smaller relative radius of curvature. As a result, from the viewpoint of the relative radius of curvature, we discover once more that higher-order SSS beams with low coherence are impacted by turbulence to a lesser extent.

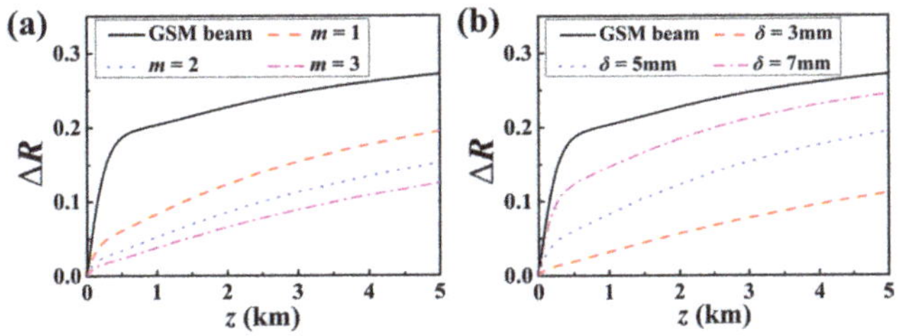

Figure 8. Relative radius of curvature of SSS beams traversing through oceanic turbulence for various (**a**) beam orders and (**b**) coherence lengths.

5. Conclusions and Discussion

We have researched how SSS beams behave statistically in the second order while they are propagating across turbulence on the ocean. From the standpoint of degenerate evolution of the SDOC and spectral intensity, it is discovered that oceanic turbulence with low kinetic energy dissipation per unit mass of the fluid and the Kolmogorov inner scale, or strong relative temperature to salinity undulations, and the mean-square temperature dissipation rate of the turbulent ocean are likely to have a more detrimental effect on such beams. Furthermore, compared to smaller-ordered structured beams with strong coherence, larger-ordered beams with low coherence are influenced by ocean turbulence to a lesser extent.

It is important to briefly discuss the physics behind the unique statistical properties of the SSS beams and their resilience to ocean turbulence. Such beams can be viewed as a superposition of coherent Gaussian modes according to prescribed weights, and the fundamental phase elements of each coherent Gaussian modes contribute to their distinct slanted propagation. Furthermore, each mode simultaneously transverses through distinct non-interfering channels, which reduces the interference-based fluctuations and averages out turbulent fluctuations. These results could have many future application in the arena of free-space optical communication. Therefore, based on the above analysis, we can design more partially coherent light sources with other novel propagation characteristics by modifying the coherent modes and weight distributions that constitute new class of partially coherent light sources, and find the appropriate coherence modes and weight distribution that most effectively reduce the negative effects of turbulence.

Author Contributions: Writing —original draft preparation, L.L. and J.H.; writing—review and editing, J.Y. and X.Z.; conceptualization, H.C.; methodology, X.Z.; software, J.H.; data curation, Y.L.; supervision, Y.C.; project administration J.Y. and Y.C. All authors have read and agreed to the published version of the manuscript.

Funding: This research was funded by National Key Research and Development Program of China (2022YFA1404800, 2019YFA0705000); National Natural Science Foundation of China (12192254, 11974218, 12004218, 92250304); China Postdoctoral Science Foundation (2020M680093, 2022M721992); Shandong Provincial Natural Science Foundation of China (ZR2020QA067); The Open Fund of the Guangdong Provincial Key Laboratory of Optical Fiber Sensing and Communications (2020GDS-GXCG08).

Institutional Review Board Statement: Not applicable.

Informed Consent Statement: Not applicable.

Data Availability Statement: Data is contained within the article.

Conflicts of Interest: The authors declare no conflict of interest.

References

1. Andrews, L.C.; Phillips, R.L. *Laser Beam Propagation through Random Media*, 2nd ed.; SPIE: Bellingham, WA, USA, 2005.
2. Tatarski, V.I. *Wave Propagation in a Turbulent Medium*; Courier Dover Publications INC: New York, NY, USA, 2016.
3. Wang, F.; Liu, X.; Cai, Y. Propagation of partially coherent beam in turbulent atmosphere: A review. *Prog. Electromagn. Res.* **2015**, *150*, 123–143. [CrossRef]
4. Korotkova, O. Light Propagation in a Turbulent Ocean. *Prog. Opt.* **2019**, *64*, 1–43.
5. Gbur, G. Partially coherent beam propagation in atmospheric turbulence. *J. Opt. Soc. Am. A* **2014**, *31*, 2038–2045. [CrossRef] [PubMed]
6. Cai, Y.; Chen, Y.; Yu, J.; Liu, X.; Liu, L. Generation of partially coherent beams. *Prog. Opt.* **2017**, *62*, 157–223.
7. Gbur, G.; Wolf, E. Spreading of partially coherent beams in random media. *J. Opt. Soc. Am. A* **2002**, *19*, 1592–1598. [CrossRef] [PubMed]
8. Yu, J.; Chen, Y.; Liu, L.; Liu, X.; Cai, Y. Splitting and combining properties of an elegant Hermite-Gaussian correlated Schell-model beam in Kolmogorov and non-Kolmogorov turbulence. *Opt. Express* **2015**, *23*, 13467–13481. [CrossRef]
9. Shakir, S.A.; Clark, T.T.; Cargill, D.S.; Carreras, R. Far-field propagation of partially coherent laser light in random mediums. *Opt. Express* **2018**, *26*, 15609–15622. [CrossRef]
10. Yu, J.; Wang, F.; Liu, L.; Cai, Y.; Gbur, G. Propagation properties of Hermite non-uniformly correlated beams in turbulence. *Opt. Express* **2018**, *26*, 16333–16343. [CrossRef]
11. Yu, J.; Huang, Y.; Wang, F.; Liu, X.; Gbour, G.; Cai, Y. Scintillation properties of a partially coherent vector beam with vortex phase in turbulent atmosphere. *Opt. Express* **2019**, *27*, 26676–26688. [CrossRef]
12. Wang, J.; Wang, M.; Lei, S.; Tan, Z.; Wang, C.; Wang, Y. Influence of Source Parameters and Non-Kolmogorov Turbulence on Evolution Properties of Radial Phased-Locked Partially Coherent Vortex Beam Array. *Photonics* **2021**, *8*, 512. [CrossRef]

13. Sun, B.; Lü, H.; Wu, D.; Wang, F.; Cai, Y. Propagation of a Modified Complex Lorentz-Gaussian-Correlated Beam in a Marine Atmosphere. *Photonics* **2021**, *8*, 82. [CrossRef]
14. Shirai, T.; Dogariu, A.; Wolf, E. Mode analysis of spreading of partially coherent beams propagating through atmospheric turbulence. *J. Opt. Soc. Am. A* **2003**, *20*, 1094–1102. [CrossRef]
15. Gori, F.; Santarsiero, M. Devising genuine spatial correlation functions. *Opt. Lett.* **2007**, *32*, 3531–3533. [CrossRef] [PubMed]
16. Chen, Y.; Wang, F.; Cai, Y. Partially coherent light beam shaping via complex spatial coherence structure engineering. *Adv. Phys.-X* **2022**, *7*, 2009742. [CrossRef]
17. Wei, D.; Wang, K.; Xu, Y.; Du, Q.; Liu, F.; Liu, J.; Dong, Y.; Zhang, L.; Yu, J.; Cai, Y.; et al. Propagation of a Lorentz Non-Uniformly Correlated Beam in a Turbulent Ocean. *Photonics* **2023**, *10*, 49. [CrossRef]
18. Liu, Y.; Zhao, Y.; Liu, X.; Liang, C.; Liu, L.; Wang, F.; Cai, Y. Statistical characteristics of a twisted anisotropic Gaussian Schell-model beam in turbulent ocean. *Photonics* **2020**, *7*, 37. [CrossRef]
19. Lajunen, H.; Saastamoinen, T. Propagation characteristics of partially coherent beams with spatially varying correlations. *Opt. Lett.* **2011**, *36*, 4104–4106. [CrossRef]
20. Gu, Y.; Gbur, G. Scintillation of nonuniformly correlated beams in atmospheric turbulence. *Opt. Lett.* **2013**, *38*, 1395–1397. [CrossRef]
21. Chen, Y.; Ponomarenko, S.A.; Cai, Y. Experimental generation of optical coherence lattices. *Appl. Phys. Lett.* **2016**, *109*, 061107. [CrossRef]
22. Liu, Y.; Chen, Y.; Wang, F.; Cai, Y.; Liang, C.; Korotkova, O. Robust far-field imaging by spatial coherence engineering. *Opto-Electron. Adv.* **2021**, *4*, 210027. [CrossRef]
23. Thorpe, S.A. *The Turbulent Ocean*; Cambridge University: Cambridge, MA, USA, 2005; pp. 1–37.
24. Ye, F.; Xie, J.; Hong, S.; Zhang, J.; Deng, D. Propagation properties of a controllable rotating elliptical Gaussian optical coherence lattice in oceanic turbulence. *Results Phys.* **2019**, *13*, 102249. [CrossRef]
25. Liu, X.; Zhou, G.; Shen, Y. Effect of oceanic turbulence with anisotropy on the propagation of multi-sinc Schell-model beams. *Results Phys.* **2022**, *36*, 105447. [CrossRef]
26. Chen, Y.; Gu, J.; Wang, F.; Cai, Y. Self-splitting properties of a Hermite-Gaussian correlated Schell-model beam. *Phys. Rev. A* **2015**, *91*, 013823. [CrossRef]
27. Nikishov, V.V.; Nikishov, V.I. Spectrum of Turbulent Fluctuations of the Sea-Water Refraction Index. *Int. J. Fluid Mech. Res.* **2000**, *27*, 82–98. [CrossRef]
28. Dan, Y.; Zhang, B. Beam propagation factor of partially coherent flat-topped beams in a turbulent atmosphere. *Opt. Express* **2008**, *16*, 15563–15575. [CrossRef]
29. Martínez-Herrero, R.; Mejías, P.M.; Piquero, G. Quality improvement of partially coherent symmetric-intensity beams caused by quartic phase distortions. *Opt. Lett.* **1992**, *17*, 1650–1651. [CrossRef]
30. Weber, H. Propagation of higher-order intensity moments in quadratic-index media. *Opt. Quantum. Electron.* **1922**, *24*, S1027–S1049. [CrossRef]
31. Wu, Y.; Zhang, Y.; Li, Y.; Hu, Z. Beam wander of Gaussian-Schell model beams propagating through oceanic turbulence. *Opt. Commun.* **2016**, *371*, 59–66. [CrossRef]

photonics

Communication

Influence of Off-Axis Noncanonical Vortex on the Dynamics of Energy Flux

Xinying Zhao [1], Huijian Liang [2], Gaofeng Wu [3,*] and Xiaoyan Pang [4,*]

1 School of Physics and Information Technology, Shaanxi Normal University, Xi'an 710061, China
2 China Academy of Space Technology-Xi'an, Xi'an 710199, China
3 School of Physics, Northwest University, Xi'an 710127, China
4 School of Electronics and Information, Northwestern Polytechnical University, Xi'an 710072, China
* Correspondence: gfwu@nwu.edu.cn (G.W.); xypang@nwpu.edu.cn (X.P.)

Abstract: In this article, we analyze the impact of the off-axis noncanonical vortex on the dynamics of 3D energy flux. The results reveal that the properties of energy flux are significantly influenced by the anisotropy parameter σ_c of the off-axis noncanonical vortex. It is demonstrated that by adjusting the anisotropy parameter, we can control the transverse energy flux, from the position of the transverse energy tornado and the distribution of the transverse energy strength to the rotation direction of the transverse energy flux as the beam propagates. The interesting phenomenon, the reversed energy flux, is also closely related to the anisotropy parameter. The position and size of the reversed energy flux region, as well as the production of two energy tornadoes with opposite rotating directions, resembling a Chinese "Taiji" pattern, can be controlled by varying the anisotropy parameter. This result will potentially provide a new freedom for tailoring the 3D optical field and be useful in optical manipulation and processing.

Keywords: energy flux; off-axis vortex; noncanonical vortex; 3D optical field; reversed energy flux

1. Introduction

A traditional vortex beam, which is called a canonical vortex beam, usually indicates intensity zero in its beam center and has a constant phase gradient around the center [1,2], arousing tremendous study on its unique characteristics. Numerous efforts have been devoted to the manipulation of the intensity distribution [3,4], polarization [5,6], coherence [7,8] and energy flux [9,10] of vortex beams and find its wide applications on optical trapping [11], optical tweezers [12] and wireless communications [13]. Additionally, another vortex beam, called the noncanonical vortex beam [14], where the phase gradient and the amplitude were spatially anisotropic, has attracted more attention. The anisotropy parameter "σ_c" describes the nonlinearity of the phase gradient and exhibits a strong influence on tailoring rich intensity and phase patterns, showing a potential application on the beam shaping [15,16]. Meanwhile, previous research studies concentrate on the properties of the on-axis noncanonical vortex beam, including the intensity, polarization and the energy flow [17–19]. In practice, the off-axis vortex is often achieved experimentally [20–23]. Hence, in this article, we will focus on the off-axis noncanonical vortex beam in a 3D-structured field and its effect on the energy flux.

Besides the intensity and phase, the energy flow is another fundamental quantity offering advantages in its application to 3D optical manipulation. This is where an anomaly phenomenon, energy back propagation, i.e., negative values of the longitudinal Poynting vectors, has attracted researchers. As early as 1919, the backward energy flow near the focus was first revealed by Ignatowskii [24]. Later, Richards and Wolf found the region of the negative propagation light of a strongly focused linearly polarized field in 1959 [25]. From that time and onwards, the condition of the negative energy flow of a sharp focused field with homogeneous (linear, circular and elliptical) polarization [26–28] and inhomogeneous

Citation: Zhao, X.; Liang, H.; Wu, G.; Pang, X. Influence of Off-Axis Noncanonical Vortex on the Dynamics of Energy Flux. *Photonics* **2023**, *10*, 346. https://doi.org/10.3390/photonics10030346

Received: 27 February 2023
Revised: 16 March 2023
Accepted: 17 March 2023
Published: 22 March 2023

(radial and azimuthal) polarization [29–31] was studied. Moreover, the negative Poynting vector acting as the "tractor force" on the particle was observed [32]. On the other hand, the transverse energy flow as an informative characteristic of an optical field can usually be decomposed into a spin part and orbit part, inspiring numerous studies on it, such as the spin-to-orbital energy flow conversion [33,34], spin–orbit coupling [35,36], and spin Hall effects of light (SHEL) [37,38]. Here, we will demonstrate that adjusting the anisotropy parameter σ_c of the off-axis noncanonical vortex can provide a simple way to control the position of the energy flux tornado, the rotation direction of the energy flux and the region of reversed energy flux, and produce a traditional Chinese "Taiji" pattern. This may be used in many applications, such as when tailoring and enriching the controlled means of particles.

2. Theoretical Background

2.1. Noncanonical Vortex

A traditional noncanonical vortex embedded in a beam at location (a, b) (in the exit pupil plane of an aplanatic system) can be described as [17]

$$V^{\text{in}}(x,y) = A(x,y)[x - a + \mathrm{i}\sigma_c(y - b)]^n, \tag{1}$$

where $A(x,y)$ is the amplitude, and, in this article, the amplitude is assumed to be Gaussian, i.e., $A(x,y) = e^{(x^2+y^2)/w_0^2}$ with w_0 denoting the waist size of the Gaussian beam. σ_c determines the noncanonical strength of the vortex, and as σ_c varies, the distributions of both the phase and amplitude change. This is the reason for its name, the anisotropy parameter. In principle, the anisotropy parameter σ_c can be a complex number, and when $\mathrm{Re}[\sigma_c] > 0$, the topological charge of the vortex expressed by Equation (1) is n (note n is positive), while for $\mathrm{Re}[\sigma_c] < 0$, the charge is $-n$. In this article, we only focus on the case that σ_c is a real number and $n = 1$, which means that the vortex is fundamental, and its topological charge is ± 1. In addition, in order to simplify the discussion, here, we also set $b = 0$, i.e., the vortex only can move along the x-axis.

The phase structures of the noncanonical vortex are depicted in Figure 1, where the inserts are their corresponding intensity distributions. In the left two columns, the anisotropy parameter is $\sigma_c = \pm 1$, and these phase-plots demonstrate that the phase gradient around the vortex is ± 1. It means that the vortex is actually the canonical type. This also can be found from Equation (1), i.e., when $\sigma_c = 1$, the noncanonical vortex degenerates into the canonical vortex. By comparing the phase structures and the intensity distributions for $\sigma_c = +1$ and $\sigma_c = -1$, we can observe that when σ_c changes its sign, only the topological charge is reversed (i.e., the phase gradient changes its sign) but the intensity distribution remains the same. It is easy to demonstrate that this rule is also true for the case $|\sigma_c| \neq 1$. Furthermore, from Figure 1, we also can find that for $|\sigma_c| < 1$, the phase changes faster near the y-axis, or the (absolute value of) phase gradient becomes bigger near the y-axis; while for $|\sigma_c| > 1$, the phase changes faster near the x-axis, or the (absolute value of) phase gradient is bigger there. The effect of the 'off-axis' is to push the intensity to the other side of the beam, i.e., to break the symmetry distribution of the intensity.

2.2. Electric Field, Magnetic Field and Energy Flux

Here, we will use the noncanonical vortex discussed above to construct a three-dimensional (3D) vector field and derive the expressions of the electric field, magnetic field and energy flux.

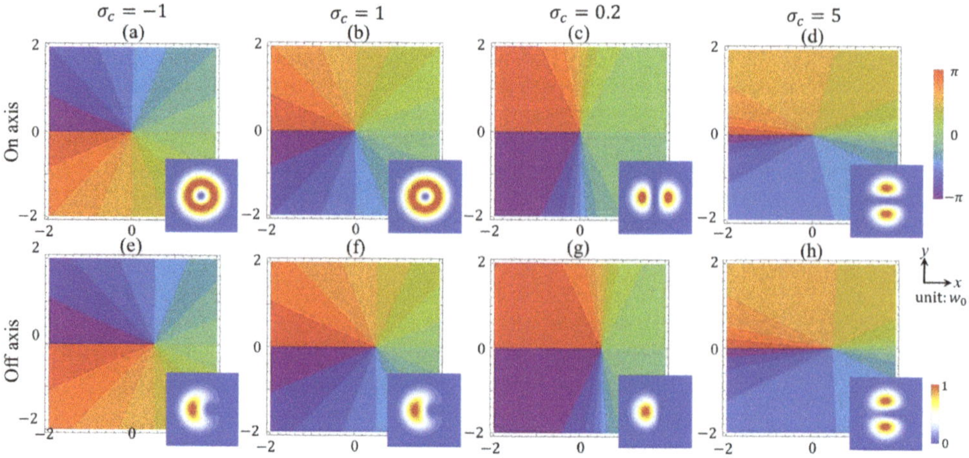

Figure 1. Structure of noncanonical vortex. The big plots are phase distributions and the inserts are their corresponding intensity distributions. In the first row, the vortex is located at the beam center, and in the second row, it is off-axis with $a = 0.5w_0$. In the left two columns $\sigma_c = \pm 1$ [(**a**,**b**,**e**,**f**)], i.e., the vortex degenerates into the canonical type.

Assume that there is an aplanatic, high numerical aperture system with the focal length f and the semi-aperture angle α, and the incident field is an off-axis noncanonical vortex beam described by Equation (1). According to the Richards–Wolf diffraction theory, the electric/magnetic field in the focal region of this structured field at an observation point $P(\rho_s, \phi_s, z_s)$ takes the form:

$$U(\rho_s, \phi_s, z_s) = -\frac{\mathrm{i}f}{\lambda}\int_0^{\alpha}\int_0^{2\pi} \sin\theta\sqrt{\cos\theta}V^{\mathrm{in}}(\theta,\phi)W_p(\theta,\phi)e^{\mathrm{i}kz_s\cos\theta} \times e^{\mathrm{i}k\rho_s\sin\theta\cos(\phi-\phi_s)}\mathrm{d}\phi\mathrm{d}\theta, \quad (2)$$

where $U(\rho_s, \phi_s, z_s)$ represents the electric or magnetic field strength and (ρ_s, ϕ_s, z_s) are the cylindrical coordinates. $k = 2\pi/\lambda$ denotes the wave number. Note that here, in order to calculate Equation (2), $V^{\mathrm{in}}(x, y)$ was changed into $V^{\mathrm{in}}(\theta, \phi)$ by applying the coordinates transform $x = \rho\cos\phi$, $y = \rho\sin\phi$ and the Abbe sine condition $\rho = f\sin\theta$. The polarization matrix $W_p(\theta, \phi)$ of the electric or magnetic field in Equation (2) is:

$$W_p(\theta,\phi) = \begin{bmatrix} 1+\cos^2\phi(\cos\theta-1) & \sin\phi\cos\phi(\cos\theta-1) \\ \sin\phi\cos\phi(\cos\theta-1) & 1+\sin^2\phi(\cos\theta-1) \\ -\sin\theta\cos\phi & -\sin\theta\sin\phi \end{bmatrix}\begin{bmatrix} w_x(\theta,\phi) \\ w_y(\theta,\phi) \end{bmatrix} \quad (3)$$

where $w_x(\theta, \phi)$ and $w_y(\theta, \phi)$ are the polarization weight functions of the x component and y component of the incident field. Here, we assume that in this article, the incident field is linearly polarized at a x direction. Thus, for the electric field, the weight functions are $w_x = 1$ and $w_y = 0$, while for the magnetic field, they are $w_x = 0$ and $w_y = 1$. By substituting the weight functions into Equation (2), the electric field and the magnetic field in the focal region can be obtained (for more details, one may refer to Appendix A).

For the electric field (e_x, e_y, e_z), after some calculations, their expressions are:

$$e_x(\rho_s, z_s, \phi_s) = -ik \int_0^\alpha P(\theta)\,(I_{x0} + I_{x1} + I_{x2} + I_{x3})\, e^{ikz_s\cos\theta} d\theta, \tag{4}$$

$$e_y(\rho_s, z_s, \phi_s) = -ik \int_0^\alpha P(\theta)\,(I_{y1} + I_{y2} + I_{y3})\, e^{ikz_s\cos\theta} d\theta, \tag{5}$$

$$e_z(\rho_s, z_s, \phi_s) = -ik \int_0^\alpha P(\theta)\,(I_{z0} + I_{z1} + I_{z2})\, e^{ikz_s\cos\theta} d\theta, \tag{6}$$

where

$$P(\theta) = f\, e^{-f^2\sin^2\theta/w_0^2}\sqrt{\cos\theta}\sin\theta, \tag{7}$$

and

$$I_{x0}(\theta;\rho_s,\phi_s) = -\frac{1}{2}a(1+\cos\theta)\, J_0(k\rho_s\sin\theta), \tag{8}$$

$$I_{x1}(\theta;\rho_s,\phi_s) = \frac{1}{4}f\sin\theta[-\sigma_c(3+\cos\theta)\sin\phi_s + i(3\cos\theta+1)\cos\phi_s]\, J_1(k\rho_s\sin\theta), \tag{9}$$

$$I_{x2}(\theta;\rho_s,\phi_s) = \frac{1}{2}a(\cos\theta-1)\cos 2\phi_s\, J_2(k\rho_s\sin\theta), \tag{10}$$

$$I_{x3}(\theta;\rho_s,\phi_s) = \frac{1}{4}f\sin\theta[\sigma_c(\cos\theta-1)\sin 3\phi_s - i(\cos\theta-1)\cos 3\phi_s]\, J_3(k\rho_s\sin\theta), \tag{11}$$

$$I_{y1}(\theta;\rho_s,\phi_s) = \frac{1}{4}f\sin\theta[\sigma_c(1-\cos\theta)\cos\phi_s - i(1-\cos\theta)\sin\phi_s]\, J_1(k\rho_s\sin\theta), \tag{12}$$

$$I_{y2}(\theta;\rho_s,\phi_s) = \frac{1}{2}a(\cos\theta-1)\sin 2\phi_s\, J_2(k\rho_s\sin\theta), \tag{13}$$

$$I_{y3}(\theta;\rho_s,\phi_s) = \frac{1}{4}f\sin\theta[\sigma_c(1-\cos\theta)\cos 3\phi_s + i(1-\cos\theta)\sin 3\phi_s]\, J_3(k\rho_s\sin\theta), \tag{14}$$

$$I_{z0}(\theta;\rho_s,\phi_s) = -\frac{1}{2}f\sin^2\theta\, J_0(k\rho_s\sin\theta), \tag{15}$$

$$I_{z1}(\theta;\rho_s,\phi_s) = ia\sin\theta\cos\phi_s\, J_1(k\rho_s\sin\theta), \tag{16}$$

$$I_{z2}(\theta;\rho_s,\phi_s) = \frac{1}{2}f\sin^2\theta[\cos 2\phi_s + i\sigma_c\sin 2\phi_s]\, J_2(k\rho_s\sin\theta). \tag{17}$$

Similarly, the expressions for the components of magnetic field (h_x, h_y, h_z) can be derived as

$$h_x(\rho_s, z_s, \phi_s) = -ik \int_0^\alpha P(\theta)\,(T_{x1} + T_{x2} + T_{x3})\, e^{ikz_s\cos\theta} d\theta, \tag{18}$$

$$h_y(\rho_s, z_s, \phi_s) = -ik \int_0^\alpha P(\theta)\,(T_{y0} + T_{y1} + T_{y2} + T_{y3})\, e^{ikz_s\cos\theta} d\theta, \tag{19}$$

$$h_z(\rho_s, z_s, \phi_s) = -ik \int_0^\alpha P(\theta)\,(T_{z0} + T_{z1} + T_{z2})\, e^{ikz_s\cos\theta} d\theta, \tag{20}$$

where

$$T_{x1}(\theta;\rho_s,\phi_s) = \frac{1}{4}f\sin\theta[\sigma_c(1-\cos\theta)\cos\phi_s - i(1-\cos\theta)\sin\phi_s]\, J_1(k\rho_s\sin\theta), \tag{21}$$

$$T_{x2}(\theta;\rho_s,\phi_s) = \frac{1}{2}a(\cos\theta-1)\sin 2\phi_s\, J_2(k\rho_s\sin\theta), \tag{22}$$

$$T_{x3}(\theta;\rho_s,\phi_s) = \frac{1}{4}f\sin\theta[\sigma_c(1-\cos\theta)\cos 3\phi_s + i(1-\cos\theta)\sin 3\phi_s]\, J_3(k\rho_s\sin\theta), \tag{23}$$

$$T_{y0}(\theta;\rho_s,\phi_s) = -\frac{1}{2}a(1+\cos\theta)\, J_0(k\rho_s\sin\theta), \tag{24}$$

$$T_{y1}(\theta;\rho_s,\phi_s) = \frac{1}{4}f\sin\theta[-\sigma_c(1+3\cos\theta)\sin\phi_s + \mathrm{i}(\cos\theta+3)\cos\phi_s]\, J_1(k\rho_s\sin\theta), \tag{25}$$

$$T_{y2}(\theta;\rho_s,\phi_s) = \frac{1}{2}a(1-\cos\theta)\cos 2\phi_s\, J_2(k\rho_s\sin\theta), \tag{26}$$

$$T_{y3}(\theta;\rho_s,\phi_s) = \frac{1}{4}f\sin\theta[\sigma_c(1-\cos\theta)\sin 3\phi_s - \mathrm{i}(1-\cos\theta)\cos 3\phi_s]\, J_3(k\rho_s\sin\theta), \tag{27}$$

$$T_{z0}(\theta;\rho_s,\phi_s) = -\frac{1}{2}\mathrm{i}\sigma_c f\sin^2\theta\, J_0(k\rho_s\sin\theta), \tag{28}$$

$$T_{z1}(\theta;\rho_s,\phi_s) = \mathrm{i}a\sin\theta\sin\phi_s\, J_1(k\rho_s\sin\theta), \tag{29}$$

$$T_{z2}(\theta;\rho_s,\phi_s) = \frac{1}{2}f\sin^2\theta[\sin 2\phi_s - \mathrm{i}\sigma_c\cos 2\phi_s]\, J_2(k\rho_s\sin\theta). \tag{30}$$

Obviously, both the electric field and the magnetic field in the focal region are strongly influenced by the anisotropy parameter σ_c, as well as the off-axis distance a.

The energy flux, by definition, is expressed by the time-averaged Poynting vector **S**. Since we have derived all the components of the electric field and magnetic field, now the energy flux can be obtained from the following expression:

$$\mathbf{S} = \frac{c}{8\pi}\mathbf{Re}(\mathbf{E}\times\mathbf{H}^*) = \begin{pmatrix} S_x \\ S_y \\ S_z \end{pmatrix} = \begin{pmatrix} e_y h_z^* - e_z h_y^* \\ e_z h_x^* - e_x h_z^* \\ e_x h_y^* - e_y h_x^* \end{pmatrix}, \tag{31}$$

where c is the speed of light in the free space. The superscript $*$ denotes the complex conjugate and **Re** is the real part of a complex number. Thus, the longitudinal energy flux is represented by S_z, which is dependent on the transverse components of both the electric field and the magnetic field; meanwhile, the transverse energy flux is (S_x, S_y), and from here on we use $\mathbf{s}_{xy} = (S_x, S_y)$ to represent this transverse vector, which relies on both the transverse and the longitudinal components of the electromagnetic field.

3. Results and Discussions

In this section, we will analyze the energy flux in the focal region. In order to obtain the effects of the off-axis noncanonical vortex on the dynamics of the energy flux, we will first analyze and summarize the behaviors of the energy flux influenced by the canonical vortex in this section.

3.1. Energy Flux: The Canonical Vortex Case $|\sigma_c| = 1$

For $|\sigma_c| = 1$, the incident vortex beam is a canonical type with its topological charge being $+1$ (when $\sigma_c = +1$) or -1 (when $\sigma_c = -1$). Let us first observe the transverse energy flux $\mathbf{s}_{xy} = (S_x, S_y)$ in this focused field with $|\sigma_c| = 1$. By imputing $\sigma_c = \pm 1$ into Equations (18)–(30), we can obtain the electric field and the magnetic field, and thereby the expression of the transverse energy flux is obtained according to Equation (31).

Figure 2 shows the transverse energy flux $\mathbf{s}_{xy}$ including the flow lines (or the lines of the transverse Poynting vectors) and the strength distribution in the focal plane ($z_s = 0$). It can be found, firstly, that there is a (transverse) Poynting singularity, i.e., a transverse energy flux tornado created in the beam center for the off-axis distance $a = 0$ or along the y_s-axis for $a \neq 0$. Secondly, the rotation direction of the tornado is counter-clockwise when the anisotropy parameter σ_c is positive, whereas it is clockwise when σ_c is negative. Thirdly and more interestingly, the sign of σ_c determines the location of the energy flux tornado, i.e., the tornado is along the $-y_s$-axis if $\sigma_c = 1$ and along the $+y_s$-axis if $\sigma_c = -1$. In other words, the strength maximum of the transverse energy flux or its strength distribution pattern depends on the sign of the anisotropy parameter σ_c. This phenomenon is actually caused by both the topological charge of the incident beam and the effect of strongly focusing, which will be discussed in the following.

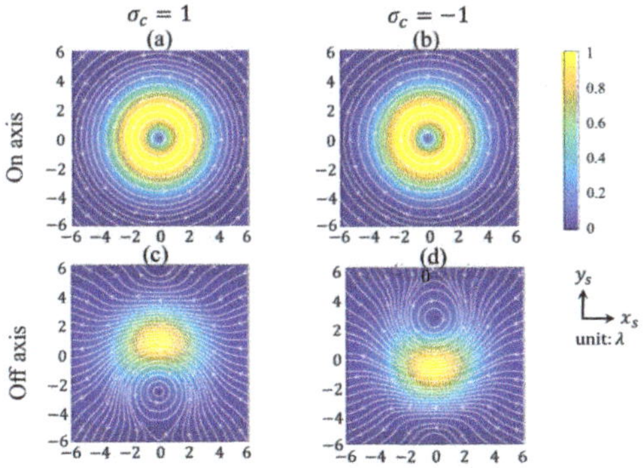

Figure 2. The transverse energy flux $\mathbf{s}_{xy}$ with flow lines (in white) and strength $|\mathbf{s}_{xy}|$ (color-coded) on the focal plane. Here (**a**) $\sigma_c = 1$, $a = 0$ (**b**) $\sigma_c = -1$, $a = 0$, (**c**) $\sigma_c = 1$, $a = 0.5w_0$ and (**d**) $\sigma_c = -1$, $a = 0.5w_0$. In all plots: $f = 2$ mm, $w_0 = 2$ mm and $\alpha = 60°$.

Propagation of the transverse energy flux $\mathbf{s}_{xy}$ in the focal region with $z_s = -4\lambda$ to $z_s = +4\lambda$ is depicted in Figure 3, where the first row shows the case of $\sigma_c = +1$ and the second row is the case of $\sigma_c = -1$. It can be observe that when $\sigma_c > 0$, the energy strength patten (and the energy flux tornado) rotates contour-clockwise as the beam propagates, while for $\sigma_c < 0$, it rotates clockwise. In order to illustrate this rotation behavior more clearly, the azimuthal angle of the strength maximum is depicted in Figure 4. (Here, in order to find both the maximum (possible) rotation angle and the tendency of this rotation behavior, the scope which is bigger than the depth of the field, $-10\lambda \leq z_s \leq 10\lambda$, is chosen.) It is demonstrated that when the beam propagates from $z_s = -10\lambda$ to 10λ, the maximum point (representing the strength distribution pattern) of the transverse energy flux rotates from near 0° ($z_s = -10\lambda$), 90° ($z_s = 0$) to almost 180° ($z_s = 10\lambda$) for $\sigma_c = +1$ and from near 0° ($z_s = -10\lambda$), −90° ($z_s = 0$) to almost −180° for $\sigma_c = -1$. That is to say, that as the beam propagates if $\sigma_c > 0$, the strength distribution pattern has a counterclockwise rotation and the total rotation angle is near 180°, while if $\sigma_c < 0$, the pattern has a clockwise rotation with the total rotation angle near −180°.

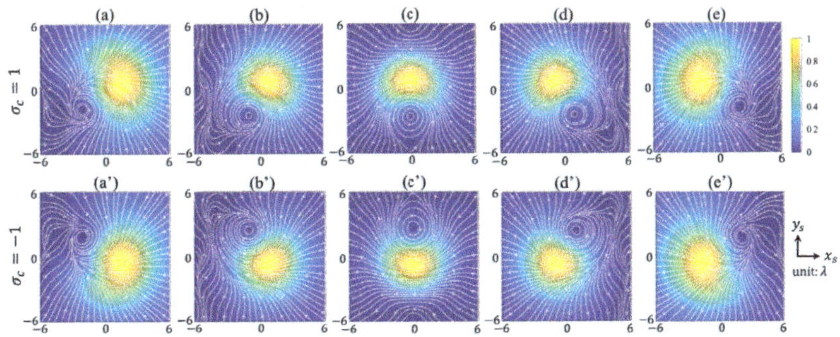

Figure 3. The transverse energy flux $\mathbf{s}_{xy}$ along the propagation direction. Here, (**a**,**a'**) $z_s = -4\lambda$, (**b**,**b'**) $z_s = -2\lambda$, (**c**,**c'**) $z_s = 0$, (**d**,**d'**) $z_s = 2\lambda$ and (**e**,**e'**) $z_s = 4\lambda$. In all plots: $a = 0.5w_0$, $f = 2$ mm, $w_0 = 2$ mm and $\alpha = 60°$.

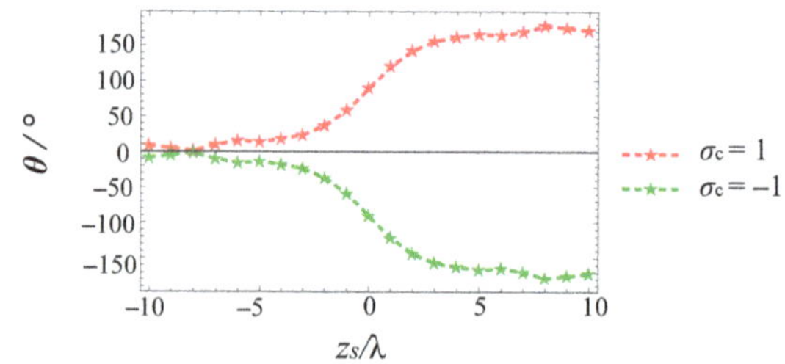

Figure 4. The rotation angle for the transverse energy flux $\mathbf{s}_{xy}$ with the beam propagation. Here, $a = 0.5w_0$, $f = 2$ mm, $w_0 = 2$ mm and $\alpha = 60°$.

After the discussion on the transverse energy flux, we will examine the longitudinal energy flux. For a common light beam, its longitudinal energy flux usually goes towards the propagation direction, i.e., $S_z > 0$. The reversed energy flux actually means the longitudinal energy flux $S_z < 0$, which has special applications in optical trappings. Here, we will observe that in our constructed field, the reserved energy flux can be found.

Figure 5 depicts the strength of the longitudinal energy flow S_z on the xOy-plane (in the first column), on the xOz-plane (in the second column) and on the yOz-plane (in the third column), separately. In this figure, the first row shows the on-axis vortex case, i.e., $a = 0$, and the second row is the case of $a = 0.5w_0$. In all plots $\sigma_c = 1$. The white lines on the xOy-plane denotes $S_z = 0$ and the reverse energy flux exists in the inner region enclosed by the white lines. It is observed that the negative energy flux shows a torus-shape distribution around the beam center for $a = 0$, while in the case of $a = 0.5w_0$, the negative energy can distribute near the beam center. In order to illustrate the energy flux with more details, here instead of the flux lines, the energy flux vectors are depicted. The Poynting vectors of $\mathbf{s}_{zx} = (S_z, S_x)$ and $\mathbf{s}_{zy} = (S_z, S_y)$ denoted by arrows are drawn on the xOz-plane and the yOz plane, respectively. The red ones represent the reversed longitudinal energy flux (the black region with reversed energy flow in Figure 5c is enlarged as an example), i.e., $S_z < 0$. From Figure 5, it can be found that the off-axis vortex changes the location of the negative energy flux region, while the size of the negative energy flux region is quite modest and almost not influenced by the off-axis vortex.

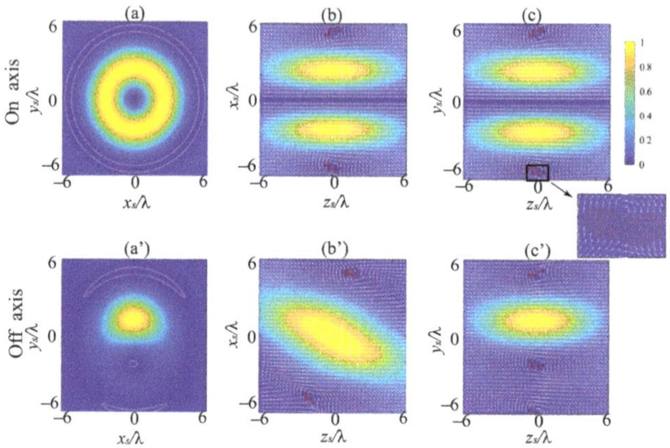

Figure 5. The longitudinal energy flow S_z on the (**a**,**a'**) xOy-plane, (**b**,**b'**) xOz-plane and (**c**,**c'**) yOz-plane. In (**a**–**c**) $\sigma_c = 1, a = 0$, (**a'**–**c'**) $\sigma_c = 1, a = 0.5w_0$. In all plots, $f = 2$ mm, $w_0 = 2$ mm and $\alpha = 60°$.

3.2. Energy Flux: The Noncanonical Vortex Case $|\sigma_c| \neq 1$

In this part, we will generalize the behaviors of the energy flux influenced by the noncanonical vortex. For $|\sigma_c| \neq 1$, the incident vortex beam is the noncanonical type with not only the phase but also the amplitude distributing anisotropically. However, its topological charge remains the same, that is $+1$ (when $\sigma_c > 0$) or -1 (when $\sigma_c < 0$).

Similarly, let us first discuss the transverse energy flux in the focused field. The transverse energy flux $\mathbf{s}_{xy}$ (including the flow lines and the strength distribution) on the focal plane for different values of the anisotropy parameter σ_c is shown in Figure 6. Similar results can be obtained in the case of the noncanonical vortex when compared to the case of the canonical vortex: (a) There is a transverse energy flux tornado (transverse Poynting singularity) along the y_s axis for $|\sigma_c| \neq 1$ with the off-axis vortex $a \neq 0$. (b) The rotation direction of the transverse Poynting singularity is counter-clockwise when σ_c is positive and clockwise when σ_c is negative. (c) The location of the transverse Poynting singularity depends on the sign of σ_c. The tornado locates at the $-y_s$ axis if $\sigma_c > 0$ and the y_s axis if $\sigma_c < 0$. Meanwhile, the difference in the results between the canonical and noncanonical cases is that the location of Poynting singularity moves closer to the center as the $|\sigma_c|$ increases. More interestingly, the strength distribution of the transverse energy flux becomes more uniform around the Poynting singularity (near focus) when the $|\sigma_c|$ is much bigger than 1, (see plots (d) and (d') in Figure 6, $\sigma_c = \pm 10$). That is to say, through only adjusting the parameter σ_c, we can control the position of the transverse energy tornado and the distribution of the transverse energy strength.

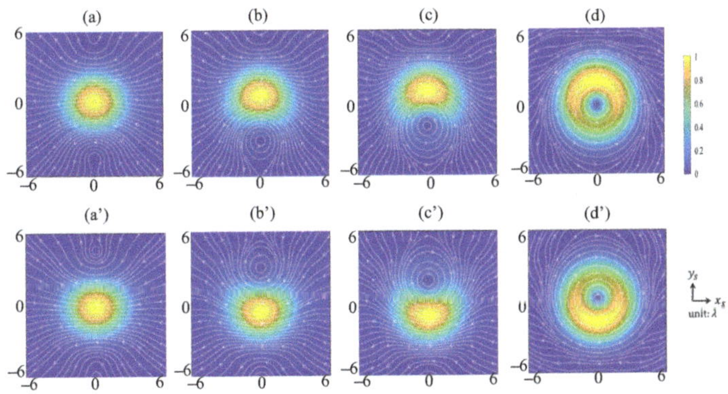

Figure 6. The transverse energy flux $\mathbf{s}_{xy}$ with flow lines and strength $|\mathbf{s}_{xy}|$ on the focal plane for $|\sigma_c| \neq 1$. Here, **(a)** $\sigma_c = 0.2$, **(a')** $\sigma_c = -0.2$, **(b)** $\sigma_c = 0.8$, **(b')** $\sigma_c = -0.8$, **(c)** $\sigma_c = 1.5$, **(c')** $\sigma_c = -1.5$ and **(d)** $\sigma_c = 10$, **(d')** $\sigma_c = -10$. In all plots, $a = 0.5w_0$, $f = 2$ mm, $w_0 = 2$ mm and $\alpha = 60°$.

The transverse energy flux $\mathbf{s}_{xy}$ for $|\sigma_c| \neq 1$ along the propagation direction from $z_s = -4\lambda$ to $z_s = 4\lambda$ is shown in Figure 7, where the anisotropy parameter $\sigma_c = 0.2$ in the first row and $\sigma_c = 10$ in the second row. It is quite obvious that the propagation behavior of the transverse energy flux for $|\sigma_c| \neq 1$ is very different from that shown in Figure 3 (i.e., $|\sigma_c| = 1$). One can observe that for $\sigma_c = 0.2, 10$, the transverse energy flux $\mathbf{s}_{xy}$ seems to not rotate when the beam propagates. The energy strength pattern for $\sigma_c = 0.2$ has one maximum on the y_s-axis on the focal plane, and this maximum laterally moves from the right side (near x_s axis) to the left side (near $-x_s$ axis) as z_s increases. When $\sigma_c = 10$, the maximum of the energy strength pattern locates at the y_s-axis on the focal plane as well, while as the beam propagates, the strength pattern splits into two parts and the maximum moves along the y_s-axis. This phenomenon can be physically explained by the joint effect of the anisotropic phase and intensity distribution, as well as the polarization of the incident beam and the strongly focusing. This joint effect results in the energy accumulated along the x_s-axis or y_s-axis when $|\sigma_c|$ is greatly larger or smaller than 1.

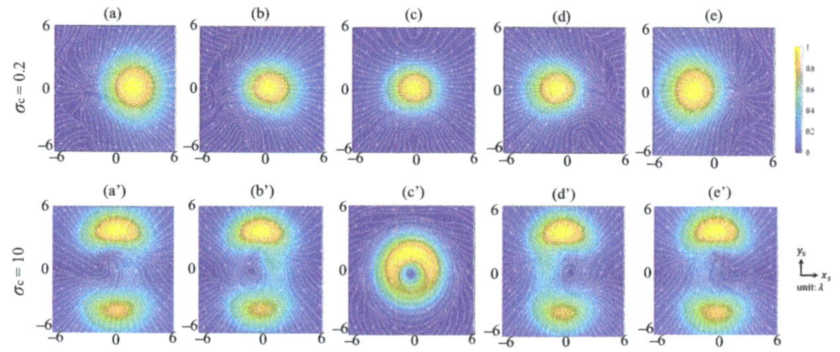

Figure 7. The transverse energy flux $\mathbf{s}_{xy}$ along the propagation direction. Here, (**a**,**a′**) $z_s = -4\lambda$, (**b**,**b′**) $z_s = -2\lambda$, (**c**,**c′**) $z_s = 0$, (**d**,**d′**) $z_s = 2\lambda$ and (**e**,**e′**) $z_s = 4\lambda$. In all plots: $a = 0.5w_0$, $f = 2$ mm, $w_0 = 2$ mm and $\alpha = 60°$.

Now, we examine some situations with $|\sigma_c|$ near 1. Figure 8 depicts the transverse energy flux $\mathbf{s}_{xy}$ with $\sigma_c = 0.8$ and 1.5 as the beam propagates. It can be observed that the rotation behavior of $\mathbf{s}_{xy}$ is very obvious, which is more like the case for the canonical vortex (see Figure 3). The variations of the azimuthal angles of the strength maximum with the propagation distance for these four cases are shown in Figure 9. From this figure, the rotational behavior and its counterclockwise manner of the transverse energy flux when $\sigma_c > 0$ can be observed as being more straight.

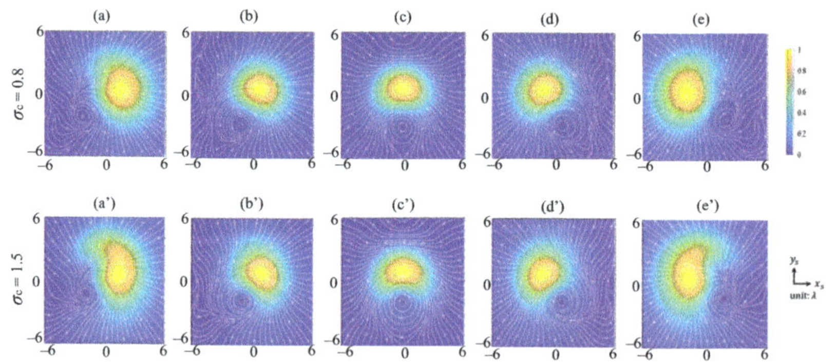

Figure 8. The transverse energy flux $\mathbf{s}_{xy}$ along the propagation direction. Here, (**a**,**a′**) $z_s = -4\lambda$, (**b**,**b′**) $z_s = -2\lambda$, (**c**,**c′**) $z_s = 0$, (**d**,**d′**) $z_s = 2\lambda$ and (**e**,**e′**) $z_s = 4\lambda$. In all plots: $a = 0.5w_0$, $f = 2$ mm, $w_0 = 2$ mm and $\alpha = 60°$.

One can find that the maximum point of the transverse energy flux stays close to 90°, as the beam propagates from $z_s = -10\lambda$ to $z_s = 10\lambda$ in the case of $\sigma_c = 10$. Meanwhile, for the case of $\sigma_c = 0.2$, it is demonstrated that the maximum point of the transverse energy strength stays near 0° when the beam propagates from $z_s = -10\lambda$ to $z_s = -3\lambda$ and keeps near 180° when the beam propagates from $z_s = 3\lambda$ to $z_s = 10\lambda$. Note that despite the azimuthal angle of the maximum point not being close to 0° or 180° in the focal region (z_s from -2λ to 2λ), it still changes dramatically from 14° ($z_s = -2\lambda$) to 166° ($z_s = 2\lambda$) compared to the canonical vortex case. Moreover, the rotation angle and behavior of the case $\sigma_c = 0.8$ and 1.5 in Figure 9 confirms the results in Figure 8. It should be noted that the trends of these two curves are not completely the same. There exist differences between $\sigma_c = 0.8$ and 1.5 in the region $z_s = -10\lambda$ to -5λ and $z_s = 5\lambda$ to 10λ. This behavior is generated by the anisotropic distribution of the phase and intensity when $|\sigma_c| \neq 1$.

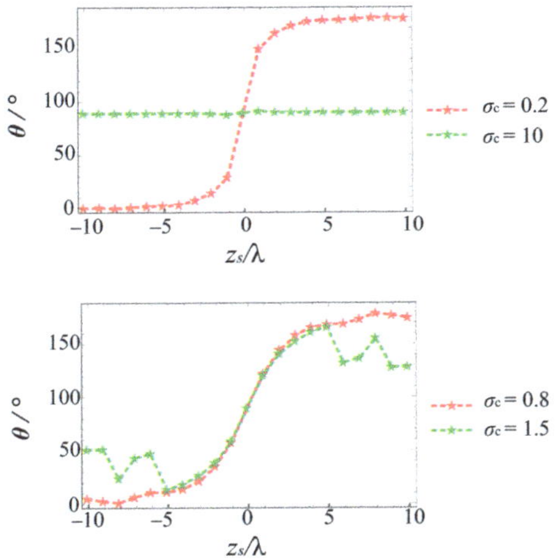

Figure 9. The rotation angle for the transverse energy flux $\mathbf{s}_{xy}$ with the beam propagation. Here, $a = 0.5w_0$, $f - 2$ mm, $w_0 = 2$ mm and $\alpha = 60°$.

The longitudinal energy flow S_z of the canonical vortex case has been investigated in detail in Figure 5. Obviously, the longitudinal energy flow becomes more complicated when $|\sigma_c| \neq 1$. The strength of longitudinal energy flow and the reverse energy flux region (denoted by the white lines) of $\sigma_c = 0.2, 0.8, 10, 15$ at the focal plane are shown in Figure 10. Through observing this figure, we can obtain that: (a) For any value of σ_c, the strength S_z and negative energy flux region is asymmetry distributed on the focal plane. (b) When $|\sigma_c| < 1$, the reverse energy flux region, which is near to the focus, becomes bigger and away from the focus as $|\sigma_c|$ decreases. Meanwhile, when $|\sigma_c| > 1$, the reverse energy flux region, which is near to the focus, becomes smaller and close to the focus as $|\sigma_c|$ increases. (c) The whole reverse energy flux region on the focal plane becomes bigger as $|\sigma_c|$ increases.

More details of $\sigma_c = 0.2$ and 15 are shown in Figure 11. The Poynting vectors of $\mathbf{s}_{zx}$ and $\mathbf{s}_{zy}$ on the xOz-plane and yOz-plane, respectively, are presented by arrows, where the red arrows still denote the reversed longitudinal energy flux. It can be found that not only the location of the negative energy flux region changes with different σ_c, but also the size of the negative energy flux region changes as well. Particularly, when $\sigma_c = 15$, the negative energy flux region grows in size on the xOz-plane. In order to show energy flux more clearly, we add the energy flow lines of $\mathbf{s}_{zx}$ and $\mathbf{s}_{zy}$ in Figure 11, which are denoted by white thick lines with arrows. It is very surprising to find that the energy flow lines exhibit an "*S*" shape and exist two energy flux tornadoes (i.e. Poynting singularities) with opposite topological charges (i.e., opposite rotating directions) on the xOz-plane in the case of $\sigma_c = 15$. Even more intriguing is that this "*S*" shape with two tornadoes is very like the Chinese "Taiji" pattern, where the two tornadoes with opposite charges exactly resemble the two eyes of the "Yin fish" and "Yang fish", respectively. As the "Yin fish" and "Yang fish" can rotate and be annihilated in the traditional "Taiji" theory, these two energy tornadoes (generated by the reversed energy flux) can also behave in the same way. These two opposite rotating energy tornadoes may provide a simple way to trap and rotate two particles with opposite rotation directions. Moreover, it is obvious that the adjusting anisotropy parameter σ_c can manipulate the size and location of the negative longitudinal energy flow region.

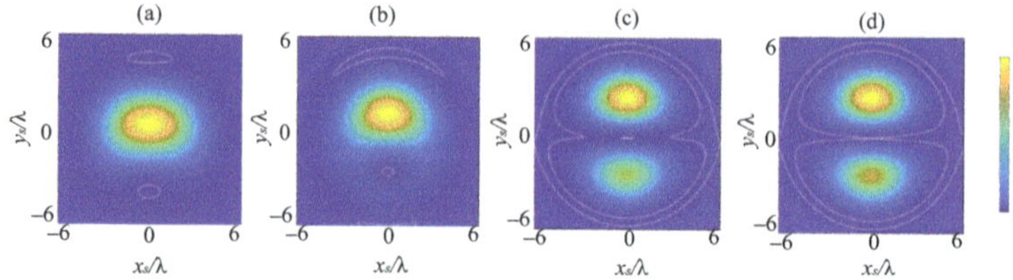

Figure 10. The longitudinal energy flow S_z on the xOy-plane, where (**a**) $\sigma_c = 0.2$, (**b**) $\sigma_c = 0.8$, (**c**) $\sigma_c = 10$ and (**d**) $\sigma_c = 15$. In all plots, $a = 0.5w_0, f = 2$ mm, $w_0 = 2$ mm and $\alpha = 60°$.

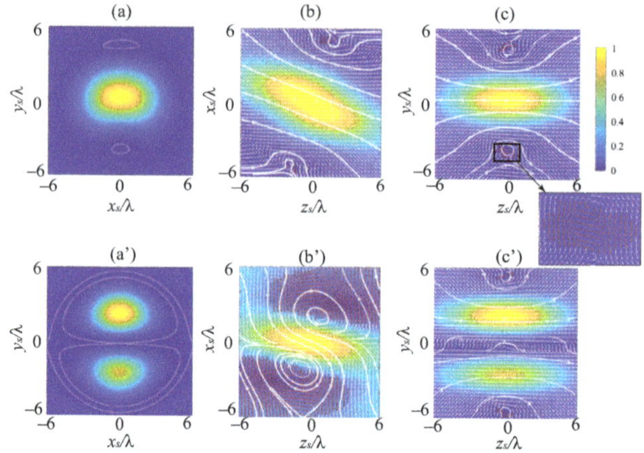

Figure 11. The longitudinal energy flow S_z on the (**a**,**a′**) xOy-plane, (**b**,**b′**) xOz-plane and (**c**,**c′**) yOz-plane. Where (**a**–**c**) $\sigma_c = 0.2$ and (**a′**–**c′**) $\sigma_c = 15$. In all plots, $a = 0.5w_0, f = 2$ mm, $w_0 = 2$ mm and $\alpha = 60°$.

4. Conclusions

In this article, an off-axis noncanonical vortex is introduced and its effect on the energy flux of a strongly focused field are discussed. For the transverse energy flux, we find that by adjusting the anisotropy parameter of the nonanonical vortex σ_c, we can control the position of the transverse energy tornado and the distribution of the transverse energy strength. Moreover, the rotation direction of the transverse energy flux $\mathbf{s}_{xy}$ is determined by the sign of σ_c as the beam propagates. There is a special case that when $|\sigma_c|$ is much smaller or bigger than 1, the transverse energy flux $\mathbf{s}_{xy}$ seems to not rotate but moves along the x_s axis or y_s axis, respectively. This phenomenon is physically caused by the joint effect of the anisotropic phase and intensity distribution, the polarization of the incident beam and the strongly focusing. In addition, for the longitudinal energy flux, it is also demonstrated that by changing the anisotropy parameter, σ_c can control the location of the reversed energy flux region, the size of the reversed energy flux region and produce two energy tornadoes with opposite rotating directions, resembling a Chinese "Taiji" pattern. Our findings not only examine the fundamental physical properties of an off-axis noncanonical vortex beam, but also provide a theoretical guidance to explore particle control methods in a 3D vector field.

Author Contributions: Conceptualization, X.P.; software, X.Z.; validation, X.Z. and X.P.; formal analysis, X.P. and X.Z.; investigation, X.P. and X.Z.; writing—original draft preparation, X.Z., H.L. and G.W.; writing—review and editing, X.Z and X.P.; visualization, H.L. and G.W. All authors have read and agreed to the published version of the manuscript.

Funding: This research was funded by the National Natural Science Foundation of China (NSFC) (12104283, 11974281, 12074310).

Institutional Review Board Statement: Not applicable.

Informed Consent Statement: Not applicable.

Data Availability Statement: The data presented in this study are available on request from the corresponding author.

Conflicts of Interest: The authors declare no conflict of interest.

Appendix A

Here, we will show some essential steps for obtaining Equations (8)–(30). By substituting the polarization matrix Equation (3) into (2), the electric field in the focal region is obtained:

$$E(\rho_s,\phi_s,z_s)=\begin{bmatrix} e_x \\ e_y \\ e_z \end{bmatrix}=-\frac{if}{\lambda}\int_0^{\alpha}\int_0^{2\pi}\sin\theta\sqrt{\cos\theta}V^{\mathrm{in}}(\theta,\phi)\begin{bmatrix} 1+\cos^2\phi(\cos\theta-1) \\ \sin\phi\cos\phi(\cos\theta-1) \\ -\sin\theta\cos\phi \end{bmatrix} \times e^{ikz_s\cos\theta}e^{ik\rho_s\sin\theta\cos(\phi-\phi_s)}\mathrm{d}\phi\mathrm{d}\theta, \tag{A1}$$

Similarly, the magnetic field in the focal region is derived as:

$$H(\rho_s,\phi_s,z_s)=\begin{bmatrix} h_x \\ h_y \\ h_z \end{bmatrix}=-\frac{if}{\lambda}\int_0^{\alpha}\int_0^{2\pi}\sin\theta\sqrt{\cos\theta}V^{\mathrm{in}}(\theta,\phi)\begin{bmatrix} \sin\phi\cos\phi(\cos\theta-1) \\ 1+\sin^2\phi(\cos\theta-1) \\ -\sin\theta\sin\phi \end{bmatrix} \times e^{ikz_s\cos\theta}e^{ik\rho_s\sin\theta\cos(\phi-\phi_s)}\mathrm{d}\phi\mathrm{d}\theta, \tag{A2}$$

where the incident field $V^{\mathrm{in}}(\theta,\phi)$ in this article is written as follows:

$$V^{\mathrm{in}}(\theta,\phi)=e^{-f^2\sin^2\theta/w_0^2}[f\sin\theta\cos\phi+i\sigma_c f\sin\theta\sin\phi-a]. \tag{A3}$$

Taking the x component of the electric field as an example, the key step for deriving the expression of e_x is to calculate the integral with respect to ϕ, i.e.,

$$\begin{aligned}&\int_0^{2\pi}(f\sin\theta\cos\phi+i\sigma_c f\sin\theta\sin\phi-a)(1+\cos^2\phi(\cos\theta-1))e^{ik\rho_s\sin\theta\cos(\phi-\phi_s)}\mathrm{d}\phi\\ &=\int_0^{2\pi}\left\{\frac{1}{4}f\sin\theta[i\sigma_c(3+\cos\theta)\sin\phi+(3\cos\theta+1)\cos\phi]+\frac{1}{4}f\sin\theta[i\sigma_c(\cos\theta-1)\sin3\phi\right.\\ &\left.+(\cos\theta-1)\cos3\phi]-\frac{1}{2}a(1+\cos\theta)-\frac{1}{2}a(\cos\theta-1)\cos2\phi\right\}e^{ik\rho_s\sin\theta\cos(\phi-\phi_s)}\mathrm{d}\phi.\end{aligned} \tag{A4}$$

Then by applying the following identities:

$$\int_0^{2\pi}\cos n\phi e^{ik\rho_s\sin\theta\cos(\phi-\phi_s)}\mathrm{d}\phi=2\pi i^n\cos n\phi_s J_n(k\rho_s\sin\theta), \tag{A5}$$

and

$$\int_0^{2\pi}\sin n\phi e^{ik\rho_s\sin\theta\cos(\phi-\phi_s)}\mathrm{d}\phi=2\pi i^n\sin n\phi_s J_n(k\rho_s\sin\theta), \tag{A6}$$

with J_n being the first kind Bessel function of order n, Equations (8)–(11) can be obtained. The other components of the electric field and magnetic field can be derived in the same way. Note that the electric and the magnetic field in the focal region in Equations (18)–(20) are consistent with the electric and magnetic field in Ref. [25] when $\sigma_c = 1$ and $a = 0$

(except the vortex factor). In addition, the electric field in Equations (8)–(11) in this article and in Ref. [39] are identical for $\sigma_c = 1$.

References

1. Soskin, M.S.; Vasnetsov, M.V. Singular Optics. In *Progress in Optics*; Wolf, E., Ed.; Elsevier: Amsterdam, The Netherlands, 2001; Volume 42, pp. 219–276. [CrossRef]
2. Gbur, G.J. *Singular Optics*; Chemical Rubber Company: Boca Raton, United States 2017. [CrossRef]
3. Otte, E.; Tekce, K.; Denz, C. Tailored intensity landscapes by tight focusing of singular vector beams. *Opt. Express* **2017**, *25*, 20194–20201. [CrossRef]
4. Zhao, X.; Pang, X.; Zhang, J.; Wan, G. Transverse Focal Shift in Vortex Beams. *IEEE Photon. J.* **2018**, *10*, 1–17. [CrossRef]
5. Kotlyar, V.V.; Kovalev, A.A.; Stafeev, S.S.; Nalimov, A.G.; Rasouli, S. Tightly focusing vector beams containing V-point polarization singularities. *Opt. Laser. Technol.* **2022**, *145*, 107479. . [CrossRef]
6. Kotlyar, V.V.; Stafeev, S.S.; Nalimov, A.G. Sharp Focusing of a Hybrid Vector Beam with a Polarization Singularity. *Photonics* **2021**, *8*, 227. [CrossRef]
7. Dong, M.; Zhao, C.; Cai, Y.; Yang, Y. Partially coherent vortex beams: Fundamentals and applications. *Sci. China Phys. Mech.* **2021**, *64*, 224201. [CrossRef]
8. Chen, J.; Liu, X.; Yu, J.; Cai, Y. Simultaneous determination of the sign and the magnitude of the topological charge of a partially coherent vortex beam. *Appl. Phys. B-Lasers O* **2016**, *122*, 201. [CrossRef]
9. Angelsky, O.V.; Bekshaev, A.Y.; Maksimyak, P.P.; Maksimyak, A.P.; Hanson, S.G.; Zenkova, C.Y. Orbital rotation without orbital angular momentum: Mechanical action of the spin part of the internal energy flow in light beams. *Opt. Express* **2012**, *20*, 3563–3571. [CrossRef]
10. Angelsky, O.V.; Bekshaev, A.Y.; Maksimyak, P.P.; Maksimyak, A.P.; Zenkova, C.Y.; Gorodynska, N.V. Circular motion of particles by the help of the spin part of the internal energy flow. In Proceedings of the ROMOPTO 2012: Tenth Conference on Optics: Micro- to Nanophotonics III, Bucharest, Romania, 3–6 September 2012; Vlad, V.I., Ed.; International Society for Optics and Photonics, SPIE: Bucharest, Romania, 2013; Volume 8882, p. 88820A. [CrossRef]
11. Taylor, M.A.; Waleed, M.; Stilgoe, A.B.; Rubinsztein-Dunlop, H.; Bowen, W.P. Enhanced optical trapping via structured scattering. *Nature Photon.* **2015**, *9*, 669–673. [CrossRef]
12. Padgett, M.J.; Molloy, J.; McGloin, D. *Optical Tweezers: Methods and Applications*; CRC Press: Boca Raton, FL, USA, 2010. [CrossRef]
13. Bozinovic, N.; Yue, Y.; Ren, Y.; Tur, M.; Kristensen, P.; Huang, H.; Willner, A.E.; Ramachandran, S. Terabit-scale orbital angular momentum mode division multiplexing in fibers. *Science* **2013**, *340*, 1545–1548. [CrossRef]
14. Freund, I.; Shvartsman, N.; Freilikher, V. Optical dislocation networks in highly random media. *Opt. Commun.* **1993**, *101*, 247–264. [CrossRef]
15. Tong, R.; Dong, Z.; Chen, Y.; Wang, F.; Cai, Y.; Setälä, T. Fast calculation of tightly focused random electromagnetic beams: Controlling the focal field by spatial coherence. *Opt. Express* **2020**, *28*, 9713–9727. [CrossRef] [PubMed]
16. Pang, X.; Xiao, W.; Zhang, H.; Feng, C.; Zhao, X. X-type vortex and its effect on beam shaping. *J. Opt.* **2021**, *23*, 125604. [CrossRef]
17. Molina-Terriza, G.; Wright, E.M.; Torner, L. Propagation and control of noncanonical optical vortices. *Opt. Lett.* **2001**, *26*, 163–165. [CrossRef]
18. Roux, F.S. Coupling of noncanonical optical vortices. *J. Opt. Soc. Am. B* **2004**, *21*, 664–670. [CrossRef]
19. Kim, G.H.; Lee, H.J.; Kim, J.U.; Suk, H. Propagation dynamics of optical vortices with anisotropic phase profiles. *J. Opt. Soc. Am. B* **2003**, *20*, 351–359. [CrossRef]
20. Lopez-Mago, D.; Perez-Garcia, B.; Yepiz, A.; Hernandez-Aranda, R.I.; Gutiérrez-Vega, J.C. Dynamics of polarization singularities in composite optical vortices. *J. Opt.* **2013**, *15*, 044028. [CrossRef]
21. Gong, L.; Wang, X.; Zhu, Z.; Lai, S.; Feng, H.; Wang, J.; Gu, B. Transversal energy flow of tightly focused off-axis circular polarized vortex beams. *Appl. Opt.* **2022**, *61*, 5076–5082. [CrossRef] [PubMed]
22. Zhang, X.; Li, P.; Liu, S.; Wei, B.; Qi, S.; Fan, X.; Wang, S.; Zhang, Y.; Zhao, J. Autofocusing of ring Airy beams embedded with off-axial vortex singularities. *Opt. Express* **2020**, *28*, 7953–7960. [CrossRef] [PubMed]
23. Lee, D.H.; Kim, H.S.; Han, I.; Bae, J.Y.; Yeo, W.J.; Jeong, S.K.; Jeon, M.; Choi, H.J.; Kim, D.U.; Lee, K.S.; et al. Generation of wavelength-tunable optical vortices using an off-axis spiral phase mirror. *Opt. Lett.* **2021**, *46*, 4216–4219. [CrossRef]
24. Ignatowsky, V.S. Diffraction by a lens of arbitrary aperture. *Trans. Opt. Inst.* **1919**, *1*, 1–36.
25. Richards, B.; Wolf, E. Electromagnetic Diffraction in Optical Systems. II. Structure of the Image Field in an Aplanatic System. *Proc. R. Soc. Lond. A* **1959**, *253*, 358–379. [CrossRef]
26. Monteiro, P.B.; Neto, P.A.M.; Nussenzveig, H.M. Angular momentum of focused beams: Beyond the paraxial approximation. *Phys. Rev. A* **2009**, *79*, 033830. [CrossRef]
27. Novitsky, A.V.; Novitsky, D.V. Negative propagation of vector Bessel beams. *J. Opt. Soc. Am. A* **2007**, *24*, 2844–2849. [CrossRef] [PubMed]
28. Kotlyar, V.V.; Kovalev, A.A.; Nalimov, A.G. Energy density and energy flux in the focus of an optical vortex: Reverse flux of light energy. *Opt. Lett.* **2018**, *43*, 2921–2924. [CrossRef]
29. Youngworth, K.S.; Brown, T.G. Focusing of high numerical aperture cylindrical-vector beams. *Opt. Express* **2000**, *7*, 77–87. [CrossRef]

30. Zhan, Q.; Leger, J.R. Focus shaping using cylindrical vector beams. *Opt. Express* **2002**, *10*, 324–331. [CrossRef]
31. Zhan, Q. Cylindrical vector beams: From mathematical concepts to applications. *Adv. Opt. Photon.* **2009**, *1*, 1–57. [CrossRef]
32. Dogariu, A.; Sukhov, S.; Sáenz, J. Optically induced 'negative forces'. *Nat. Photon.* **2013**, *7*, 24–27. [CrossRef]
33. Petersen, J.; Volz, J.; Rauschenbeutel, A. Chiral nanophotonic waveguide interface based on spin-orbit interaction of light. *Science* **2014**, *346*, 67–71. [CrossRef]
34. Huang, S.Y.; Zhang, G.L.; Wang, Q.; Wang, M.; Tu, C.; Li, Y.; Wang, H.T. Spin-to-orbital angular momentum conversion via light intensity gradient. *Optica* **2021**, *8*, 1231–1236. [CrossRef]
35. Geng, T.; Li, M.; Guo, H. Orbit-induced localized spin angular momentum of vector circular Airy vortex beam in the paraxial regime. *Opt. Express* **2021**, *29*, 14069–14077. [CrossRef]
36. Otte, E.; Rosales-Guzmán, C.; Ndagano, B.; Denz, C.; Forbes, A. Entanglement beating in free space through spin–orbit coupling. *Light Sci. Appl.* **2018**, *7*, 18009. [CrossRef] [PubMed]
37. Onoda, M.; Murakami, S.; Nagaosa, N. Hall effect of light. *Phys. Rev. Lett.* **2004**, *93*, 083901. [CrossRef] [PubMed]
38. Zhu, W.; Zheng, H.; Zhong, Y.; Yu, J.; Chen, Z. Wave-vector-varying Pancharatnam-Berry phase photonic spin Hall effect. *Phys. Rev. Lett.* **2021**, *126*, 083901. [CrossRef] [PubMed]
39. Zhao, X.; Zhang, J.; Pang, X.; Wan, G. Properties of a strongly focused Gaussian beam with an off-axis vortex. *Opt. Commun.* **2017**, *389*, 275–282. [CrossRef]

photonics

MDPI

Article

Effect of Twisting Phases on Linear–Circular Polarization and Spin–Orbital Angular Momentum Conversions in Tightly Focused Vector and Scalar Beams

Shu-Dan Wu, Khian-Hooi Chew and Rui-Pin Chen *

Key Laboratory of Optical Field Manipulation of Zhejiang Province, Department of Physics, Zhejiang Sci-Tech University, Hangzhou 310018, China
* Correspondence: chenrp@zstu.edu.cn

Abstract: We theoretically investigated the effect of a new type of twisting phase on the polarization dynamics and spin–orbital angular momentum conversion of tightly focused scalar and vector beams It was found that the existence of twisting phases gives rise to the conversion between the linear and circular polarizations in both scalar and vector beams during focusing. The linear–circular polarization conversion further leads to an optical spin–orbital angular momentum transformation in the longitudinal component (LC). Therefore, even in a scalar optical field with a uniform linear polarization distribution, a circular polarization (spin angular momentum), and an orbital angular momentum (OAM) can appear in the cross-section and the longitudinal component, respectively, while being tightly focused. The novel distributions of the optical field, state of polarization (SOP) and OAM in the focal region are sensitively dependent on the twisted strength of the twisting phase These results provide a more flexible manipulation of a structured optical field in the aspects of the optical field, SOP, and OAM.

Keywords: twisting phase; vector beams; state of polarization (SOP); orbital angular momentum (OAM)

check for updates

Citation: Wu, S.-D.; Chew, K.-H.; Chen, R.-P. Effect of Twisting Phases on Linear–Circular Polarization and Spin–Orbital Angular Momentum Conversions in Tightly Focused Vector and Scalar Beams. *Photonics* **2023**, *10*, 151. https://doi.org/10.3390/photonics10020151

Received: 1 January 2023
Revised: 29 January 2023
Accepted: 30 January 2023
Published: 1 February 2023

1. Introduction

The tightly focusing properties of an optical field are extensively studied due to their fundamental interests and potential applications, especially for the vector beam with a non-uniform state of polarization (SOP) in the beam cross-section [1–6]. There are two landmark findings about the tightly focused optical field. Firstly, the spin–orbital angular momentum conversion in the longitudinal component (LC) of a tightly focused circular polarized light [7,8] has found corresponding applications such as particle manipulation [9] and information processing [10]. On the other hand, a strong longitudinal electric field generated in a tightly focused radial vector optical field can be manipulated by the initial SOP distribution [1,6]. The super-diffraction phenomenon based on the LC of a tightly focused radial vector optical field has led to many technology breakthroughs in super-resolution imaging [11,12], optical needle [9,13,14], focal field modulation [15], and optical micro-processing [16].

Recently, the modulation of various spatial phases, including the azimuthal phase [17], radial phase [18], and caustic phase [19] on the vector beam, has generated many novel phenomena and extended the applications of a vector beam [1]. A conventional twisted phase discovered by Simon and Mukunda in 1993 is a secondary phase that exists only in the partially coherent beams and can cause the beam to rotate during transmission with non-separability [20]. The effects of a conventional twisted phase on the beam focusing [21], the orbital angular momentum (OAM) [22], and the propagation properties in the linear medium [23] and non-linear media [24] have been extensively explored. More recently, a new type of twisting phase with an expression consistent with the image dispersion term in the Zernike polynomial has been proposed based on the conventional twisted phase [25]. The advantage of the new twisting phase over the conventional twisted phase is that it can exist in a fully coherent beam and the rotation angle can be adjusted by the twisting phase [25]. Research on beams carrying the twisting phase is still in its infancy but has found promising applications in optical detection and particle manipulation [26]. As a spatially non-uniform phase distribution, the effect of the new type of twisting phase on a tightly focused vector vortex beam may result in novel phenomena including the polarization conversion and the redistribution of spin angular momentum (SAM) and OAM, especially the manipulation of spin–orbital angular momentum conversion in the LC.

In this work, the effect of a twisting phase on the tightly focused properties of scalar and vector beams is theoretically demonstrated. The distribution characteristics of the optical field, SOP, and OAM of a twisted vector optical field in the focal region are analyzed. Both the SAM and OAM appear in the transverse section at the focal region of the twisted beam, regardless of the scalar or vector beams. In particular, the circular polarization appears in the focal region of a scalar twisted beam with uniform linear polarization due to the linear–circular polarization conversion, leading to an optical spin–orbital angular momentum transformation in the LC during the focusing. The interaction among the twisting phase, vortex, and SOP in a vector vortex beam during focusing induces the novel distributions of SOP, OAM, and optical field in the focal region. These results can extend our understanding of the twisted vector optical field and provide a new approach to manipulate a structured optical field by the twisting phase.

2. Intensity Distribution and SOP of a Twisted Vector Beam in the Focal Region

The vector vortex beams with a twisting phase in the field cross-section in the Cartesian coordinate system can be expressed as [17,27,28]

$$E(\rho,\phi) = e^{i(u\rho^2 \cos\phi \sin\phi + n\phi)} \times \left[\cos(m\phi+\phi_0)e_x + e^{i\Delta\phi}\sin(m\phi+\phi_0)e_y\right] \tag{1}$$

where u is the twisted strength of the twisting phase, m is the polarization topological charge number, n is the vortex topological charge number. ρ and ϕ are the parameters that describe the polar radius and azimuthal angle of the initial field in polar coordinates, respectively. e_x and e_y are the unit vectors in the x-direction and y-direction, respectively. The only parameter that determines the initial SOP distribution is the azimuthal angle ϕ. When $\Delta\phi = 0$, the x-directional components are in the same phase as the y-directional components, and the initial SOP of the vector optical field is a locally linear polarization. The directions of the linear polarization in the cross-section of the optical field are determined by the azimuthal angle ϕ. When $m = 1$ with the initial phase $\phi_0 = 0$ or $\pi/2$, the vector optical field described by Equation (1) is the radially or azimuthally polarized field [17,19,27,28] with a twisting phase, respectively. When $m = 0$, the optical field is a scalar linear polarized light field. On the other hand, if $\Delta\phi = \pi/2$, there is a phase difference $\pi/2$ between the x component and the y component, resulting in a hybrid SOP distribution with the linear, elliptical, and circular polarizations located at different positions in the cross-section of the optical field. When $m = 0$ with the initial phase $\phi_0 = -\pi/4$ or $\pi/4$, the vector optical field described by Equation (1) degrades into the right- or left-handed circular polarized fields with twisting phases, respectively.

Based on the Richards and Wolf vectorial diffraction theory [29], the three-dimensional optical fields of the vector optical beam with a twisting phase focused by the high numerical aperture (NA) lens in the focal region are derived in the integral formula as

$$
\begin{aligned}
E(r,\varphi,z) = & -\tfrac{ikf}{2\pi}\int_0^{\alpha}\sin\theta d\theta\int_0^{2\pi}\sqrt{\cos\theta}\cdot\exp(iu\rho^2\cos\phi\sin\phi+in\phi)\times\exp[ik(z\cos\theta+r\sin\theta\cos(\varphi-\phi))] \\
& \times\left\{\left[\cos(m\phi+\phi_0)\cos\phi+e^{i\Delta\phi}\sin(m\phi+\phi_0)\sin\phi\right]\begin{pmatrix}\cos\theta\cos\phi\\ \cos\theta\sin\phi\\ \sin\theta\end{pmatrix}\right. \\
& \left.+\left[e^{i\Delta\phi}\sin(m\phi+\phi_0)\cos\phi-\cos(m\phi+\phi_0)\sin\phi\right]\begin{pmatrix}-\sin\phi\\ \cos\phi\\ 0\end{pmatrix}\right\}d\phi
\end{aligned} \tag{2}
$$

where $k = 2\pi/\lambda$ is the wave number, λ is the wavelength and f is the focal length of the high NA objective. $\alpha = \arcsin(NA/\eta)$ is the maximal angle describing the ratio of the pupil radius to the beam waist radius determined by the NA of the objective lens, and $NA = 0.9$ in this work. η is the refractive index of the surrounding medium. r is the polar radius, φ and z are the azimuthal angle and longitudinal positions of the observation point in the focal region in the cylindrical coordinate system, as shown in Figure 1.

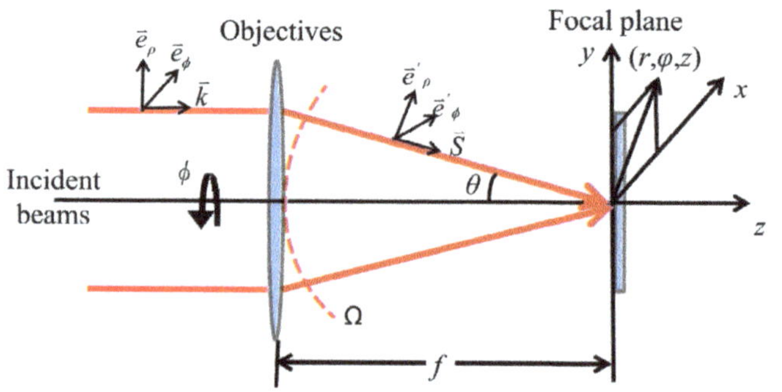

Figure 1. Schematic diagram of a vector vortex beam under a tightly focused condition, where the focal region is located at $z = 0$.

Without the modulation of the twisting phase, the total intensity distribution of a vector beam in the focal region is a petal-like annular spot with $|2m-2|$ petals. In the LC distribution, the number of petal-like annular spots is $|2m-2|$ [30], while in distributions of the transverse x- or y- components, this number equals $|2m|$. In addition, the ring distributions are formed for the focused profiles of the transverse component by combining the x- and y- components, the number of a petal-like annular spots of the total intensity distribution of a vector beam in the focal region is mainly attributed to the contribution of the longitudinal component [31]. However, the effect of the twisting phase on the intensity distribution of the transverse component (TC) and LC of a vector beam in the focal region causes the gradual bending of the $|2m-2|$ petal-like spots and the intensity redistribution in the diagonal (45° and 135°) directions, as shown in Figures 2 and 3. Figure 2 shows the effect of different twisting intensity u on intensity and SOP distributions of a vector beam ($n = 0$, $m = 3$) in the focal region. When the twisted strength $u < 2\times10^{12}$ mm^{-2}, the influence of twisting phase on the vortex phase (or polarization topological charge) can be negligible. With the increasing values of twisting strength, the intensity distributions in the diagonal (45° and 135°) directions become obvious with different SOP due to the modulation of the twisting phase. Figure 3 shows the effect of a twisting strength ($u = 10^{13}$ mm^{-2}) on different vector beams in the focal region. The intensity concentration in the diagonal (45° and 135°) directions becomes more evident with the increase in polarization topological charge. In addition, the intensity distributions of different components in the focal region are also sensitively dependent on the modulation of the twisting phase. In particular, the effect of a twisting phase on the tightly focused vector beams with locally linear polarization and hybrid SOP results in the different intensity distributions of the LC, as shown in Figure 3.

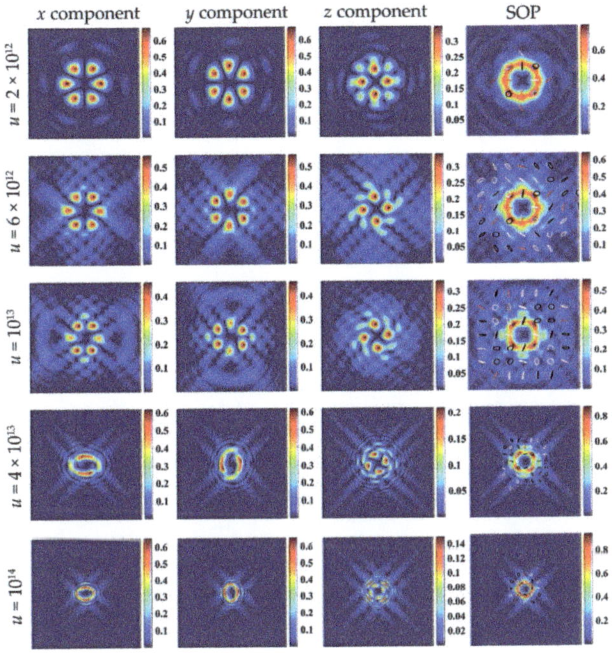

Figure 2. Intensity distributions of the TC (transverse component, x-direction, and y-direction), LC (longitudinal component), and SOP of a vector optical beam ($n = 0$, $m = 3$, $\phi_0 = \pi/2$ in Equation (1)) with different twisted strengths in the focal region. Here the size of each plot is $3\lambda \times 3\lambda$, the red, black, and white lines in the plots in the last column indicate the linear, left-, and right-circular polarizations, respectively.

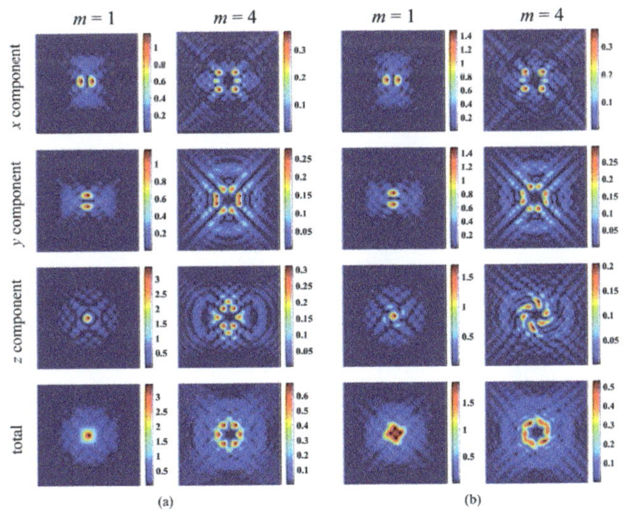

Figure 3. Intensity distributions of the TC (transverse component, x-direction, and y-direction), LC (longitudinal component), and total field of vector optical beams with a twisting phase ($u = 10^{13}$ mm^{-2}) and different polarization topological charges m in the focal region: (**a**) locally linear polarizations ($\Delta\phi = 0$ in Equation (1)); (**b**) hybrid SOP ($\Delta\phi = \pi/2$ in Equation (1)). Here the size of each plot is $3\lambda \times 3\lambda$ and the vortex topological charge $n = 0$.

3. The Effect of a Twisting Phase on Linear–Circular Polarization and SAM-OAM Conversions of the Scalar and Vector Beams in the Focal Region

The linear–circular polarization and SAM-OAM conversions of a structured optical field are important topics in optical field manipulation due to their fundamental interest and practical applications [32–40]. The SAM is dependent on circular polarization [32–34], whereas the OAM is related to the gradient phase distribution [35,36]. The SAM and OAM of a structured beam at the focal region can be calculated as [32–36]

$$S \propto \mathrm{Im}[E^{*} \times E] \tag{3}$$

$$L \propto \mathrm{Im}[r \times (E^{*} \cdot (\nabla) E)] \tag{4}$$

where Im[.] represents the imaginary part, and the asterisk denotes the complex conjugate. S and L describe the SAM and OAM densities, respectively. The distributions of SAM at the focal region confirm the occurrence of linear–circular polarization conversion during focusing, as shown in the third column in Figure 4. The appearance and modulation of SAM in the focal region are closely related to the twisting phase and the initial SOP, as a comparison between the first and third columns in Figure 4 shows. In particular, when the initial field is a scalar optical field with a uniform linear polarization distribution, there is a circular polarization appearance in the cross-section in the focal region due to the linear–circular polarization conversion, resulting from the modulation of the spatially non-uniform phase distribution of a twisting phase uxy, as shown in the first row in Figure 4a. If the initial field is a vector optical field with a non-uniform SOP, the conversion of linear–circular polarization gives rise to a novel SOP distribution in the focal region. When the initial SOP is a locally linear polarization ($\Delta\phi = 0$ in Equation (1)), the hybrid polarization state, including linear and circular polarizations, appears with the central symmetry distribution, as shown in Figure 4. If the initial SOP of the optical field is the hybrid polarizations ($\Delta\phi = \pi/2$ in Equation (1)), the SOP distributions of the focused and twisted vector optical field become more complicated because of the linear–circular polarization conversions during propagation. However, the SOP distribution in the focal region retains the central symmetrical shape if the initial SOP is central symmetric, as shown in Figure 4.

The OAM has attracted extensive interest and intensive investigation due to its fundamental interest and applications [35,36]. The spin–orbital angular momentum conversion in LC in the tightly focused circularly polarized scalar optical field has been demonstrated [8,32]. Moreover, the OAM resulting from a twisting phase has been reported [22]. Therefore, the effect of a twisting phase on the optical OAM of a twisted vector beam in the focal region will result in a more complicated spin–orbital angular momentum conversion and redistributions of OAM, especially in the LC.

The optical angular momentum of a structured optical field has been extensively studied in recent years [41–45]. The OAM appears in linearly polarized beams with optical vortices [42–44], in particular, the specific SAM orthogonal to the propagation direction of a linearly polarized vortex beam during propagation [42–45]. Many interesting results about angular momentum have been analyzed and reported such as the spin–orbit interaction [37,38], and the angular momentum of focused beams [39,40]. Different from the vortex phase with a spiral distribution, the twisting phase uxy can compress the optical field in the $\pm 45^{\circ}$azimuthal angle [46]. The SAM appears in the focal region due to the linear–circular polarization conversion (see the first row in Figure 4a and the first rows in Figure 5a,b), which further results in the appearance of OAM in LC during focusing, as shown in the third rows in Figure 5a,b for the twisting strength $u = 10^{13}$ mm^{-2} and 8×10^{13} mm^{-2}. In addition, the existence of a twisting phase can not only induce the OAM of TC, but also redistributes the OAM of the vortex in the field cross-section, as shown in the second rows in Figure 5a,b. For a scalar optical field with left or right circular polarization, the OAM of the LC and TC in the focused vortex field has been extensively explored [47]. There are no OAM, either for a focused non-vortex with circular polarizations in the TC, or for a focused 1-order vortex ($n = 1$) with the right circular polarization in the LC, as shown in Figure 6a,b [7]. Interestingly, the OAM appears in both the two cases if the initial field is embedded with a twisting phase, as shown in Figure 6c,d. By comparing Figures 5 and 6, it is evident that even in a scalar field, the influences of the twisting phase on the OAM of the LC are different for linear and circular polarizations. For a scalar vortex beam with circular polarization, both the OAM in TC and LC are affected by the presence of the twisting phase. In particular, they are redistributed in the focal region due to the modulation of the twisting phase, as shown in Figure 6c,d. With the increase in the twisting strength u, the modulation effect of the twisting phase becomes more obvious. The OAM distributions in the diagonal (45° and 135°) directions which are induced by the twisting phase become obvious, and the OAM distributions related to the polarization topological charges are compressed toward the center, as shown in Figure 6e,f for $u = 8 \times 10^{13}$ mm^{-2}.

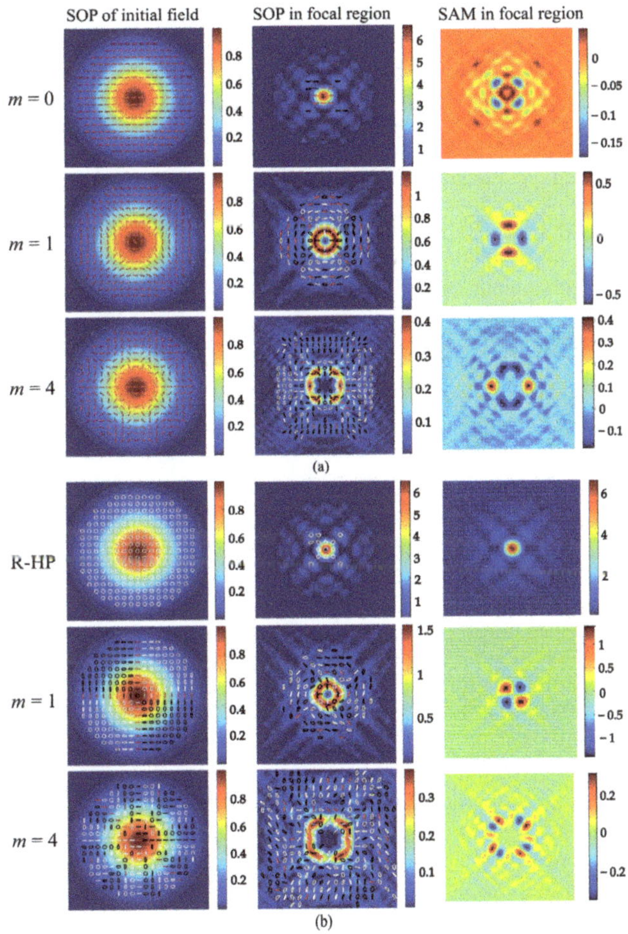

Figure 4. The SOP of scalar and vector beams in the initial field and focal region, and the corresponding SAM distribution in the focal region. The first column is the initial SOP of the input optical field, and the second and third columns are the SOP and SAM distributions with a twisting phase ($u = 10^{13}$ mm^{-2}) in the focal region: (**a**) the locally linear polarization ($\Delta\phi = 0$ in Equation (1)), (**b**) the hybrid SOP ($\Delta\phi = \pi/2$ in Equation (1)). The size of each plot in the first column is $5500\lambda \times 5500\lambda$, and the size of each plot in the second and third columns is $3\lambda \times 3\lambda$. Here the vortex topological charge $n = 0$. The red, black, and white lines indicate the linear, left-, and right-polarizations, respectively.

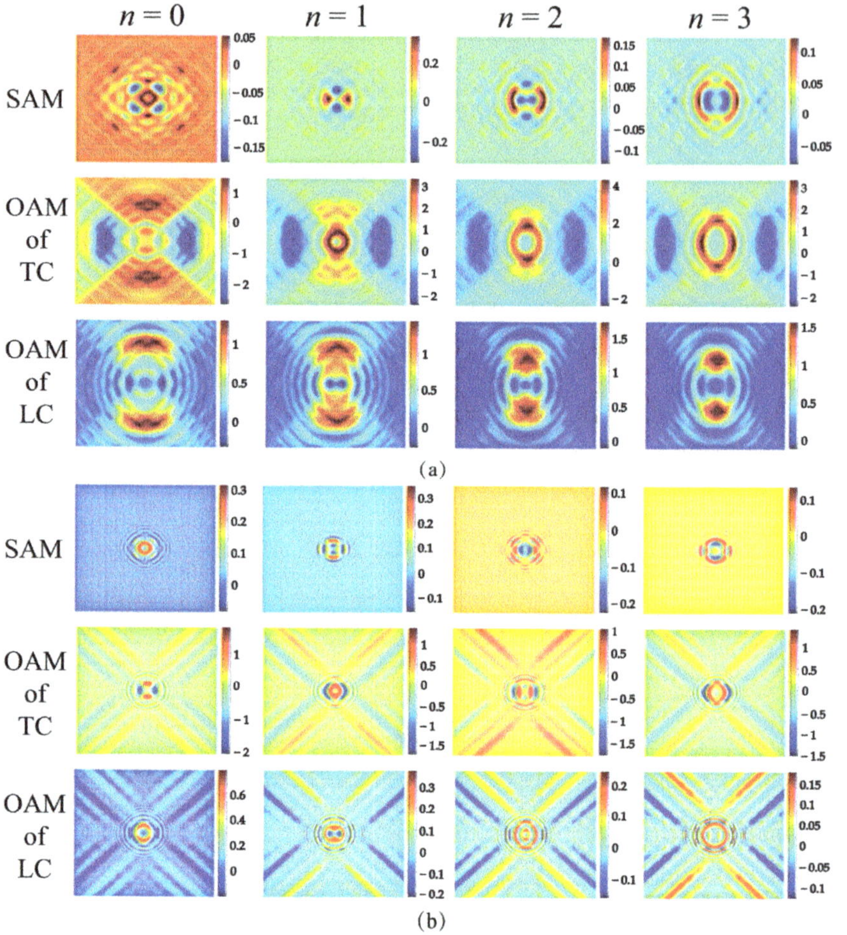

Figure 5. The SAM and OAM distributions of a focused scalar linear polarized beam (the initial field with $m = 0$ and $\phi_0 = 0$ in Equation (1)) in the focal region with different vortex topological charge numbers and the modulation of a twisting phase: (**a**) the twisted strength $u = 10^{13}$ mm^{-2}, (**b**) the twisted strength $u = 8 \times 10^{13}$ mm^{-2}. Here the size of each plot is $3\lambda \times 3\lambda$.

The SAM-OAM conversion in a focused vector vortex beam [48] and a high non-paraxial vector vortex beam [49] have been reported. The effect of a twisting phase also plays an important role in the linear–circular polarization conversion and SAM-OAM conversion of a tightly focused vector vortex beam with non-uniform SOP in the focal region. For the vector vortex beam with locally linear polarization or hybrid SOP ($\Delta\phi = 0$ or $\pi/2$ in Equation (1)) and the same topological charge number, the TC OAM distributions are similar, as shown in the first column in Figure 7. It can be explained as the OAM of two corresponding orthogonal polarization components being equal and incoherent, although the OAM is modulated by the twisting phases. However, the OAM distributions of the LC are different for the focused twisted vector beam with the same topological charge number but dissimilar SOP (locally linear polarization or hybrid SOP), as shown in the second column in Figure 7. The distinct OAM distributions of the LC can be explained by the variations in the linear–circular polarization conversions of the two SOP during the focusing, leading to the differences in the spin–orbital angular momentum transformations and OAM distributions in the LC. Thus, the effect of a twisting phase on the tightly focused scalar and vector optical field can induce the generation and redistribution of OAM in both TC and LC, as shown in Figure 7c,d. These results provide a new way to further vectorially manipulate the structured optical field in the aspects of the optical field, SOP, and OAM via the twisting phase.

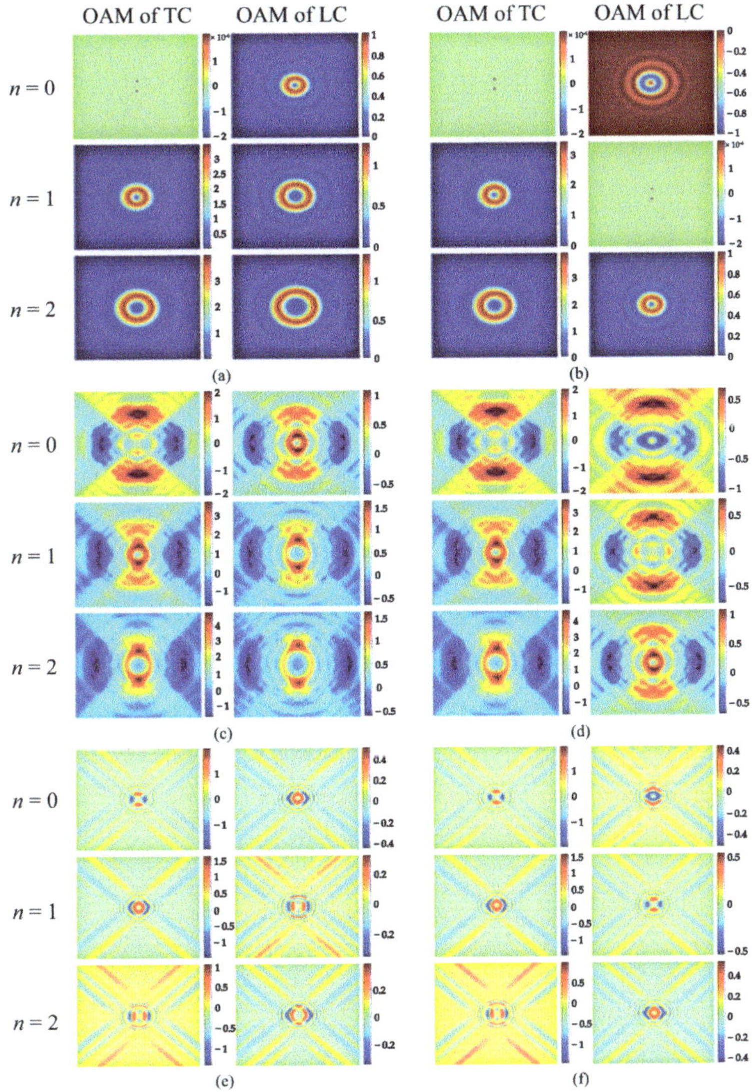

Figure 6. The OAM distribution of a focused scalar circularly polarized beam in the focal region with different vortex topological charges without (**a**,**b**) or with (**c**,**d**) the modulation of a twisting phase: (**a**) the left circularly polarized light (the initial field with $m = 0$ and $\phi_0 = \pi/4$ in Equation (1)), (**b**) the right circularly polarized light (the initial field with $m = 0$ and $\phi_0 = -\pi/4$ in Equation (1)), (**c**) the left circularly polarized light modulated by a twisting phase with $u = 10^{13}$ mm^{-2}, (**d**) the right circularly polarized light modulated by a twisting phase with $u = 10^{13}$ mm^{-2}, (**e**) the left circularly polarized light modulated by a twisting phase with $u = 8 \times 10^{13}$ mm^{-2}, (**f**) the right circularly polarized light modulated by a twisting phase with $u = 8 \times 10^{13}$ mm^{-2}. Here the size of each plot is $3\lambda \times 3\lambda$.

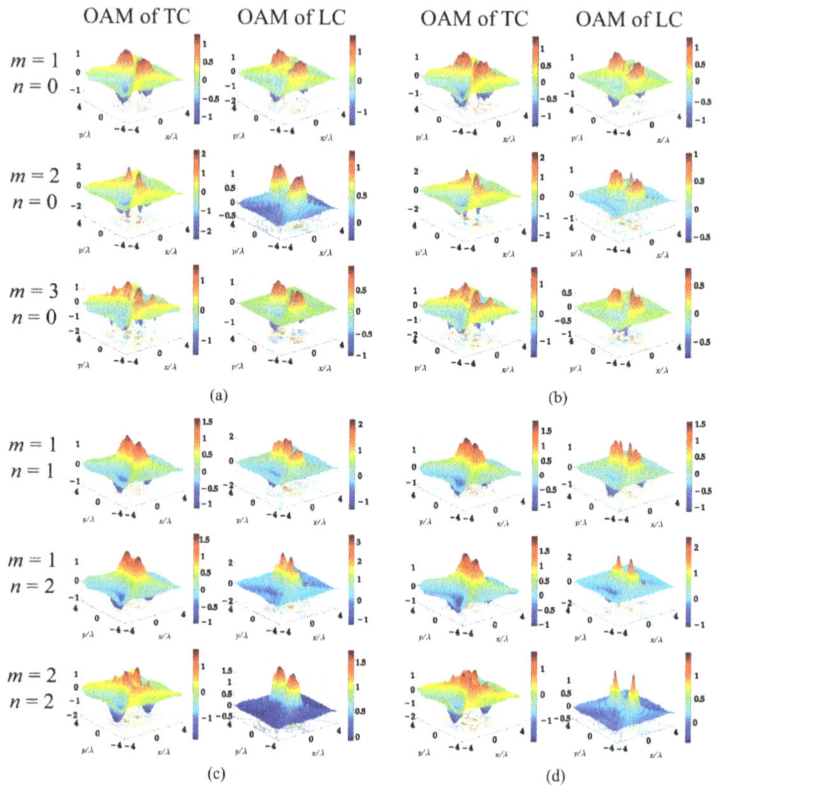

Figure 7. The OAM distribution of a focused twisted vector vortex beam in the focal region: (**a**,**c**) the locally linear polarization ($\Delta\phi = 0$ in Equation (1)); (**b**,**d**) the hybrid SOP ($\Delta\phi = \pi/2$ in Equation (1)). Here, the size of each plot is set as $4\lambda \times 4\lambda$. The initial phase is $\phi_0 = 0$ and the twisted strength is $u = 10^{13}$ mm^{-2}.

4. Discussion

The conventional non-separable quadratic twisting phase $u(x_1y_2 - x_2y_1)$ only exists in a partially coherent light field [20]. As a new type of twisting phase uxy (or $u\rho^2\cos\phi\sin\phi$ in Equation (1)) since being introduced into coherent light, some novel properties have been revealed, such as the rotation of a twisted anisotropic beam [46], measurement of vortex topological charge number [50], and the optical angular momentum resulting from the twisting phase [26]. In this work, the effect of a twisting phase on the polarization dynamics and spin–orbital angular momentum conversion of a vector vortex optical field in the focal region is studied. The results indicate that the linear–circular polarization conversion and the SAM-OAM conversion in LC occur in both scalar and vector twisted beams in the focal region due to the modulation of the twisting phase.

The underlying physics of these novel phenomena can be explained based on the non-uniform spatial phase distribution of a twisting phase uxy, which plays an essential role in the linear–circular polarization conversion during focusing. This is because the modulation and distribution of the non-uniform twisting phase uxy in the field cross-section induce a phase difference between the orthogonal polarization components in the focal region, as well as the linear–circular polarization conversion that occurs during focusing. In particular, even in a scalar optical field with a uniform linear polarization, there will be a circular polarization component (SAM) during propagation, further inducing the appearance of OAM in the LC due to the SAM-OAM conversion during tightly focusing. The OAM of a longitudinal component in a tightly focusing beam is related to the distributions of transverse orbital and spin angular momentum, as recognized in Equation (1). In a tightly focused scalar optical field with linear polarizations, the transverse OAM is influenced by the twisting phase due to the non-uniform phase distribution of the twisting phase whereas the SAM depends on the linear–circular polarization conversion. Therefore, the longitudinal component of a tightly focused scalar optical field with linear polarizations can be manipulated by different twisting strengths u because the transverse OAM distribution and the linear–circular polarization conversion are sensitively dependent on the twisting strength of a twisting phase. These results provide a new way to manipulate the SOP and optical angular momentum.

5. Conclusions

The effect of a twisting phase on the focused scalar and vector optical field has been studied. Unlike other phase modulations, the twisting phase uxy is spatially non-uniform in distribution that results in the linear–circular polarization conversion during focusing. Since the linear–circular conversion can also occur in a scalar beam with uniform linear polarization, the conversion further induces the appearance of OAM in the LC due to the spin–orbital angular momentum transformation. The spatial distributions of the optical field, SOP, and OAM of a twisted scalar or vector beam in the focused region are sensitively dependent on the twisted strength of a twisting phase. These results provide a new way to vectorially manipulate the spatial distribution of the optical field, SOP, and OAM of a structured optical field by a twisting phase.

Author Contributions: Conceptualization, R.-P.C.; methodology, R.-P.C. and S.-D.W.; validation, R.-P.C. and K.-H.C.; investigation, S.-D.W. and R.-P.C.; data curation, S.-D.W. and R.-P.C.; supervision, R.-P.C.; writing—original draft preparation, S.-D.W.; writing—review and editing, R.-P.C. and K.-H.C. All authors have read and agreed to the published version of the manuscript.

Funding: This research was funded by the Zhejiang Provincial Key Research and Development Program (No. 2022C04007); National Natural Science Foundation of China (No. 11874323).

Institutional Review Board Statement: Not applicable.

Informed Consent Statement: Not applicable.

Data Availability Statement: Not applicable.

Conflicts of Interest: The authors declare no conflict of interest.

References

1. Youngworth, K.; Brown, T. Focusing of high numerical aperture cylindrical-vector beams. *Opt. Express* **2000**, *7*, 77–87. [CrossRef] [PubMed]
2. Zhan, Q.; Leger, J. Focus shaping using cylindrical vector beams. *Opt. Express* **2002**, *10*, 324–331. [CrossRef] [PubMed]
3. Pohl, D. Operation of a Ruby laser in the purely transverse electric mode TE01. *Appl. Phys. Lett.* **1972**, *20*, 266–267. [CrossRef]
4. Mushiake, Y.; Matsumura, K.; Nakajima, N. Generation of radially polarized optical beam mode by laser oscillation. *Proc. IEEE* **1972**, *9*, 1107–1109. [CrossRef]
5. Tidwell, S.C.; Kim, G.H.; Kimura, W.D. Efficient radially polarized laser beam generation with a double interferometer. *Appl. Opt.* **1993**, *27*, 5222. [CrossRef]
6. Dorn, R.; Quabis, S.; Leuchs, G. Sharper focus for a radially polarized light beam. *Phys. Rev. Lett.* **2003**, *23*, 233901. [CrossRef]
7. Bomzon, Z.E.; Gu, M.; Shamir, J. Angular momentum and geometrical phases in tight-focused circularly polarized plane waves. *Appl. Phys. Lett.* **2006**, *89*, 241104. [CrossRef]
8. Zhao, Y.; Edgar, J.S.; Jeffries, G.D.M.; McGloin, D.; Chiu, D.T. Spin-to-orbital angular momentum conversion in a strongly focused optical beam. *Phys. Rev. Lett.* **2007**, *7*, 73901. [CrossRef]
9. Maragò, O.M.; Jones, P.H.; Gucciardi, P.G.; Volpe, G.; Ferrari, A.C. Optical trapping and manipulation of nanostructures. *Nat. Nanotechnol.* **2013**, *8*, 807–819. [CrossRef]
10. Barreiro, J.T.; Wei, T.; Kwiat, P.G. Remote preparation of single-photon "hybrid" entangled and vector-polarization states. *Phys. Rev. Lett.* **2010**, *3*, 30407. [CrossRef]
11. Sheppard, C.J.R.; Choudhury, A. Annular pupils, radial polarization, and superresolution. *Appl. Opt.* **2004**, *43*, 4322–4327. [CrossRef] [PubMed]

12. Hao, X.; Kuang, C.; Wang, T.; Liu, X. Phase encoding for sharper focus of the azimuthally polarized beam. *Opt. Lett.* **2010**, *35*, 3928–3930. [CrossRef] [PubMed]
13. Zhan, Q. Trapping metallic Rayleigh particles with radial polarization. *Opt. Express* **2004**, *12*, 3377–3382. [CrossRef] [PubMed]
14. Zhong, M.; Gong, L.; Di, L.; Zhou, J.; Wang, Z.; Li, Y. Optical trapping of core-shell magnetic microparticles by cylindrical vector beams. *Appl. Phys. Lett.* **2014**, *105*, 181112.
15. Chen, Z.; Zeng, T.; Ding, J. Reverse engineering approach to focus shaping. *Opt. Lett.* **2016**, *41*, 1929–1932. [CrossRef] [PubMed]
16. Bialynicki-Birula, I. Local and nonlocal observables in quantum optics. *New J. Phys.* **2014**, *16*, 113056. [CrossRef]
17. Wang, H.; Wang, X.; Li, Y.; Chen, J.; Guo, C.; Ding, J. A new type of vector fields with hybrid states of polarization. *Opt. Express* **2010**, *18*, 10786. [CrossRef] [PubMed]
18. Pan, Y.; Li, Y.; Li, S.; Ren, Z.; Si, Y.; Tu, C.; Wang, H. Vector optical fields with bipolar symmetry of linear polarization. *Opt. Lett.* **2013**, *38*, 3700–3703. [CrossRef]
19. Chen, R.P.; Chen, Z.; Gao, Y.; Ding, J.; He, S. Flexible manipulation of the polarization conversions in a structured vector field in free space. *Laser Photonics Rev.* **2017**, *11*, 1700165. [CrossRef]
20. Simon, R.; Mukunda, N. Twisted Gaussian Schell-model beams. *J. Opt. Soc. Am. A* **1993**, *1*, 95. [CrossRef]
21. Friberg, A.T.; Tervonen, E.; Turunen, J. Focusing of twisted Gaussian Schell-model beams. *Opt. Commun.* **1994**, *4–6*, 127–132. [CrossRef]
22. Serna, J.; Movilla, J.M. Orbital angular momentum of partially coherent beams. *Opt. Lett.* **2001**, *26*, 405–407. [CrossRef] [PubMed]
23. Friberg, A.T.; Tervonen, E.; Turunen, J. Interpretation and experimental demonstration of twisted Gaussian Schell-model beams. *J. Opt. Soc. Am. A* **1994**, *11*, 1818. [CrossRef]
24. Ponomarenko, S.A. Twisted Gaussian Schell-model solitons. *Phys. Rev. E* **2001**, *64*, 36618. [CrossRef] [PubMed]
25. Wan, L.; Zhao, D. Controllable rotating Gaussian Schell-model beams. *Opt. Lett.* **2019**, *44*, 735–738. [CrossRef] [PubMed]
26. Liu, Z.; Wan, L.; Zhou, Y.; Zhang, Y.; Zhao, D. Progress on studies of beams carrying twist. *Photonics* **2021**, *8*, 92. [CrossRef]
27. Chen, H.; Hao, J.; Zhang, B.; Xu, J.I.; Ding, J.; Wang, H. Generation of vector beam with space-variant distribution of both polarization and phase. *Opt. Lett.* **2011**, *36*, 3179–3181. [CrossRef]
28. Chen, R.P.; Chen, Z.; Chew, K.H.; Li, P.G.; Yu, Z.; Ding, J.; He, S. Structured caustic vector vortex optical field: Manipulating optical angular momentum flux and polarization rotation. *Sci. Rep.* **2015**, *5*, 10628. [CrossRef]
29. Richards, B.; Wolf, E. Electromagnetic diffraction in optical systems. II. Structure of the image field in an aplanatic system. *Proc. R. Soc. Lond. Ser. A* **1959**, *1274*, 358–379.
30. Huang, K.; Shi, P.; Cao, G.W.; Li, K.; Zhang, X.B.; Li, Y.P. Vetor-vortex Bessel-Gauss beam and their tightly focusing properties. *Opt. Lett.* **2011**, *36*, 888–890. [CrossRef]
31. Dai, X.; Li, Y.; Liu, L. Tight focusing properties of hybrid-order Poincaré sphere beams. *Opt. Commun.* **2018**, *426*, 46–53. [CrossRef]
32. Berry, M. Paraxial beams of spinning light. *Proc. SPIE* **1998**, *3847*, 6–11.
33. Wang, X.L.; Chen, J.; Li, Y.; Ding, J.; Guo, C.S.; Wang, H.T. Optical orbital angular momentum from the curl of polarization. *Phys. Rev. Lett.* **2010**, *25*, 253602. [CrossRef] [PubMed]
34. Bialynicki-Birula, I.; Bialynicka-Birula, Z. Canonical separation of angular momentum of light into its orbital and spin parts. *J. Opt.* **2011**, *6*, 64014. [CrossRef]
35. Padgett, M.; Bowman, R. Tweezers with a twist. *Nat. Photonics* **2011**, *5*, 343–348. [CrossRef]
36. Liu, J.; Li, Z. Controlled mechanical motions of microparticles in optical tweezers. *Micromachines* **2018**, *9*, 232. [CrossRef]
37. Bliokh, K.Y.; Alonso, M.A. Angular momenta and spin-orbit interaction of non-paraxial light in free space. *Phys. Rev. A* **2010**, *82*, 063825. [CrossRef]
38. Bliokh, K.Y.; Rodríguez-Herrera, O.; Lara, D.; Ostrovskaya, E.A.; Dainty, C. Optical nanoprobing via spin-orbit interaction of light. *Phys. Rev. Lett.* **2010**, *104*, 253601.
39. Bekshaev, A.Y. A simple analytical model of the angular momentum transformation in strongly focused light beams. *Cent. Eur. J. Phys.* **2010**, *8*, 947–960. [CrossRef]
40. Monteiro, P.B.; Neto, P.A.M.; Nussenzveig, H.M. Angular momentum of focused beams: Beyond the paraxial approximation. *Phys. Rev. A* **2009**, *79*, 033830. [CrossRef]
41. Bekshaev, A.Y.; Bliokh, K.Y.; Soskin, M. Internal flows and energy circulation in light beams. *J. Opt.* **2011**, *13*, 053001. [CrossRef]
42. Angelsky, O.V.; Bekshaev, A.Y.; Hanson, S.G.; Zenkova, C.Y.; Mokhun, I.I.; Zheng, J. Structured light: Ideas and concepts. *Front. Phys.* **2020**, *8*, 00114. [CrossRef]
43. Bliokh, K.Y.; Nori, F. Transverse and longitudinal angular momenta of light. *Phys. Rep.* **2015**, *592*, 1–38. [CrossRef]
44. Aiello, A.; Banzer, P.; Neugebauer, M.; Leuchs, G. From transverse angular momentum to photonic wheels. *Nat. Photonics* **2015**, *9*, 789–795. [CrossRef]
45. Bekshaev, A.Y. Transverse spin and the hidden vorticity of propagating light fields. *J. Opt. Soc. Am. A* **2022**, *39*, 1577–1583. [CrossRef] [PubMed]
46. Liu, Q.; Chew, K.H.; Huang, Y.; Liu, C.; Hu, X.; Li, Y.; Chen, R.P. Effect of twisting phases on the polarization dynamics of a vector optical field. *Photonics* **2022**, *9*, 722. [CrossRef]
47. Kotlyar, V.V.; Kovalev, A.A.; Nalimov, A.G. Energy density and energy flux in the focus of an optical vortex: Reverse flux of light energy. *Opt. Lett.* **2018**, *43*, 2921–2924. [CrossRef]

48. Pan, D.; Chew, K.H.; Wu, Y.; Chen, R.P. Conversions of linear-circular polarizations and spin-orbital angular momentums in a focused vector vortex beam with fractional topological charges. *Optik* **2022**, *252*, 168473. [CrossRef]
49. Chen, R.P.; Chew, K.H.; Dai, C.Q.; Zhou, G. Optical spin-to-orbital angular momentum conversion in the near field of a highly nonparaxial optical field with hybrid states of polarization. *Phys. Rev. A* **2017**, *96*, 053862. [CrossRef]
50. Shen, D.; Zhao, D. Measuring the topological charge of optical vortices with a twisting phase. *Opt. Lett.* **2019**, *44*, 2334–2337. [CrossRef]

photonics

MDPI

Article

Propagation of a Lorentz Non-Uniformly Correlated Beam in a Turbulent Ocean

Dongmei Wei [1], Ke Wang [1], Ying Xu [1], Qian Du [1], Fangning Liu [1], Juan Liu [1], Yiming Dong [2], Liying Zhang [2], Jiayi Yu [1,*], Yangjian Cai [1,*] and Xinlei Zhu [1,*]

1 Shandong Provincial Engineering and Technical Center of Light Manipulation and Shandong Provincial Key Laboratory of Optics and Photonic Devices, School of Physics and Electronics, Shandong Normal University, Jinan 250014, China
2 Department of Physics, Shaoxing University, Shaoxing 312000, China
* Correspondence: jiayiyu0528@sdnu.edu.cn (J.Y.); yangjiancai@sdnu.edu.cn (Y.C.); zhuxinleiphy@sdnu.edu.cn (X.Z.)

Abstract: We study the propagation characteristics (spectral intensity and degree of coherence) of a new type of Lorentz non-uniformly correlated (LNUC) beam based on the extended Huygens–Fresnel principle and the spatial power spectrum of oceanic turbulence. The effects of the oceanic turbulence parameters and initial beam parameters on the evolution propagation characteristics of LNUC beams are studied in detail by numerical simulation. The results indicate that such beams exhibit self-focusing propagation features in both free space and oceanic turbulence. Decreasing the dissipation rate of kinetic energy per unit mass of fluid and the Kolmogorov inner scale, or increasing the relative strength of temperature to salinity undulations and the dissipation rate of mean-square temperature of the turbulent ocean tends to increase the negative effects on the beams. Furthermore, we propose a strategy of increasing the beam width and decreasing the coherence length, to reduce the negative effects of the turbulence.

Keywords: non-uniformly correlated beam; self-focusing; oceanic turbulence; turbulence resistance

Citation: Wei, D.; Wang, K.; Xu, Y.; Du, Q.; Liu, F.; Liu, J.; Dong, Y.; Zhang, L.; Yu, J.; Cai, Y.; et al. Propagation of a Lorentz Non-Uniformly Correlated Beam in a Turbulent Ocean. *Photonics* **2023**, *10*, 49. https://doi.org/10.3390/photonics10010049

Received: 28 November 2022
Revised: 12 December 2022
Accepted: 28 December 2022
Published: 3 January 2023

1. Introduction

Since the invention of the laser in 1960, due to its coherence, brightness, monochromaticity and directivity, it has been widely used in industrial processing, medicine, optical communication systems and other fields, contributing to the progress of human society and economic development [1,2]. Free-space optical communication is an important application of laser; it uses the propagation of light beams to wirelessly transmit data for the receiver. This technique is useful in situations where physical connectivity is not possible. Research on the propagation of light beams in random media has attracted much attention in recent decades [3–5]. However, researchers prefer to study the interaction of light beams with the turbulent atmosphere, as well as the propagation characteristics of light beams in the turbulent atmosphere, but there seems to be a tendency to forget that turbulent ocean, which covers 71 percent of the Earth's total area, is another important random media. This may be because the power spectra of refractive index fluctuations of the atmosphere and ocean have similar structures [3,5]. We know temperature, humidity, salinity fluctuations and the irregular motion of water are the main drivers of oceanic turbulence. Apart from absorption and scattering, the main disturbance of light beams in the ocean is random vibrations in refraction [5,6]. Relative to turbulent atmosphere, the situation of light beams propagating in turbulent ocean is more complex. Therefore, with the development of underwater technologies, it is necessary and meaningful to study the influence of oceanic turbulence on light beam propagation characteristics and how to reduce the negative effects of oceanic turbulence on light beams.

It is known that light beams with low spatial coherence, known as partially coherent beams (PCBs), display better propagation performance than comparable coherent light beams [7], such as better resistance to turbulence [4,8], which has been physically explained from the perspective of the coherent mode representation [9,10]. With the development of the unified theory of coherence and polarization and the control of optical coherent structures [11], more and more light beams with novel physical features caused by different correlation structures have been customized [7,12]. PCBs with prescribed correlation structures not only exhibit novel propagation features, but also have better resistance to turbulence. Heretofore, researchers have investigated the propagation characteristics of some PCBs with prescribed correlation structures in oceanic turbulence, such as multi-sinc Schell-model beams [13], partially coherent radially and azimuthally polarized rotating elliptical Gaussian beams [14], radially polarized multi-Gaussian Schell-models [15], etc. These works all confirm the excellent turbulence resistance capability of PCBs with prescribed correlation structures in oceanic turbulence. However, the studied target beams are confined to the Schell-model correlated type, i.e., uniformly correlated beams. One distinct and broad class of beams with a spatially variant correlation function, referring to non-uniformly correlated beams, exhibit special self-focusing properties and possess not only lower scintillation but also higher intensity than Gaussian–Schell model beams over certain propagation ranges. This type of beam shows potential for free-space optical communication applications [16–20]. Ding et al. introduced a partially coherent source with circular coherence and showed the propagation behavior of this beam in turbulent ocean [21]. They only discussed the influence of the oceanic turbulence parameters on the propagation characteristics, and did not give a strategy to reduce the negative effects of oceanic turbulence.

In this paper, we introduce a new class of non-uniform correlated beams, termed Lorentz non-uniformly correlated (LNUC) beam, and investigate its propagation characteristics in oceanic turbulence, and discuss the effects of oceanic turbulence parameters on the evolution of spectral intensity and degree of coherence (DOC). We also give a strategy on how to adjust the initial parameters, beam width and coherence length of the beam to reduce the negative effects of oceanic turbulence.

2. Theoretical Model of a LNUC Beam

We consider a scalar, monochromatic, beam-like field generated by a statistical stationary source propagating along the z-axis. In the source plane ($z = 0$), the second-order statistics of such source is characterized by the two-point cross-spectral density (CSD) in the space-frequency domain,

$$W(\mathbf{r}_1, \mathbf{r}_2) = \langle E^*(\mathbf{r}_1) E(\mathbf{r}_2) \rangle, \tag{1}$$

where $\mathbf{r} = (x, y)$ represents the spatial position in the source plane, $E(\mathbf{r})$ is a component of the electric-field vector, the asterisk represents the complex conjugate, and the angular brackets represent the ensemble mean. For brevity, we will work with a single frequency and suppress its explicit description in the arguments of the electric filed and the CSD.

To be physically implemented, the CSD must be Hermitian and correspond to a non-negative definite kernel. The condition of non-negative definiteness will be satisfied if it may be given as a one-dimensional superposition integral as [22]

$$W(\mathbf{r}_1, \mathbf{r}_2) = \int p(v) H^*(\mathbf{r}_1, v) H(\mathbf{r}_2, v) dv. \tag{2}$$

here v is a one-dimensional position in Fourier space, $p(v)$ is a weight function with a Fourier transform, and $H(\mathbf{r}, v)$ is an arbitrary kernel function. To generate partially coherent sources with non-uniform correlation structures, we consider a kernel of the form

$$H(\mathbf{r}, v) = \tau(\mathbf{r}) \exp\left(-iv\mathbf{r}^2\right), \tag{3}$$

where $\tau(\mathbf{r}) = \exp(-\mathbf{r}^2/2w_0^2)$ with w_0 as the beam width and $k = 2\pi/\lambda$ as the wave number with λ being the wavelength. Then, we set a new weight function $p(v)$ to satisfy the following specific distribution

$$p(v) = \frac{\delta^2}{2} \exp\left(-\delta^2 |v|\right). \tag{4}$$

Substituting Equations (3) and (4) into Equation (2), we obtain

$$W(\mathbf{r}_1, \mathbf{r}_2) = \exp\left(-\frac{\mathbf{r}_1^2 + \mathbf{r}_2^2}{2w_0^2}\right) \mu(\mathbf{r}_1, \mathbf{r}_2), \tag{5}$$

where $\mu(\mathbf{r}_1, \mathbf{r}_2)$ represents the spectral DOC of the proposed partially coherent source, given by a Lorentz form

$$\mu(\mathbf{r}_1, \mathbf{r}_2) = \frac{\delta^4}{\delta^4 + \left(\mathbf{r}_1^2 - \mathbf{r}_2^2\right)^2}, \tag{6}$$

where δ denotes the spatial coherence length. Equation (6) shows that the spectral DOC is not only related to the separation, but also to the specific positions of two points, that is, not a homogeneous function, it is non-uniformly correlated and it takes the form of a Lorentz function. Therefore, this new type of partially coherent source, whose CSD is given by Equations (5) and (6), is referred to as a LNUC source.

The density plot of the absolute value of the spectral DOC of a LNUC source is shown in Figure 1. The beam parameters are set to w_0 = 1 cm and δ = 1 cm. It is found that the distribution of the spectral DOC displays circlar symmetry in the $x_1 - y_1$ plane and rectangular symmetry in the $x_1 - x_2$ plane. In the $x_1 - x_2$ plane, the high-coherence region of the LNUC beam is confined to the center and two diagonal lines of the beam in the source plane, which confirms that the correlation structure of such a source is of non-uniform type. We therefore obtain a new type of non-uniformly correlated source with a prescribed correlation function.

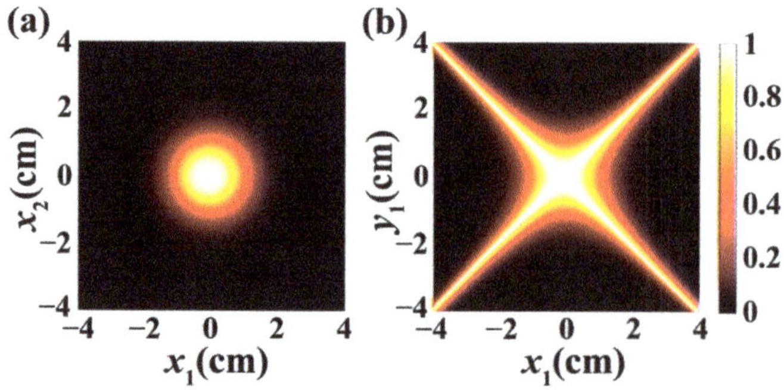

Figure 1. Density plot of the absolute value of the spectral DOC of a LNUC beam (**a**) in the $x_1 - x_2$ plane with $y_1 = y_2 = 0$, (**b**) in the $x_1 - y_1$ plane with $x_2 = y_2 = 0$.

3. CSD of a LNUC Beam Propagating in Oceanic Turbulence

When a PCB propagates in a turbulence medium from the source plane along the positive half plane $z > 0$, under the paraxial approximation, its CSD at an arbitrary cross-section perpendicular to the z-axis can be described by the extended Huygens–Fresnel principle [3]

$$W(\boldsymbol{\rho}_1,\boldsymbol{\rho}_2,z)=\frac{k^2}{4\pi^2z^2}\int\int_{-\infty}^{\infty}W(\mathbf{r}_1,\mathbf{r}_2)\exp\left\{-\frac{ik}{2z}\left[(\boldsymbol{\rho}_1-\mathbf{r}_1)^2-(\boldsymbol{\rho}_2-\mathbf{r}_2)^2\right]\right\}\times\langle\exp[\Psi(\mathbf{r}_1,\boldsymbol{\rho}_1,z)+\Psi^*(\mathbf{r}_2,\boldsymbol{\rho}_2,z)]\rangle d^2\mathbf{r}_1d^2\mathbf{r}_2, \tag{7}$$

where $\boldsymbol{\rho}_1=(\rho_{1x},\rho_{1y})$ and $\boldsymbol{\rho}_2=(\rho_{2x},\rho_{2y})$ represent two spatial positions in the target plane, z is the propagation distance, and $\Psi(\mathbf{r},\boldsymbol{\rho})$ is the complex perturbation phase caused by the refractive index fluctuation of the random transmission medium between the source and target planes. The last term in the above equation, the ensemble average, can be expressed as the following formula [3]

$$\langle\exp[\Psi(\mathbf{r}_1,\boldsymbol{\rho}_1,z)+\Psi^*(\mathbf{r}_2,\boldsymbol{\rho}_2,z)]\rangle=\exp\left\{-\frac{\pi^2k^2zT}{3}\left[(\mathbf{r}_1-\mathbf{r}_2)^2+(\mathbf{r}_1-\mathbf{r}_2)\cdot(\boldsymbol{\rho}_1-\boldsymbol{\rho}_2)+(\boldsymbol{\rho}_1-\boldsymbol{\rho}_2)^2\right]\right\}, \tag{8}$$

where $T=\int_0^\infty\kappa^3\Phi_n(\kappa)d\kappa$ with $\Phi_n(\kappa)$ is the power spectrum of refractive-fluctuations of the oceanic turbulence, it is defined as [5,23]

$$\Phi_n(\kappa)=C_0\varepsilon^{-1/3}\kappa^{-11/3}\chi_T\left[1+2.35(\kappa\eta)^{2/3}\right]\left[\exp(-A_T\sigma)+\omega^{-2}\exp(-A_S\sigma)-2\omega^{-1}\exp(-A_{TS}\sigma)\right], \tag{9}$$

where $C_0=0.388\times10^{-8}$, ε is the dissipation rate of kinetic energy per unit mass of fluid, ranging from 10^{-1} m^2/s^3 to 10^{-10} m^2/s^3, χ_T is the dissipation rate of mean-square temperature, and it varies from 10^{-4} K^2/s to 10^{-10} K^2/s, η is the Kolmogorov inner scale. $\omega\in[-5,0]$ is a non-dimensional parameter indicating the relative strength of temperature to salinity undulations, -5 and 0 mean turbulences are dominantly induced by temperature and by salinity, respectively. The quantities are $A_T=1.863\times10^{-2}$, $A_S=1.9\times10^{-4}$, $A_{TS}=9.41\times10^{-3}$ and $\sigma=8.284(\kappa\eta)^{4/3}+12.978(\kappa\eta)^2$ [23]. Thus, T can be expressed in the following form

$$T=C_0\varepsilon^{-1/3}\chi_T\eta^{-1/3}\left(47.5708\omega^{-2}-17.6701\omega^{-1}+6.78335\right). \tag{10}$$

We can now calculate the CSD of LNUC beams in the target plane using the above equations. For the convenience of operation, we substitute Equations (2) and (3) into Equation (7), and after exchanging the integration order, we obtain the formula

$$W(\boldsymbol{\rho}_1,\boldsymbol{\rho}_2,z)=\frac{k^2}{4\pi^2z^2}\int_{-\infty}^{\infty}p(v)P(\boldsymbol{\rho}_1,\boldsymbol{\rho}_2,v,z)dv, \tag{11}$$

where $P(\boldsymbol{\rho}_1,\boldsymbol{\rho}_2,v,z)$ can be obtained after integrating over $\mathbf{r}_1$ and $\mathbf{r}_2$,

$$\begin{aligned}P(\boldsymbol{\rho}_1,\boldsymbol{\rho}_2,v,z)&=\frac{w_0^2}{w_z^2}\exp\left[-\frac{ik}{2z}(\boldsymbol{\rho}_1^2-\boldsymbol{\rho}_2^2)\right]\exp\left[-\left(\frac{w_0^2k^2}{4z^2}+\frac{1}{3}\pi^2k^2zT\right)(\boldsymbol{\rho}_1-\boldsymbol{\rho}_2)^2\right]\\&\times\exp\left\{-\frac{1}{w_z^2}\left[-i\left[\frac{kw_0^2}{2z}\left(1-\frac{2vz}{k}\right)-\frac{1}{3}\pi^2kz^2T\right](\boldsymbol{\rho}_1-\boldsymbol{\rho}_2)+\left(\frac{\boldsymbol{\rho}_1+\boldsymbol{\rho}_2}{2}\right)\right]^2\right\},\end{aligned} \tag{12}$$

where

$$w_z^2=w_0^2\left(1-\frac{2vz}{k}\right)^2+\left(\frac{z}{kw_0}\right)^2+\frac{4\pi^2z^3}{3}T. \tag{13}$$

Therefore, we can obtain the CSD of LNUC beams in the target plane by Equations (4), (11) and (12), and we can analyze the spectral intensity of such beams by using the diagonal elements of the CSD, i.e.,

$$S(\boldsymbol{\rho})=W(\boldsymbol{\rho},\boldsymbol{\rho}), \tag{14}$$

and the spectral DOC of such beams using the normalized CSD, i.e.,

$$\mu(\boldsymbol{\rho}_1, \boldsymbol{\rho}_2) = \frac{W(\boldsymbol{\rho}_1, \boldsymbol{\rho}_2)}{\sqrt{W(\boldsymbol{\rho}_1, \boldsymbol{\rho}_1)W(\boldsymbol{\rho}_2, \boldsymbol{\rho}_2)}}. \tag{15}$$

4. Numerical Calculation and Analysis of Propagation Characteristics of LNUC Beams

In this section, we study the propagation features of LNUC beams propagating in free space and in oceanic turbulence by using the above derived formulas. In the following examples, the initial beam parameters and turbulence parameters are set as λ = 632.8 nm, $w_0 = 1$ cm, $\delta = 1$ cm, $\varepsilon = 10^{-3}$ m^2/s^3, $\chi_T = 10^{-8}$ K^2/s, $\eta = 1$ mm, and $\omega = -3.0$. If the parameter is used as a variable, it will be marked in the figures.

4.1. In Free Space

The propagation characteristics of LNUC beams in free space can be obtained by using Equations (11)–(15) with the setting $T = 0$. The intensity evolution of the LNUC beams in free space is shown in Figure 2. Figure 2a presents normalized density plots of transverse intensities at different propagation distances, and Figure 2b illustrates normalized spectral density distribution of the LNUC beam at $\boldsymbol{\rho}_x - z$ plane. We clearly observe that, as the propagation distance increases, the beam spot size undergoes a decreasing process and the on-axis intensity undergoes an increasing process, implying that such beam exhibits self-focusing propagation characteristics during propagation. Figure 2c shows how the on-axis intensity depends on both propagation distance and coherence length in free-space. One confirms that the on-axis intensity of LNUC beams increases to a maximum over short distances and then gradually decreases. Furthermore, the peak value increases with decreasing coherence length, which means that the self-focusing property of the low-coherence LNUC beams is more dramatic.

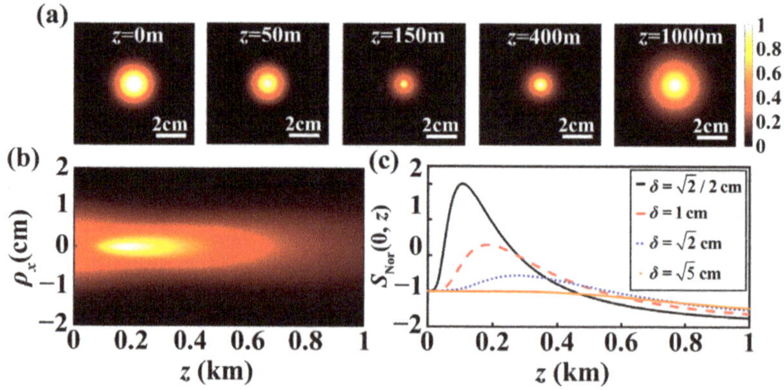

Figure 2. Evolution of the normalized spectral intensity on propagation in free space, (**a**) in $\boldsymbol{\rho}_x - \boldsymbol{\rho}_y$ cross-section; (**b**) at $\boldsymbol{\rho}_x - z$ cross-section; (**c**) on-axis.

For clarity and without loss of generality, we choose two spatial positions on the x-axis, i.e., $\rho_{1y} = \rho_{2y} = 0$, to discuss the evolution of the spectral DOC of such a beam on propagation in free space. Figure 3 shows the evolution of the absolute value of the spectral DOC of a LNUC beam. From this, we can draw the following conclusions: firstly, the high-coherence area is confined to center and diagonal lines of the beam; secondly, the high-coherence area increases with the propagation distance; thirdly, the spectral DOC distribution is symmetrical in the horizontal, vertical and diagonal directions.

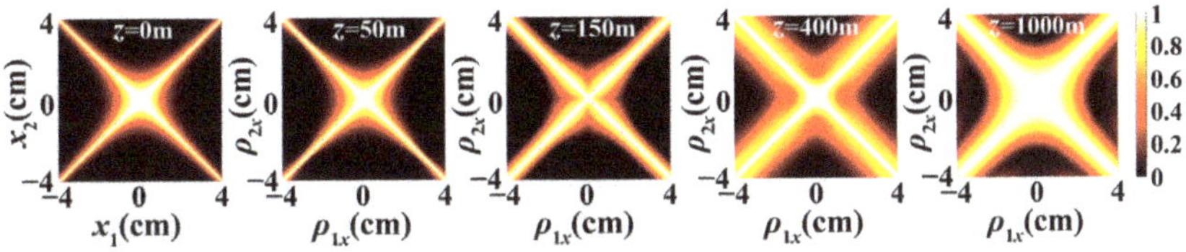

Figure 3. Evolution of the normalized spectral DOC on propagation in free space.

4.2. In Turbulent Ocean

Figure 4 illustrates the evolving behavior of the spectral intensity and DOC of a LNUC beam in oceanic turbulence, and we find that such a beam still displays a self-focusing feature in oceanic turbulence. We find from Figure 4b that as the propagation distance increases, the coverage of the high-coherence area gradually decreases along the diagonal lines (around $\rho_{1x} = -\rho_{2x}$). With further increases in propagation distance, the distribution of the spectral DOC becomes spatially quasi-homogeneous, almost constant along the diagonal lines ($\rho_{2x} - \rho_{1x}$ = constant), which means that such beam is negatively affected by the oceanic turbulence. Comparing with Figures 2–4 one confirms that with the increase in propagation distance, the influence of oceanic turbulence gradually accumulates and plays a dominant role, and the LNUC beam is negatively affected, which is realized as the spectral intensity evolution is greatly broadened, and the spectral DOC gradually degenerates into spatial quasi-homogeneous.

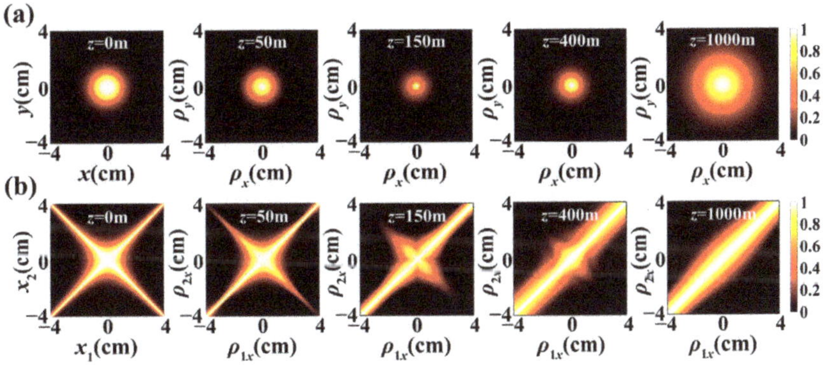

Figure 4. Evolution of the (**a**) normalized spectral intensity and (**b**) spectral DOC on propagation in oceanic turbulence.

Next, we will discuss the effect of turbulence parameters on the spectral intensity and the DOC of the proposed beams, and discuss the role of the initial beam parameters in resisting the negative effects of turbulence.

4.2.1. Turbulence Parameters

We first focus on the effect of the dissipation rate of kinetic energy per unit mass of fluid. The effects of different values of this turbulence parameter on the spectral intensity and DOC degradation at a propagation distance z = 150 m in oceanic turbulence are compared in Figure 5.

We find that with a small rate of dissipation of kinetic energy per unit mass of fluid, the size of the beam spot increases significantly and the distribution of spectral DOC degenerates more quickly to spatially quasi-homogeneous. It can convince the readers that oceanic turbulence with a small rate of dissipation of kinetic energy per unit mass of fluid has a large effect on the evolution of the spectral intensity and DOC of the proposed beams. To demonstrate this conclusion more favorably, we plot the normalized spectral intensity on-axis on propagation in oceanic turbulence and the cross-line ($\rho_{2x} = 0$) of the absolute value of the spectral DOC at $z = 150$ m for different rates of dissipation of kinetic energy per unit mass of fluid in Figure 6.

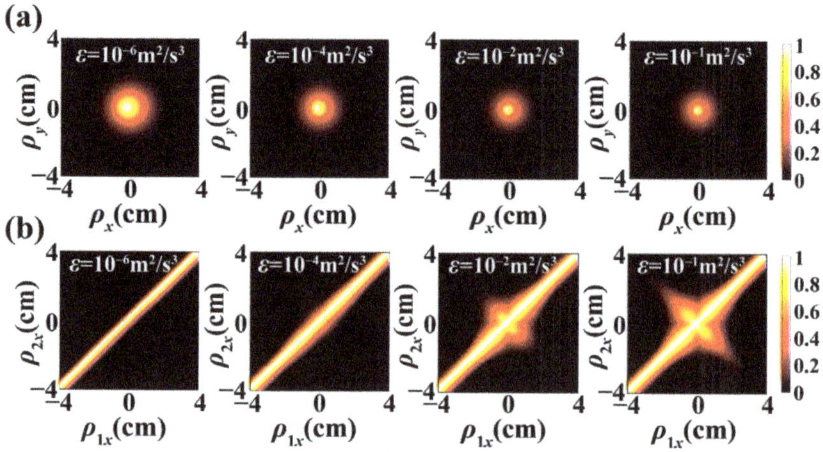

Figure 5. Density plot of (**a**) the normalized spectral intensity and (**b**) the absolute value of the spectral DOC of a LNUC beam in oceanic turbulence with different values of the dissipation rate of kinetic energy per unit mass of fluid at propagation distance $z = 150$ m.

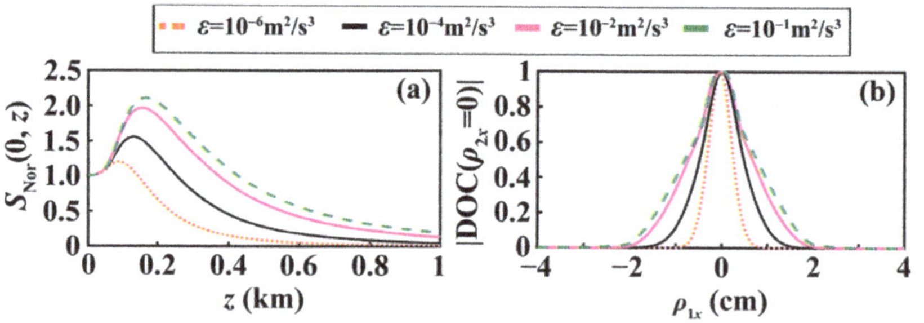

Figure 6. (**a**) Normalized on-axis intensity of LNUC beams propagating in oceanic turbulence and (**b**) the cross-line ($\rho_{2x} = 0$) of the absolute value of the spectral DOC at $z = 150$ m for different rates of dissipation of kinetic energy per unit mass of fluid.

We confirm from Figure 6a that the LNUC beam also displays a self-focusing propagation property in oceanic turbulence. The peak value of the spectral intensity decreases with the decrease in the rate of dissipation of kinetic energy per unit mass of fluid, which means a weaker self-focusing property of the LNUC beam in the oceanic turbulence with a low rate of dissipation of kinetic energy per unit mass of fluid. Figure 6b illustrates the same conclusion from the perspective of the degradation of the absolute value of the spectral DOC. It degrades more rapidly to a Gaussian distribution as the rate of dissipation of kinetic energy per unit mass of fluid decreases. Therefore, we can say the oceanic turbulence with low rate of dissipation of kinetic energy per unit mass of fluid has a large negative influence on the evolution of the spectral intensity and DOC of LNUC beams.

We next discuss the effect of the other three turbulence parameters, the Kolmogorov inner scale, the relative strength of temperature to salinity undulations and the dissipation rate of the mean-square temperature of oceanic turbulence on the evolution of the spectral intensity and DOC. For brevity, we only plot the normalized on-axis intensity of LNUC beams on propagation and the cross-line ($\rho_{2x} = 0$) of the absolute value of the DOC at

$z = 150$ m for different values of these three oceanic turbulence parameters in Figure 7. We confirm that the oceanic turbulence with small Kolmogorov inner scale, or the large relative strength of temperature to salinity undulations and large dissipation rate of mean-square temperature, has a large negative influence on propagation characteristics of the proposed beams.

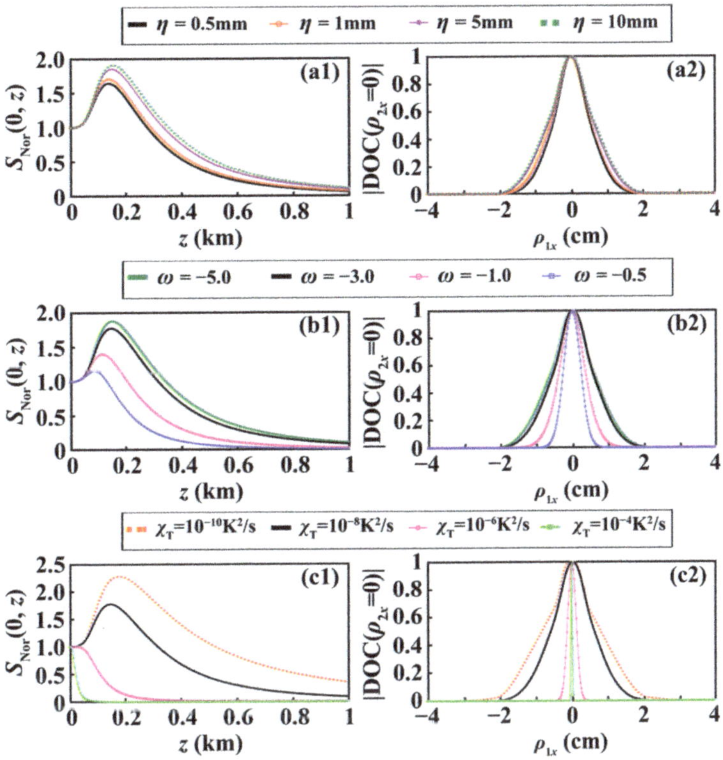

Figure 7. Normalized on-axis intensity of the LNUC beam propagating in oceanic turbulence and the cross-line ($\rho_{2x} = 0$) of the absolute value of the spectral DOC at $z = 150$ m for different oceanic turbulence parameters: (**a**) the Kolmogorov inner scale; (**b**) the relative strength of temperature to salinity undulations; (**c**) the dissipation rate of mean-square temperature.

4.2.2. Beam Parameters

Based on the above analysis, we understand the effect of different turbulence parameters on the spectral intensity and DOC of LNUC beams propagation in oceanic turbulence. In practice, it is difficult for us to change the parameters of oceanic turbulence. Therefore, it is necessary to explore how to maintain the excellent propagation characteristics of the proposed beams in oceanic turbulence, i.e., how to improve the oceanic turbulence resistance by adjusting its initial beam parameters.

Figure 8 illustrates how the normalized on-axis intensity depends on the beam width and coherence length during propagating in oceanic turbulence. One confirms from Figure 8a that as the beam width increases, the peak value of the intensity becomes larger, which means that the self-focusing propagation feature of the beam with a large beam width is more dramatic. We also plot the normalized on-axis intensity of LNUC beams in oceanic turbulence for different coherence lengths. We find that LNUC beams with low coherence exhibit more pronounced self-focusing propagation characteristics. Therefore, adjusting both the beam width and the coherence length of LNUC beams can be used to improve the oceanic turbulence resistance.

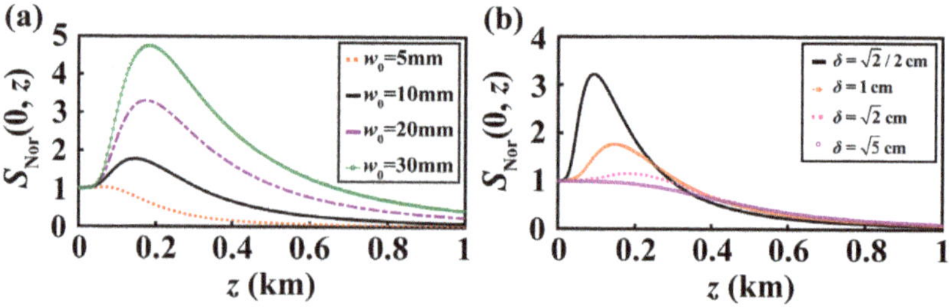

Figure 8. Normalized on-axis intensity of LNUC beams propagating in oceanic turbulence for different (**a**) beam widths and (**b**) coherence lengths.

Figure 9 shows the distribution of the absolute value of the spectral DOC of LNUC beams at propagation distance z = 150 m for different beam widths and coherence lengths From Figure 9a, we confirm that the spectral DOC of the LNUC beam with large beam width has small bumps (side lobes) on both sides of the main peak at z = 150 m. As the beam width decreases, the small bumps (side lobes) gradually disappear and the spectral DOC finally degenerates into a Gaussian distribution. Furthermore, the distribution of the spectral DOC is closely related to the coherence length. We find the conversion from the Gaussian to non-Gaussian distribution as the coherence length decreases, which implies that LNUC beams with low coherence are less affected by the oceanic turbulence from the perspective of the spectral DOC. Therefore, we conclude that LNUC beams with large beam width and low coherence are less affected by oceanic turbulence.

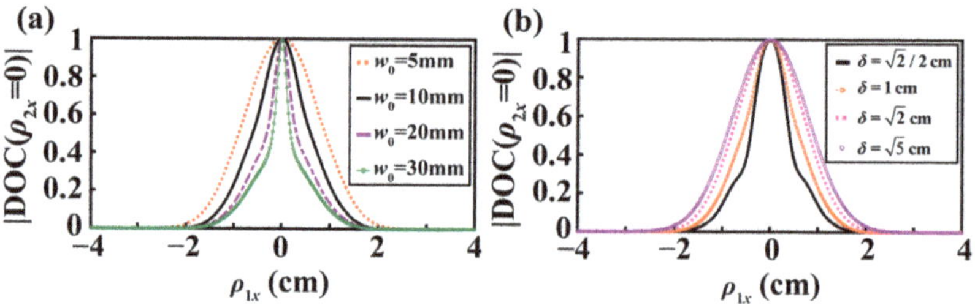

Figure 9. The cross-line (ρ_{2x} = 0) of the absolute value of the spectral DOC of LNUC beams at z = 150 m for different (**a**) beam widths and (**b**) coherence lengths.

5. Discussion

With regard to the numerical results we present, it is worth discussing and explaining the physical nature that leads to LNUC beams having their novel propagation features. From Equation (2), one finds that a partially coherent source can be regarded as an integral/superposition over the beam coherence mode according to a prescribed weight. Equation (3) shows that each mode being superimposed has a quadratic phase factor, which causes them to converge individually, and the focal distance is determined by the expression $z = 1/2v$. At different propagation distances, then, different modes will dominate. Therefore, designing a prescribed weight $p(v)$ (see Equation (4)) and constructing a new type of non-uniform correlated beam by it is the key. By superimposing different modes with the prescribed weights, it is possible to tailor beams with extraordinary propagation properties along the z-direction.

Next, let us explain the physics of how adjusting the coherence length can improve the resistance of turbulence of such beams. We know, the partially coherent source synthetical by the superposition of more coherent modes has better resistance to turbulence [9,10]. From Equation (4), we confirm that the coherence length δ determines the effective value range of v, i.e., the waist of the weight function. By decreasing the coherence of the beam, the width of the weight function can be increased, thereby increasing the number of the modes. Therefore, LNUC beams with low coherence will be less affected by the oceanic turbulence.

6. Conclusions

We have introduced a new type of non-uniformly correlated beams that named LNUC beams, and have studied the propagation of such beams in free space and oceanic turbulence. An analytical expression for the CSD of LNUC beams in oceanic turbulence is derived, and the evolutions of the spectral intensity and DOC of such beams are numerically illustrated. LNUC beams exhibit self-focusing propagation features in both free space and oceanic turbulence. We have found that such beams with large beam width and low coherence exhibit more dramatic self-focusing features in oceanic turbulence. Furthermore, adjusting the initial beam width and coherence length can mitigate negative effects from oceanic turbulence. We also investigate the effect of oceanic turbulence parameters, the dissipation rate of kinetic energy per unit mass of fluid, the dissipation rate of mean-square temperature, the Kolmogorov inner scale and the relative strength of temperature to salinity undulations on the evolution of intensity and DOC of such beams. The class of LNUC beams proposed here display novel self-focusing properties on propagation in free space and oceanic turbulence, and such beams show potential for applications such as free-space optical communications and laser radar.

Author Contributions: Writing—original draft preparation, D.W. and K.W.; writing—review and editing, J.Y. and X.Z.; conceptualization, Y.D. and L.Z.; methodology, X.Z.; software, Q.D. and J.L.; data curation, Y.X. and F.L.; supervision, Y.C.; project administration J.Y. and Y.C. All authors have read and agreed to the published version of the manuscript.

Funding: This research was funded by the National Key Research and Development Program of China (2022YFA1404800, 2019YFA0705000); the National Natural Science Foundation of China (12192254, 11974218, 12004218, 92250304); the China Postdoctoral Science Foundation (2020M680093, 2022M721992); the Shandong Provincial Natural Science Foundation of China (ZR2020QA067); and the Open Fund of the Guangdong Provincial Key Laboratory of Optical Fiber Sensing and Communications (Jinan University).

Institutional Review Board Statement: Not applicable.

Informed Consent Statement: Not applicable.

Data Availability Statement: Data is contained within the article.

Conflicts of Interest: The authors declare no conflict of interest.

References

1. Garceschavez, V.; Mcgloin, D.; Melville, H.; Sibbett, W.; Dholakia, K. Simultaneous micromanipulation in multiple planes using a self-reconstructing light beam. *Nature* **2002**, *419*, 145–147. [CrossRef] [PubMed]
2. Lavery, M.P.; Speirits, F.C.; Barnett, S.M.; Padgett, M.J. Detection of a Spinning Object Using Light's Orbital Angular Momentum. *Science* **2013**, *34*, 537–540. [CrossRef] [PubMed]
3. Andrews, L.C.; Phillips, R.L. *Laser Beam Propagation through Random Media*, 2nd ed.; SPIE: Bellingham, WA, USA, 2005; pp. 35–56.
4. Gbur, G. Partially coherent beam propagation in atmospheric turbulence. *J. Opt. Soc. Am. A* **2014**, *31*, 2038–2045. [CrossRef] [PubMed]
5. Thorpe, S.A. *The Turbulent Ocean*; Cambridge University: Cambridge, MA, USA, 2005; pp. 1–37.
6. Korotkova, O. Light Propagation in a Turbulent Ocean. *Prog. Opt.* **2019**, *64*, 1–43.
7. Cai, Y.; Chen, Y.; Yu, J.; Liu, X.; Liu, L. Generation of partially coherent beams. *Prog. Opt.* **2017**, *62*, 157–223.
8. Wang, F.; Liu, X.; Cai, Y. Propagation of partially coherent beam in turbulent atmosphere: A review (Invited review). *Prog. Electromagn. Res.* **2015**, *150*, 123–143. [CrossRef]
9. Gbur, G.; Wolf, E. Spreading of partially coherent beams in random media. *J. Opt. Soc. Am. A* **2002**, *8*, 1592–1598. [CrossRef] [PubMed]
10. Shirai, T.; Dogariu, A.; Wolf, E. Mode analysis of spreading of partially coherent beams propagating through atmospheric turbulence. *J. Opt. Soc. Am. A* **2003**, *20*, 1094–1102. [CrossRef] [PubMed]

11. Wolf, E. *Introduction to the Theory of Coherence and Polarization of Light*; Cambridge University: Cambridge, MA, USA, 2007; pp. 174–197.
12. Chen, Y.; Wang, F.; Liu, L.; Zhao, C.; Cai, Y.; Korotkova, O. Generation and propagation of partially coherent beams with nonconventional correlation functions: A review. *J. Opt. Soc. Am. A* **2014**, *31*, 2083–2096.
13. Liu, X.; Zhou, G.; Shen, Y. Effect of oceanic turbulence with anisotropy on the propagation of multi-sinc Schell-model beams. *Results Phys.* **2022**, *36*, 105447. [CrossRef]
14. Sun, C.; Lv, X.; Ma, B.; Zhang, J.; Deng, D.; Hong, W. Statistical properties of partially coherent radially and azimuthally polarized rotating elliptical Gaussian beams in oceanic turbulence with anisotropy. *Opt. Express* **2019**, *27*, A245–A256. [CrossRef] [PubMed]
15. Song, Z.; Han, Z.; Ye, J.; Liu, Z.; Liu, S.; Liu, B. Propagation properties of radially polarized multi-Gaussian Schell-model beams in oceanic turbulence. *J. Opt. Soc. Am. A* **2019**, *36*, 1719–1726. [CrossRef] [PubMed]
16. Gu, Y.; Gbur, G. Scintillation of nonuniformly correlated beams in atmospheric turbulence. *Opt. Lett.* **2013**, *38*, 1395–1397. [CrossRef] [PubMed]
17. Yu, J.; Wang, F.; Liu, L.; Cai, Y.; Gbur, G. Propagation properties of Hermite non-uniformly correlated beams in turbulence. *Opt. Express* **2018**, *26*, 16333–16343. [CrossRef] [PubMed]
18. Yu, J.; Cai, Y.; Gbur, G. Rectangular Hermite nonuniformly correlated beams and its propagation properties. *Opt. Express* **2018**, *26*, 27894–27906. [CrossRef] [PubMed]
19. Lin, S.; Wang, C.; Zhu, X.; Lin, R.; Wang, F.; Gbur, G.; Cai, Y.; Yu, J. Propagation of radially polarized Hermite non-uniformly correlated beams in a turbulent atmosphere. *Opt. Express* **2020**, *28*, 27238–27249. [CrossRef] [PubMed]
20. Lin, S.; Wu, J.; Xu, Y.; Zhu, X.; Gbur, G.; Cai, Y.; Yu, J. Analysis and experimental demonstration of propagation features of radially polarized specific non-uniformly correlated beams. *Opt. Lett.* **2022**, *47*, 305–308. [CrossRef] [PubMed]
21. Ding, C.; Koivurova, M.; Turunen, J.; Pan, L. Self-focusing of a partially coherent beam with circular coherence. *J. Opt. Soc. Am. A* **2017**, *34*, 1441–1447. [CrossRef] [PubMed]
22. Gori, F.; Santarsiero, M. Devising genuine spatial correlation functions. *Opt. Lett.* **2007**, *32*, 3531–3533. [CrossRef] [PubMed]
23. Nikishov, V.V.; Nikishov, V.I. Spectrum of Turbulent Fluctuations of the Sea-Water Refraction Index. *Int. J. Fluid Mech. Res.* **2000**, *27*, 82–98. [CrossRef]

photonics

MDPI

Article

Manipulation of Energy Flow with X-Type Vortex

Han Zhang [1], Tianhu Zhang [1], Xinying Zhao [2] and Xiaoyan Pang [1,*]

[1] School of Electronics and Information, Northwestern Polytechnical University, Xi'an 710072, China
[2] School of Physics & Information Technology, Shaanxi Normal University, Xi'an 710061, China
* Correspondence: xypang@nwpu.edu.cn

Abstract: In this study, a new method for manipulating energy flow in a 3D vector field is proposed. In this method, an azimuthally-polarized beam with a noncanonical vortex, the X-type vortex, is focused in a high-numerical aperture system. It is found that, instead of the invariance of the energy flow which is characteristic of the traditional vortex (i.e., canonical vortex), both the longitudinal and the transverse energy flows in virtue of the X-type vortex rotate around the beam center as the beam propagates, and this rotational behavior (including the maxima location and the rotational angle) can be adjusted by the anisotropic parameter and the order the X-type vortex. Through defining a complex transverse Poynting field and applying the equivalence principle, the transverse energy flow and its topological reactions are discussed in the focal plane. Our result shows that, by changing the anisotropic parameter of the X-type vortex, rich topological reactions will occur, resulting in various distribution patterns of the energy flow, such as multi vortex-type singularities around the beam center. Our research demonstrates newly-observed features of the X-type vortex and also provides a simple method to manipulate energy flows both along longitudinal and transverse directions, which will be useful in optical manipulations.

Keywords: Poynting vector; optical vortex; singularity; energy flow

check for updates

Citation: Zhang, H.; Zhang, T.; Zhao, X.; Pang, X. Manipulation of Energy Flow with X-Type Vortex. *Photonics* **2022**, *9*, 998. https://doi.org/10.3390/photonics9120998

Received: 24 November 2022
Accepted: 15 December 2022
Published: 17 December 2022

Publisher's Note: MDPI stays neutral with regard to jurisdictional claims in published maps and institutional affiliations.

1. Introduction

The energy flow of light has been studied for more than 100 years since the possibility of backward energy was revealed in the near-focus field in 1919 [1]. Later, in 1959, the energy flow was also analyzed near the Airy rings in a classical article [2]. Because of its important role in both fundamental and applied research, the study of energy flow has become more vigorous in recent decades [3–23]. In fundamental research, the energy flow provides a natural method for exploring the most intimate features of an optical field, such as the (intrinsic) energy flow being divided into a spin part and an orbital part corresponding to the two different angular momenta of an optical field [3–6], which reflects the physical nature of light; the behaviors of the energy flow can connect to the topological reactions of optical singularities, which supports a method of explaining the special features of singular optics [7,8]. On the other hand, energy flow has been utilized in optical manipulations. The absorptive particle can move along the direction of the energy flow and the velocity of the movement is proportional to the modulus of the energy flow [9,10]. Recently, beams with backward energy flow have also attracted a lot of interest for their role as a 'tractor' to exert pulling force on a particle in manipulation schemes [11–16]. Many methods have been proposed in manipulating energy flows, such as tailoring the phase structures [6,17,18] and/or polarization distributions [19–22]. Research on energy flow continues to be expanded on in fundamental and applied optics [16,23].

An optical vortex usually refers to the canonical vortex with a constant phase gradient around its center [8,24–26]. Since it has peculiar characteristics, such as carrying orbital angular momentum, the optical vortex has been studied extensively and utilized in a wide range of applications, such as in optical tweezers [27], optical communications [28], imaging [29], microscopy [30,31], etc. The vortex also plays a key role in most research on the manipulation of energy flows [12,14–20,22]. Besides the well-known canonical vortex, there also exists the noncanonical vortex, which has not received as much attention in most studies [32–36]. For a noncanonical vortex, the phase gradient is no longer constant along the azimuthal direction; thus, there will exist an 'anisotropic parameter' characterizing the phase distribution, which actually provides more freedom for beam structure [33,36]. Very recently, an X-type vortex, as one type of the noncanonical vortex, was proposed, and it was found that this noncanonical vortex could shape the intensity distribution in rich structures in 3D vector fields [36]. In this article, we will use the X-type vortex and show its effects on energy flows.

2. Materials and Methods

The X-type vortex is a type of noncanonical optical vortex, with an anisotropic phase distribution [36]. The transverse field of an optical beam embedded with an X-type vortex can be expressed as:

$$V^{(X)}(x,y) = A(x,y)\left(\sqrt{x^2+y^2}\right)^l \frac{(x+\mathrm{i}\sigma_c y)^l}{\left(\sqrt{x^2+\sigma_c{}^2y^2}\right)^l}, \tag{1}$$

where σ_c is the anisotropic parameter determining the phase distribution and l ($l \in N$) represents the order of the vortex. When $\sigma_c = \pm 1$, the X-type vortex degenerates into a conventional (canonical) vortex. It is also convenient to re-write the Equation (1) in the polar coordinates ($x = \rho cos\phi, y = \rho sin\phi$):

$$V^{(X)}(\rho,\phi) = A(\rho,\phi)\rho^l e^{\mathrm{i}l\arctan(\sigma_c \tan\phi)}, \tag{2}$$

Which shows that the phase is a nonlinear function of azimuthal direction ϕ and the phase gradient along the ϕ direction equals $l\sigma_c/\left(\cos^2\phi + \sigma_c^2\sin^2\phi\right)$. Examples of the phase distributions are depicted in Figure 1a, where the plot with $l = 1$ and $\sigma_c = 1$ represents the conventional (canonical) vortex with order 1, and one can see that the phase changes uniformly along the ϕ direction. The phases in the plots with $\sigma_c > 1$ change faster near the x-axis, where the phase in the plot with $\sigma_c < 1$ changes faster near the y-axis.

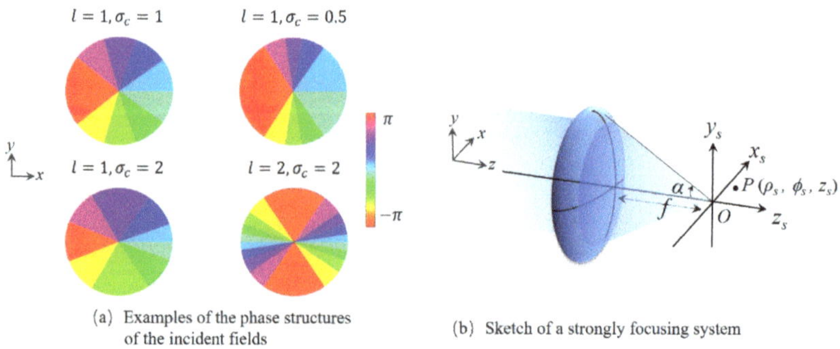

(a) Examples of the phase structures of the incident fields (b) Sketch of a strongly focusing system

Figure 1. A strongly focusing system with an X-type vortex beam as the incident field.

Next, we will use this X-type vortex to construct a 3D vector field in a high numerical aperture (NA) system. First, consider a high NA system with a semi-aperture angle α and a focal length f. The focus of this system is located at point O, the origin of the Cartesian coordinate system, see Figure 1b. Then, assume that a Gaussian beam embedded with an X-type vortex is incident upon this focusing system, i.e., the complex amplitude of the incident field can be expressed by Equation (2) with $A(\rho,\phi) = exp\left(-\frac{\rho^2}{w_0^2}\right)$, where w_0 is the waist size. This X-type vortex can be generated by using a phase-spatial light modulator or a programmable q-plate [37–39]. According to the Richards-Wolf vector diffraction theory, the electric/magnetic field at a point $P(\rho_s, \phi_s, z_s)$ in the focal region can be written as [2]

$$\mathbf{U}(\rho_s,\phi_s,z_s) = -\frac{ikf}{2\pi}\int_0^{\alpha}\int_0^{2\pi} V^{(X)}(\rho,\phi)\, \mathbf{Q}_{E(H)}(\theta,\phi)\sqrt{\cos\theta}e^{ikz_s\cos\theta} \times e^{ik\rho_s\sin\theta\cos(\phi-\phi_s)}\sin\theta d\theta d\phi, \tag{3}$$

where $\mathbf{U}(\rho_s,\phi_s,z_s)$ represents the 3D electric field ($\mathbf{E}(\rho_s,\phi_s,z_s)$) or magnetic field ($\mathbf{H}(\rho_s,\phi_s,z_s)$), and where $k = 2\pi/\lambda$ is the wave number with λ denoting the wavelength of the free space. $\mathbf{Q}_E(\theta,\phi)(\mathbf{Q}_H(\theta,\phi))$ is the polarization matrix of the electric (magnetic) field, which can be expressed as

$$\mathbf{Q}_{E(H)}(\theta,\phi) = \begin{bmatrix}\cos\theta+\sin^2\phi(1-\cos\theta)\\ (\cos\theta-1)\cos\phi\sin\phi\\ -\sin\theta\cos\phi\end{bmatrix}a(\theta,\phi) + \begin{bmatrix}(\cos\theta-1)\cos\phi\sin\phi\\ \cos\theta+\cos^2\phi(1-\cos\theta)\\ -\sin\theta\sin\phi\end{bmatrix}b(\theta,\phi), \tag{4}$$

where $a(\theta,\phi)$ and $b(\theta,\phi)$ are the weight functions for the x-polarized and y-polarized components of the incident beam, respectively. Assume this incident beam is also azimuthally polarized with order m; thus, for the electric field:

$$\begin{pmatrix} a(\theta,\phi)\\ b(\theta,\phi)\end{pmatrix} = \begin{pmatrix} -\sin m\phi\\ \cos m\phi\end{pmatrix}, \tag{5}$$

and for the magnetic field:

$$\begin{pmatrix} a(\theta,\phi)\\ b(\theta,\phi)\end{pmatrix} = \begin{pmatrix} -\cos m\phi\\ -\sin m\phi\end{pmatrix}. \tag{6}$$

For simplicity, from here on, we only consider the case of $m = 1$; the other cases can be derived in the same way. By substituting Equations (5) and (6) into Equation (4), we can obtain two simple expressions for the polarization matrix:

$$\mathbf{Q}_E(\theta,\phi) = \begin{bmatrix}-\sin\phi\\ \cos\phi\\ 0\end{bmatrix}, \tag{7}$$

$$\mathbf{Q}_H(\theta,\phi) = \begin{bmatrix}-\cos\theta\cos\phi\\ -\cos\theta\sin\phi\\ \sin\theta\end{bmatrix}. \tag{8}$$

thus, the electric field $\mathbf{E}(\rho_s,\phi_s,z_s)$ and magnetic field $\mathbf{H}(\rho_s,\phi_s,z_s)$ in the focal region can be calculated as:

$$\mathbf{E}(\rho_s,\phi_s,z_s) = \begin{bmatrix}e_x\\ e_y\\ e_z\end{bmatrix} = -\mathrm{i}k\int_0^{\alpha} L(\theta)\begin{bmatrix}I_x(\theta,\phi)\\ I_y(\theta,\phi)\\ 0\end{bmatrix}d\theta, \tag{9}$$

$$\mathbf{H}(\rho_s,\phi_s,z_s) = \begin{bmatrix}h_x\\ h_y\\ h_z\end{bmatrix} = -\mathrm{i}k\int_0^{\alpha} L(\theta)\begin{bmatrix}-\cos\theta\, I_y(\theta,\phi)\\ \cos\theta\, I_x(\theta,\phi)\\ \sin\theta\, I_z(\theta,\phi)\end{bmatrix}d\theta, \tag{10}$$

where

$$L(\theta) = \sqrt{\cos\theta}e^{-(f\sin\theta)^2/w_0^2}f^l(\sin\theta)^{l+1}e^{ikz_s\cos\theta}, \tag{11}$$

and

$$I_x(\theta,\phi) = \int_0^{2\pi}\frac{-\sin\phi(\cos\phi+\mathrm{i}\sigma_c\sin\phi)^l}{\left(\sqrt{\cos^2\phi+{\sigma_c}^2\sin^2\phi}\right)^l}e^{ik\rho_s\sin\theta\cos(\phi-\phi_s)}d\phi, \tag{12}$$

$$I_y(\theta,\phi) = \int_0^{2\pi}\frac{\cos\phi(\cos\phi+\mathrm{i}\sigma_c\sin\phi)^l}{\left(\sqrt{\cos^2\phi+{\sigma_c}^2\sin^2\phi}\right)^l}e^{ik\rho_s\sin\theta\cos(\phi-\phi_s)}d\phi, \tag{13}$$

$$I_z(\theta,\phi) = \int_0^{2\pi} \frac{(\cos\phi + i\sigma_c \sin\phi)^l}{\left(\sqrt{\cos^2\phi + \sigma_c{}^2 \sin^2\phi}\right)^l} e^{ik\rho_s \sin\theta\cos(\phi-\phi_s)} d\phi. \quad (14)$$

The energy flow, which is described by the time-averaged Poynting vector **P**, can now be written in terms of the expressions of these 3D electric and magnetic fields, as in [3,12]:

$$\mathbf{P} \propto \text{Re}[\mathbf{E} \times \mathbf{H}^*] = \begin{pmatrix} p_x \\ p_y \\ p_z \end{pmatrix} = \begin{pmatrix} \text{Re}[e_y h_z^*] \\ -\text{Re}[e_x h_z^*] \\ \text{Re}[e_x h_y^* - e_y h_x^*] \end{pmatrix}, \quad (15)$$

where Re[·] means the real part and the superscript * denotes the complex conjugate. The following discussions on the energy flow are mainly based on the equations derived in this section.

3. Results and Discussions

In this section, we first discuss the behaviors of the energy flow as the beam propagates. After that, the properties of the transverse energy flow in the focal plane will be analyzed.

3.1. Longitudinal Energy Flow along the Propagation Direction

Let us first analyze the longitudinal energy flow, which is the longitudinal component of the Poynting vector p_z along the propagation direction. When $\sigma_c = 1$, i.e., the case for the canonical vortex, the integral with azimuthal angle ϕ in Equations (12)–(14) can be calculated and the first kind of Bessel functions with trigonometric functions will be obtained. Since the relations of the transverse components of the electric field and the magnetic field (see Equations (9)–(13)), it can be calculated easily that p_z has a circular symmetry, i.e., $p_z(\rho_s, \phi_s + \Delta_\phi, z_s) = p_z(\rho_s, \phi_s, z_s)$, with Δ_ϕ being an arbitrary azimuthal angle. This means that the longitudinal energy flow is uniformly distributed along the azimuthal direction, which can be seen in Figure 2b. In addition, when $\sigma_c \neq 1$, the X-type vortex (noncanonical case), this symmetry will be broken, see Figure 2a,c).

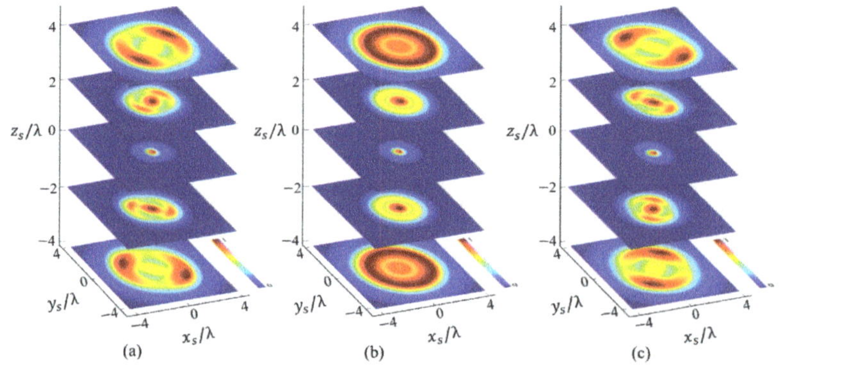

Figure 2. The longitudinal energy flow p_z along the propagation direction: (**a**) $\sigma_c = 0.5$; (**b**) $\sigma_c = 1$; (**c**) $\sigma_c = 2$. In all plots $l = 1$.

Figure 2 depicts the distribution of the longitudinal component p_z in different transverse planes along the propagation direction, where the order of the beam is chosen as $l = 1$ and the anisotropic parameter $\sigma_c = 0.5$ in plot (a), $\sigma_c = 1$ plot (b), and $\sigma_c = 2$ in plot (c). In this figure, the semi-aperture angle α is set as 60°, and, unless otherwise specified, $\alpha = 60°$ in this article. In this figure, p_z is always positive except certain points where $p_z = 0$, which implies that the longitudinal energy flow (if it exists) always points to the $+z_s$ direction. Through observing this figure, we can obtain the hypotheses that: (a) for any value of σ_c (when $l = 1$), the maximum of the energy is located at the focus on the focal plane, while, when the propagation distance is far from the focal plane, the maximum/maxima is/are gradually 'thrown out' of the beam center; (b) for $\sigma_c \neq 1$ (the X-type case), P_z has two maxima in the transverse plane (which has a distance from the focal plane). Also, for $|\sigma_c| < 1$, the two maxima in the $-z_s$ space are more likely to stay in the second and fourth quadrants and, in the $+z_s$ space, to stay in the first and third quadrants. For $|\sigma_c| > 1$, the trend is just the opposite; c) more interestingly, when $\sigma_c \neq 1$, the distribution pattern of P_z rotates in a counterclockwise manner along the beam propagation direction. This indicates that *the X-type vortex leads to a rotation of the longitudinal energy flow as the beam propagates*. In the following, this rotational behavior will be examined more deeply.

First of all, in order to characterize this rotational behavior quantitively, a rotational angle φ_r is introduced, as shown in Figure 3. φ_r is defined as the azimuthal angle of one maximum point of p_z. In this study, to remain consistent, this maximum point is always chosen as the one near the $+x_s$ axis from the $-z_s$ space. In other words, first, we fix one maximum point in the $-z_s$ space (the condition for this point is that it is near the $+x_s$ axis), then we follow the position of this point as the beam propagates and record its rotational angle φ_r. For instance, in Figure 3, the maximum point is the one in the fourth quadrant and φ_r is about $-37°$. Thus, the variation in the rotational angle φ_r with the beam propagation can be drawn, and, as shown in Figure 4, four curves of φ_r for $\sigma_c = 0.1$, 0.5, 2, and 10 are depicted. In this figure, the range is $-14\lambda < z_s < 14\lambda$ and the order $l = 1$. From this figure, the rotational behavior and its counterclockwise manner of the longitudinal energy flow p_z can be seen more clearly. The accumulated rotational angle (here denoted by $\int \varphi_r$) from -14λ to 14λ is quite big, which is more than 100° in all four of these cases. Also, this rotation can be manipulated by the propagation distance z_s and the anisotropic parameter σ_c. As $|z_s|$ increases and/or σ_c moves further from 1, the rotational behavior becomes more obvious and the accumulated rotational angle $\int \varphi_r$ becomes bigger: $\int \varphi_r = 112.6°$ for $\sigma_c = 0.1$, and 10 and $\int \varphi_r = 100.4°$ for $\sigma_c = 0.5$, 2. As $|z_s|$ approaches $+\infty$, $\int \varphi_r$ is infinitely nearer to 180°. In addition, the curves for $\sigma_c = 0.1$ and 10 are the same except for a constant difference, which is also the case for $\sigma_c = 0.5$ and 2. This is because, in each group, these two values of σ_c are reciprocal of each other, and, according to Equations (1) and (2), replacing σ_c with $1/\sigma_c$ is equivalent to exchanging the x and the y coordinates. Furthermore, one also can find that the curves which are very near focus are very steep; this is mainly caused by the dramatic changing of the wavefront spacings in the high NA system [40,41].

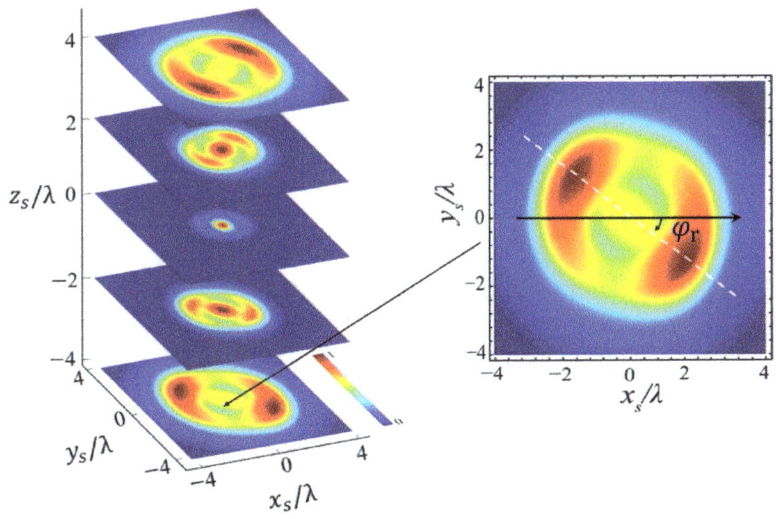

Figure 3. Definition of a rotational angle φ_r . Here, $\sigma_c = 0.5, l = 1, z_s = -4$.

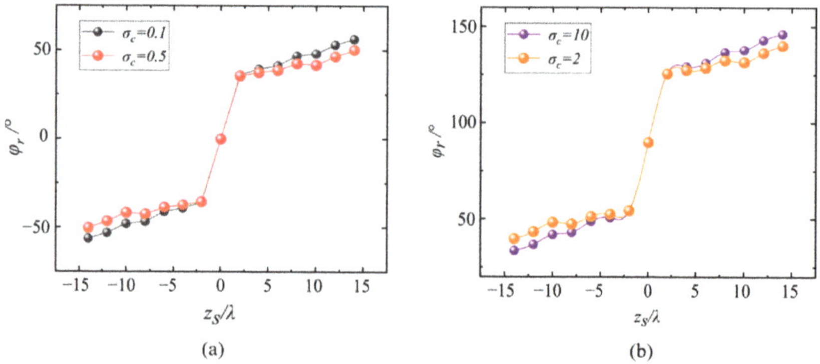

Figure 4. Variation in the rotational angle φ_r for the longitudinal energy flow p_z with the beam propagation. Here, (**a**) $\sigma_c = 0.1,\ 0.5$; (**b**) $\sigma_c = 2,\ 10$. In both plots $l = 1$.

In the above analysis, the beam order l is set as 1, and, here, we will show that changing the order l can also manipulate the rotational behavior of p_z. The distribution of p_z along the propagation direction for the X-type vortex with order 2 ($l = 2$) is illustrated in Figure 5. By observing Figure 5 with Figure 2, we can find: (a) for $l = 2$, the beam center is always hollow, i.e., the energy of p_z is mainly distributed around the beam center, which is the main difference in the distribution of p_z between $l = 2$ and $l = 1$. Further, more generally, for any $l \neq 1$, the distribution of p_z always exhibits a hollow shape. This result can be obtained directly from the expression of e_x, e_y, and P_z, using Equations (9)–(15). Consider the simplest case $\sigma_c = 1$;from Equations (12) and (13) we can obtain $I_{x(y)} = C_{x(y)}\ J_{l\pm1}(k\rho_s \sin\theta)$, with $J_{l\pm1}$ being the first kind of Bessel function. This expression implies that, only when $l = 1$, the e_x and e_y (also h_x, and h_y) will not be zero along the z_s axis ($\rho_s = 0$) (note in this article $l \in \mathbb{N}$). Therefore, p_z ($\rho_s = 0$) $= 0$ for any $l \neq 1$. Furthermore, since a property of the Bessel function is that, as l increases, the radius of the hollow part widens, which can be observed from the comparison of Figures 3 and 5. Although these two results can be derived from the properties of the Bessel function, they can be easily generalized to the case of $\sigma_c \neq 1$. (b) The longitudinal component p_z in $l = 2$ also has a similar two maximum points and rotational behaviors to those in $l = 1$. For $\sigma_c \neq 1$, p_z has two main maxima in the transverse plane instead of maximum rings of $\sigma_c = 1$, and the positions of the maxima are located in the same quadrants as in in $l = 1$. More importantly, the distribution pattern of p_z also rotates in a counterclockwise manner along the propagation direction, which is the same as it is in $l = 1$.

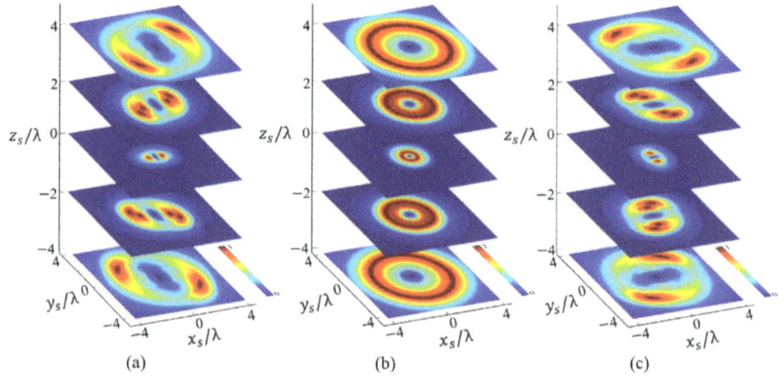

Figure 5. The longitudinal energy flow p_z along the propagation direction: (**a**) $\sigma_c = 0.5$; (**b**) $\sigma_c = 1$; (**c**) $\sigma_c = 2$. In all the plots $l = 2$.

Similarly, the rotational angle φ_r for $l = 2$ can also be drawn, as shown in Figure 6. By comparing Figure 6 with Figure 4, one can see that the rotation for $l = 2$ goes more smoothly as the beam propagates, and the accumulated rotational angle for $l = 2$ is generally bigger than its corresponding one for $l = 1$, i.e., $\int \varphi_r\ (l = 2) = 117.5^\circ > \int \varphi_r\ (l = 1) = 112.6^\circ$ for $\sigma_c = 0.1,\ 10$, and $\int \varphi_r\ (l = 2) = 108.9^\circ > \int \varphi_r\ (l = 1) = 100.4^\circ$ for $\sigma_c = 0.2,\ 5$. Also, it is not hard to calculate that, as l increases, this accumulated rotational angle $\int \varphi_r$ becomes slightly bigger.

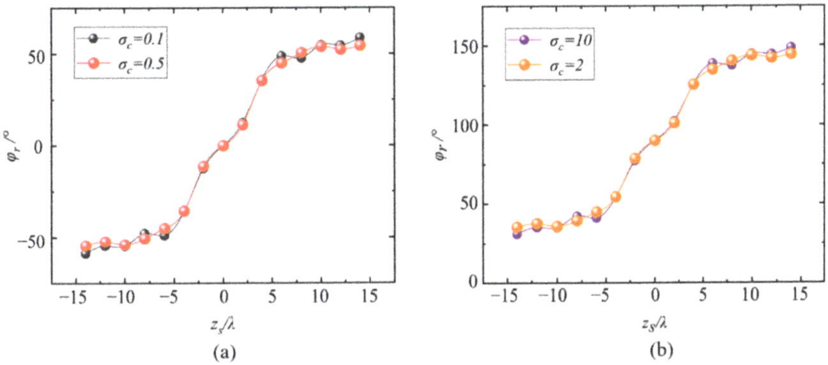

Figure 6. Variation in the rotational angle φ_r for the longitudinal energy flow p_z with the beam propagation. Here, (**a**) $\sigma_c = 0.1,\ 0.5$; (**b**) $\sigma_c = 2,\ 10$. In both plots $l = 2$.

3.2. Transverse Energy Flow along the Propagation Direction

Transverse energy flow is one of the characteristics of 3D vector optical fields. In addition to p_z, the transverse energy flow, $\mathbf{p}_{xy} = (p_x, p_y)$, as it will be shown, can also rotate with the beam propagating.

Figure 7 depicts the transverse energy flow $\mathbf{p}_{xy}$ in different transverse planes along the propagation direction. Here, the energy strength $|\mathbf{p}_{xy}|$ is presented by different colors and the direction of $\mathbf{p}_{xy}$; the flow lines are drawn by black lines with an arrow. It can be seen that, firstly, the distribution of the transverse energy flow $\mathbf{p}_{xy}$ for $\sigma_c \neq 1$ can also rotate along the propagation direction, and this rotation can be found to be counterclockwise. This indicates that *the X-type vortex can not only cause the rotation of the longitudinal flow p_z, but can also make the transverse energy rotate with the beam propagating*. Similarly, the rotation of the transverse energy flow can be measured quantitively by using the rotational angle φ_r, while φ_r, here is defined with respect to the transverse energy distribution.

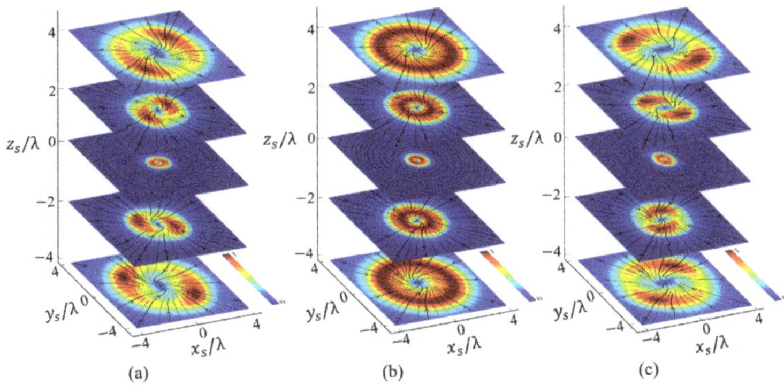

Figure 7. The transverse energy flow $\mathbf{p}_{xy}$ along the propagation direction: (**a**) $\sigma_c = 0.5$; (**b**) $\sigma_c = 1$; (**c**) $\sigma_c = 2$. In all the plots $l = 1$.

The variation in the rotational angle φ_r for the transverse energy flow along the propagation direction is drawn in Figure 8. One can see that the accumulated rotational angle $\int \varphi_r$ is slightly bigger than its corresponding longitudinal component, with $\int \varphi_r = 119.9°$ for $\sigma_c = 0.1,\ 10$ and $\int \varphi_r = 116.6°$ for $\sigma_c = 0.5,\ 2$. It is worth noting that, although the overall rotational tendency is counterclockwise, in the range of about $|z| < \lambda$, $\mathbf{p}_{xy}$ will show a short clockwise rotation. Due to the limitation of the sampling points in Figure 7, this abnormal rotation cannot be seen. In Figure 9, the clockwise rotational behavior is shown, and one can find that, in the range $-0.8\lambda < z_s < 0.8\lambda$, the accumulated clockwise rotational angle is about $-22.6°$ for $\sigma_c = 0.1,\ 10$ and $-36.9°$ for $\sigma_c = 0.5,\ 2$.

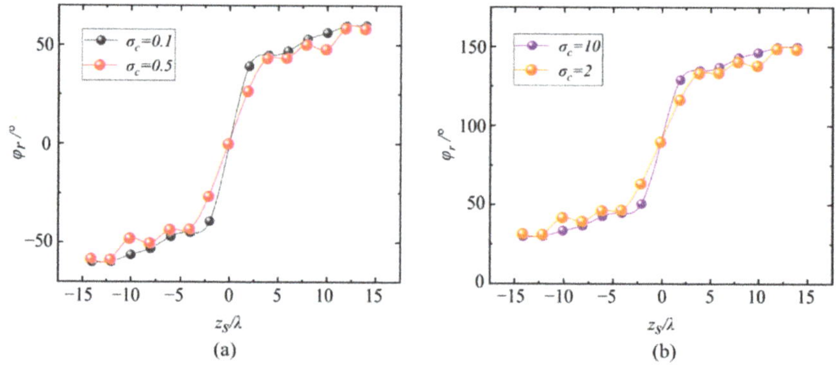

Figure 8. Variation in the rotational angle φ_r for the transverse energy flow $\mathbf{p}_{xy}$ with the beam propagation. Here, (**a**) $\sigma_c = 0.1,\ 0.5$; (**b**) $\sigma_c = 2,\ 10$. In both plots $l = 1$.

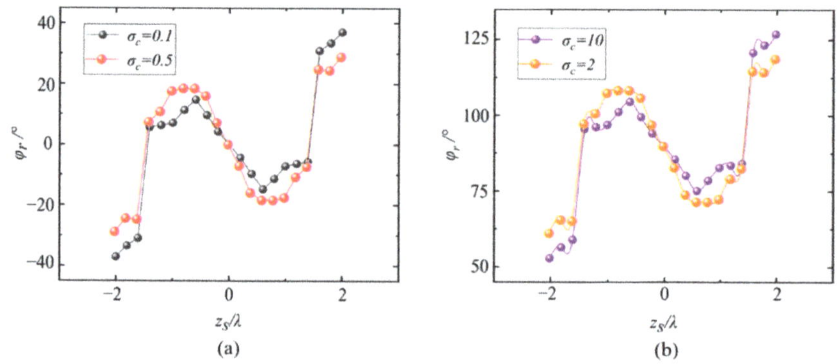

Figure 9. Same as Figure 8, but with more details in a narrower range ($-2\lambda < z_s < 2\lambda$).

Besides the effects of the propagation distance z_s and the anisotropic parameter σ_c on the transverse energy flow $\mathbf{p}_{xy}$, the beam order l, as we discussed for the longitudinal energy flow, will also influence the rotational behavior of $\mathbf{p}_{xy}$. Figure 10 depicts the distributions of the transverse energy $|\mathbf{p}_{xy}|$ along the propagation direction in the second-order case ($l = 2$). It can also be seen that, generally, the transverse energy also rotates in a counterclockwise manner as the beam propagates and, in Figure 11, the corresponding rotational angle φ_r is shown, where the accumulated rotational angle $\int \varphi_r = 123.9°$ for $\sigma_c = 0.1,\ 10$ and $\int \varphi_r = 116.8°$ for $\sigma_c = 0.5,\ 2$ in the range $-14\lambda < z_s < 14\lambda$. We should note that abnormal rotational behavior also exists in the second-order case, which can be seen in Figure 12. It can be seen that this abnormal behavior is more complicated than it is in the case of $l = 1$. When $\sigma_c = 0.1,\ 10$, the clockwise rotation is in a very narrow range, about $-0.2\lambda < z_s < 0.2\lambda$; however, the accumulated rotational angle $\int \varphi_r$ is as big as $-70.0°$, while, for $\sigma_c = 0.5, 2$, $\int \varphi_r$ is about $-18.4°$ within $-0.8\lambda < z_s < 0.8\lambda$. This abnormal rotational behavior occurs in the range closer to the focal plane, which may be caused by the redistribution of the topological structure of the transverse energy flow on account of the X-type vortex; this redistribution will be discussed in the following part.

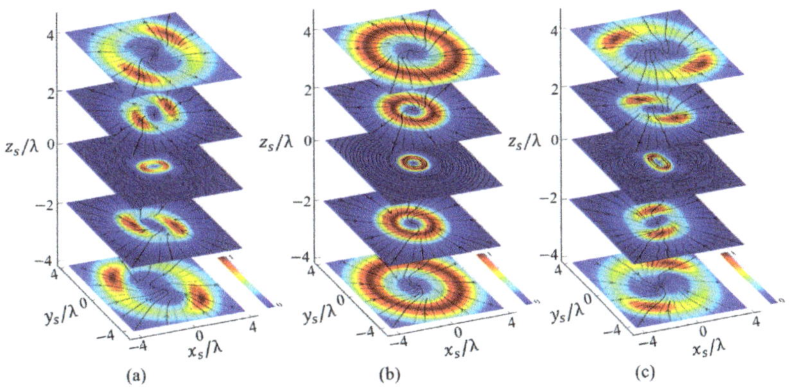

Figure 10. The transverse energy flow $\mathbf{p}_{xy}$ along the propagation direction: (**a**) $\sigma_c = 0.5$; (**b**) $\sigma_c = 1$; (**c**) $\sigma_c = 2$. In all the plots $l = 2$.

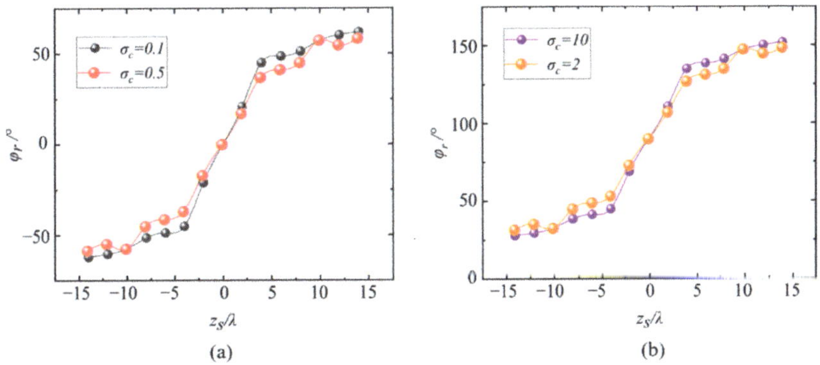

Figure 11. Variation in the rotational angle φ_r for the transverse energy $|\mathbf{p}_{xy}|$ with the beam propagation. Here, (**a**) $\sigma_c = 0.1, 0.5$; (**b**) $\sigma_c = 2, 10$. In both plots $l = 2$.

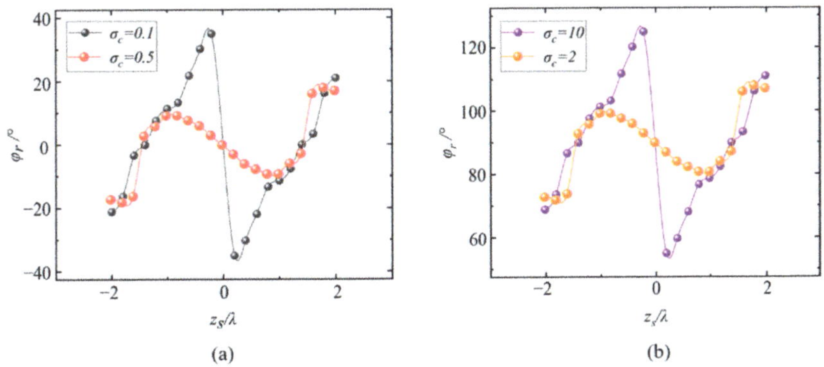

Figure 12. Same as Figure 11, but with more details in a narrower range ($-2\lambda < z_s < 2\lambda$).

In addition, here, we should note that the rotational angle φ_r describes the rotational behavior of the transverse/longitudinal energy flow distribution (i.e., the energy flow 'pattern'), which is quite different from the 'skew angle' observed in traditional vortex beams [42,43]. The rotational behavior characterized by the 'skew angle' denotes the direction of rotation of the Poynting vector reflecting the orbital angular momentum effect of the vortex beams, while the rotation behavior discussed in this article is a kind of 'energy distribution' rotation essentially coming from the inconstant phase gradient of the X-type vortex, which, as we discussed, cannot be observed in a traditional (canonical) vortex beam. Furthermore, the energy flow in all of the above cases rotates in a counterclockwise manner due to the positive charge in the X-type vortex. If the topological charge of the X-type vortex is negative, i.e., $\sigma_c < 0$, and the charge is equal to $-l$ ($l \in \mathbb{N}$), the energy flow will rotate clockwise.

3.3. *Transverse Energy Flow in the Focal Plane*

The transverse energy flow $\mathbf{p}_{xy}$, including the flow lines (or the Poynting vectors) and the energy distribution, in virtue of the X-type vortex, also can exhibit interesting structures. Here, we will focus on the focal plane to examine these structures of the transverse energy flow.

First, we will look at the case of $l = 1$. Figure 13 illustrates the transverse energy flow on the focal plane for different values of the anisotropic parameter σ_c, where the energy strength is also denoted by color, and the black flow lines and the white arrows represent the energy flow lines and the transverse Poynting vectors, respectively. It can be found, on one hand, that, when $\sigma_c < 1$ (plots (a) and (b)) there are two energy maxima located on the x_s-axis, while, as σ_c increases from 0.1, 0.5, 1, and 2 to 10, the energy maxima move from the x_s-axis to the y_s-axis, and, especially in the case of conventional vortex ($\sigma_c = 1$), the transverse energy is distributed uniformly along the azimuthal direction, i.e., there are no longer any maximum points. On the other hand, the flow lines for the conventional vortex ($\sigma_c = 1$) rotate around the beam center azimuthally and form circular shaped trajectories, while, when $\sigma_c \neq 1$, i.e., for the X-type vortex, the trajectories of these flow lines around the beam center become elliptical. More specifically, the energy flow lines near the outer sides of the maxima (such as the flow lines near $|x_s| > 0.8\lambda$ in plot (a), and the flow lines near $|y_s| > 0.8\lambda$ in plot €) derivate from the azimuthal trajectories around the beam center in the case of $\sigma_c \neq 1$, which implies that, although the topological charge of the energy flow in the beam center does not change in virtue of the X-type vortex for $l = 1$, the new Poynting singularities will be formed as σ_c changes, i.e., the topological structure of the energy flow on the focal plane is changed. This redistribution of the topological structure is more obvious and typical as l gets bigger, and, here, we adopt the case of $l = 2$ as an example for further analysis, which is depicted in Figure 14.

The transverse energy flow $\mathbf{p}_{xy}$ (with flow lines and vectors) in the focal plane with the anisotropic parameter changing from 0.1, 0.5, 1, and 2 to 10 for $l = 2$ is drawn in Figure 14. It can be seen that there exist three main singular points in the focal plane, the 'original' Poynting singularity at the beam center, 'O', and two (constructed) off-axis singularities 'A' and 'B'. When $\sigma_c = 0.1$ (plot (a)), points A and B are located on the x_s-axis with topological charge $+1$, while the point O has a charge of -1. As σ_c increases to 0.5, the two off-axis singularities A and B move closer to the beam center, and, when σ_c arrives at 1, the points A and B merge with original point O, resulting into a new singular point O_+ with topological charge $+1$, as shown in plot (c). This process obeys the conservation law of topological events. While, as σ_c continues to increase, the singular point O_+ splits into three singularities again, the off-axis singularities A and B no longer exist on the x_s-axis; instead, they are located on the y_s-axis.

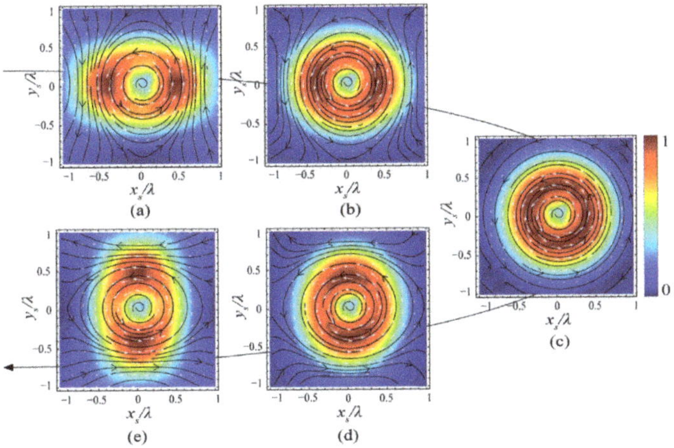

Figure 13. The transverse energy flow in the focal plane for $l = 1$: (**a**) $\sigma_c = 0.1$; (**b**) $\sigma_c = 0.5$; (**c**) $\sigma_c = 1$; (**d**) $\sigma_c = 2$; (**e**) $\sigma_c = 10$.

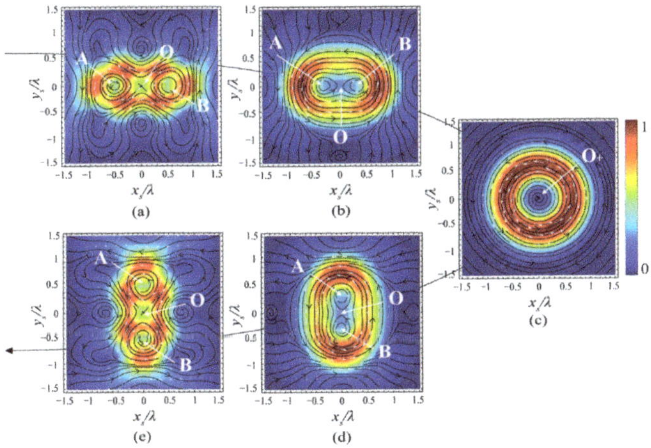

Figure 14. The transverse energy flow on the focal plane for $l = 2$: (**a**) $\sigma_c = 0.1$; (**b**) $\sigma_c = 0.5$; (**c**) $\sigma_c = 1$; (**d**) $\sigma_c = 2$; (**e**) $\sigma_c = 10$.

This topological event can be observed more obviously in a phase-type figure. Here, we define a complex transverse Poynting field $p_{xy}^{(c)}$ as:

$$p_{xy}^{(c)} = p_x + \mathrm{i}p_y, \tag{16}$$

where p_x and p_y are still the x- and y-Poynting components, respectively. By applying the equivalence of the vector field and its corresponding complex field, the topological structure of the Poynting vector field can also be described by the phase-structure of $p_{xy}^{(c)}$. Here, the phase singularity of $p_{xy}^{(c)}$ is equivalent to the vector singularity of the Poynting vector $\mathbf{p}_{xy}$, and, also, the topological charge of this phase singularity is equal to the charge of the vector singularity. The contour plot of the phase of the complex transverse Poynting field, $\arg\left[p_{xy}^{(c)}\right]$ (here $\arg[\cdot]$ means the argument or the phase of $p_{xy}^{(c)}$), is illustrated in Figure 15, where the intersections of different contours indicate the phase singularities of $p_{xy}^{(c)}$, which also means the Poynting (vector) singularity of the transverse energy flow $\mathbf{p}_{xy}$. In order to observe the topological behavior more clearly, the plots with $\sigma_c = 0.3$ and $\sigma_c = 3$

are added. The three main singular points A, B, and O are marked out in this figure, and it is easy to see that, as σ_c increases, the points A and B, both with charge $+1$, move from the x_s-axis to merge with point O with charge -1, then to be created again on the y_s-axis. In addition to these three singularities and their topological reaction, we can also find other singularities in Figure 15, i.e., the points at the contour intersections around points A, B, and O, namely 'surrounding singularities.' As σ_c increases (Figure 15a–d), it can be seen that these surrounding singularities will gradually annihilate each other and disappear when $\sigma_c = 1$ (plot (d)), and, as σ_c continues to increase (Figure 15e–g), these surrounding singularities emerge again with their positions having a 90° rotation. More specifically, it is found that, for $\sigma_c = 1$, there is no surrounding singularity; instead, there exists a 'singularity ring,' i.e., an edge-type singularity (denoted in white in Figure 15d). This means that the multi vortex-type singularities of the energy flow appearing in the case of the X-type vortex will degenerate into a simple edge-type singularity for a canonical vortex. Thus far, the topological structures and the related reactions in the focal plane have been observed and analyzed. It is found that the topological structures change greatly with the topological reaction on account of the X-type vortex. Consequently, the transverse energy and flow directions are re-distributed, which implies that, by adjusting the anisotropic parameter σ_c, one can realize the manipulation of the transverse energy flows in the focal plane. Since the particle will move along the energy flow direction, the multi-vortex-type singularities of the transverse energy flow may provide a method to trap/rotate particles in the focal plane and transport them from one axis to another. It is well-known that the energy distribution strongly depends on the singular points; therefore, the topological reactions also mean that the complicated topological structures and related the transverse energy distribution in the transverse planes move nearer to the focus, which may be the reason for the abnormal rotation discussed in previous section.

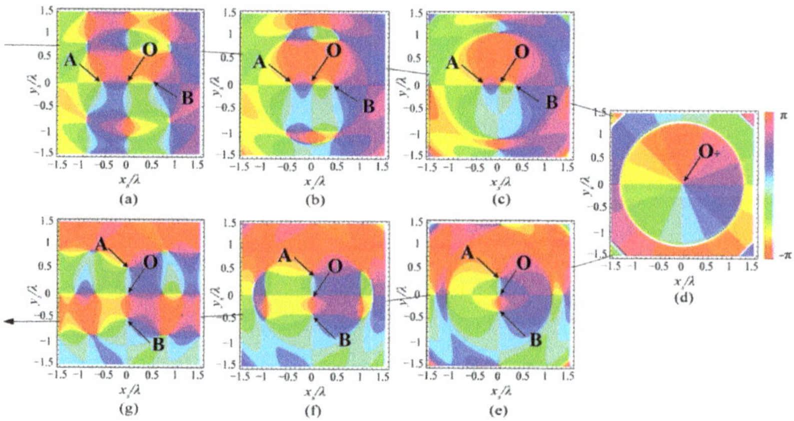

Figure 15. Contours of the phase of complex transverse Poynting field, arg$\left[p_{xy}^{(c)}\right]$ on the focal plane for $l = 2$: **(a)** $\sigma_c = 0.1$; **(b)** $\sigma_c = 0.3$; **(c)** $\sigma_c = 0$; **(d)** $\sigma_c = 1$; **(e)** $\sigma_c = 2$; **(f)** $\sigma_c = 3$; **(g)** $\sigma_c = 10$.

4. Conclusions

In this article, the energy flow in a 3D vector field constructed by strongly focusing azimuthally-polarized beams with an X-type vortex is studied. It is found that, in virtue of the X-type vortex, the energy flow, including the longitudinal component and the transverse component, can rotate along the propagation direction. By adjusting the anisotropic parameter σ_c and the vortex order l of the X-type vortex, the location of the energy maxima and the rotation angle can be manipulated. Different from the longitudinal energy flow, the transverse energy flow will rotate inversely (i.e., in a clockwise manner) in a very short propagation distance near the focus, which may be caused by the complicated topological structures in that range. The transverse energy flow in the focal plane is also discussed, and it is found that the energy distribution can show very distinguishing patterns as the anisotropic parameter σ_c varies. Through defining a complex transverse Poynting field and applying the equivalence principle in singular optics, the topological behaviors of the transverse energy flow are analyzed, which shows that, instead of the simple edge-type singularity existing in the canonical vortex case ($\sigma_c = 1$), many vortex-type singularities emerge in the X-type vortex case, and their locations and topological reactions are determined by the anisotropic parameter σ_c. This research not only explores the physical properties of the X-type vortex, but also provides a method to construct a rotating energy flow with beam propagating and to form tunable energy flows in the focal plane, which may have applications in optical manipulations, such as rotating particles along the longitudinal or transverse directions, and in controlling the chirality of nanostructures [44,45].

Author Contributions: Conceptualization, X.P.; software, H.Z.; validation, H.Z. and T.Z.; formal analysis, X.P. and X.Z.; investigation, X.P. and H.Z.; writing—original draft preparation, H.Z. and T.Z.; writing—review and editing, X.P. and H.Z.; visualization, H.Z., T.Z. and X.Z. All authors have read and agreed to the published version of the manuscript.

Funding: This research was funded by the National Natural Science Foundation of China (NSFC) (11974281, 12104283), and by the Fundamental Research Funds for the Central Universities (No. GK202103021).

Institutional Review Board Statement: Not applicable.

Informed Consent Statement: Not applicable.

Data Availability Statement: The data presented in this study are available on request from the corresponding author.

Conflicts of Interest: The authors declare no conflict of interest.

References

1. Ignatowskii, V.S. Diffraction by lens of arbitrary aperture. *Trans. Opt.* **1919**, *1*, 1–36. [CrossRef]
2. Richards, B.; Wolf, E. Electromagnetic diffraction in optical systems, II. Structure of the image field in an aplanatic system. *Proc. R. Soc. Lond. A* **1959**, *253*, 358–379. [CrossRef]
3. Berry, M.V. Optical currents. *J. Opt. A Pure Appl. Opt.* **2009**, *11*, 094001. [CrossRef]
4. Bekshaev, A.; Bliokh, K.Y.; Soskin, M. Internal flows and energy circulation in light beams. *J. Opt.* **2011**, *13*, 053001. [CrossRef]
5. Angelsky, O.V.; Bekshaev, A.Y.; Maksimyak, P.P.; Maksimyak, A.P.; Hanson, S.G.; Zenkova, C.Y. Orbital rotation without orbital angular momentum: Mechanical action of the spin part of the internal energy flow in light beams. *Opt. Express* **2012**, *20*, 3563. [CrossRef] [PubMed]
6. Otte, E.; Asché, E.; Denz, C. Shaping optical spin flow topologies by the translation of tailored orbital phase flow. *J Opt.* **2019**, *21*, 064001. [CrossRef]
7. Schouten, H.F.; Visser, T.D.; Lenstra, D. Optical vortices near sub-wavelength structures. *J. Opt. B Quantum Semiclassical Opt.* **2004**, *6*, S404. [CrossRef]
8. Gbur, G.J. *Singular Optics*; Chemical Rubber Company: Boca Raton, FL, USA, 2017.
9. Friese, M.E.J.; Enger, J.; Rubinsztein-Dunlop, H.; Heckenberg, N. Optical angular-momentum transfer to trapped absorbing particles. *Phys. Rev. A* **1996**, *54*, 1593. [CrossRef]
10. Tkachenko, G.; Brasselet, E. Optofluidic sorting of material chirality by chiral light. *Nat. Commun.* **2014**, *5*, 1–7. [CrossRef]
11. Gao, D.; Novitsky, A.; Zhang, T.; Cheong, F.C.; Gao, L.; Lim, C.T. Unveiling the correlation between non-diffracting tractor beam and its singularity in Poynting vector. *Laser Photonics Rev.* **2015**, *9*, 75–82. [CrossRef]
12. Kotlyar, V.V.; Stafeev, S.S.; Nalimov, A.G. Energy backflow in the focus of a light beam with phase or polarization singularity. *Phys. Rev. A* **2019**, *99*, 033840. [CrossRef]
13. Stafeev, S.S.; Kotlyar, V.V.; Nalimov, A.G.; Kozlova, E.S. The non-vortex inverse propagation of energy in a tightly focused high-order cylindrical vector beam. *IEEE Photon. J.* **2019**, *11*, 1–10. [CrossRef]
14. Khonina, S.N.; Ustinov, A.V. Increased reverse energy flux area when focusing a linearly polarized annular beam with binary. *Opt. Lett.* **2019**, *44*, 2008–2011. [CrossRef]
15. Wang, H.; Hao, J.; Zhang, B.; Han, C.; Zhao, C.; Shen, Z.; Xu, J.; Ding, J. Donut-like photonic nanojet with reverse energy flow. *Chin. Opt. Lett.* **2021**, *19*, 102602. [CrossRef]
16. Zhou, J.; Ma, H.; Zhang, Y.; Zhang, S.; Min, C.; Yuan, X. Energy flow inversion in an intensity-invariant focusing field. *Opt. Lett.* **2022**, *47*, 1494–1497. [CrossRef] [PubMed]

17. Man, Z.; Bai, Z.; Zhang, S.; Li, X.; Li, J.; Ge, X.; Zhang, Y.; Fu, S. Redistributing the energy flow of a tightly focused radially polarized optical field by designing phase masks. *Opt. Lett.* **2018**, *26*, 23935. [CrossRef]
18. Gong, L.; Wang, X.; Zhu, Z.; Lai, S.; Feng, H.; Wang, J.; Gu, B. Transversal energy flow of tightly focused off-axis circular polarized vortex beams. *Appl. Opt.* **2022**, *61*, 5076–5082. [CrossRef]
19. Gao, X.; Pan, Y.; Zhang, G.; Zhao, M.; Ren, Z.; Tu, C.; Li, Y.; Wang, H. Redistributing the energy flow of tightly focused ellipticity-variant vector optical fields. *Photonics Res.* **2017**, *5*, 640–648. [CrossRef]
20. Man, Z.; Li, X.; Zhang, S.; Bai, Z.; Lyu, Y.; Li, J.; Ge, X.; Sun, Y.; Fu, S. Manipulation of the transverse energy flow of azimuthally polarized beam in tight focusing system. *Opt. Commun.* **2019**, *431*, 174–180. [CrossRef]
21. Pan, Y.; Gao, X.; Zhang, G.; Li, Y.; Tu, C.; Wang, H. Spin angular momentum density and transverse energy flow of tightly focused kaleidoscope-structured vector optical fields. *APL Photonics* **2019**, *4*, 096102. [CrossRef]
22. Kotlyar, V.V.; Stafeev, S.S.; Kovalev, A.A. Reverse and toroidal flux of light fields with both phase and polarization higher-order singularities in the sharp focus area. *Opt. Express* **2019**, *27*, 16689–16702. [CrossRef] [PubMed]
23. Wang, S.; Cao, H.; Sun, J.; Qin, F.; Cao, Y.; Li, X. Subwavelength generation of orientation-unlimited energy flow in 4π microscopy. *Opt. Express* **2022**, *30*, 138–145. [CrossRef] [PubMed]
24. Soskin, M.S.; Vasnetsov, M.V. *Singular Optics Progress in Optics*; Wolf, E., Ed.; Elsevier: Amsterdam, The Netherlands, 2001; Volume 42, pp. 219–276.
25. Shen, Y.; Wang, X.; Xie, Z.; Min, C.; Fu, X.; Liu, Q.; Gong, M.; Yuan, X. Optical vortices 30 years on: OAM manipulation from topological charge to multiple singularities. *Light Sci. Appl.* **2019**, *8*, 1–29. [CrossRef]
26. Liu, X.; Sun, C.; Deng, D. Propagation properties and radiation force of circular Airy Gaussian vortex beams in strongly nonlocal nonlinear medium. *Chin. Phys. B* **2021**, *30*, 024202. [CrossRef]
27. Polimeno, P.; Magazzu, A.; Iati, M.; Patti, F.; Saija, R.; Boschi, C.; Donato, M.; Gucciardi, P.; Jones, P.; Volpe, G.; et al. Optical tweezers and their applications. *J. Quant. Spectrosc. Radiat. Transf.* **2018**, *218*, 131–150. [CrossRef]
28. Willner, A.E.; Huang, H.; Yan, Y.; Ren, Y.; Ahmed, N.; Xie, G.; Bao, C.; Li, L.; Cao, Y.; Zhao, Z.; et al. Optical communications using orbital angular momentum beams. *Adv. Opt. Photon.* **2015**, *7*, 66–106. [CrossRef]
29. Tamburini, F.; Anzolin, G.; Umbriaco, G.; Bianchini, A.; Barbieri, C. Overcoming the Rayleigh criterion limit with optical vortices. *Phys. Rev. Lett.* **2006**, *97*, 163903. [CrossRef]
30. D'ambrosio, V.; Spagnolo, N.; Re, L.D.; Slussarenko, S.; Li, Y.; Kwek, L.C.; Marrucci, L.; Walborn, S.P.; Aolita, L.; Sciarrino, F. Photonic polarization gears for ultra-sensitive angular measurements. *Nat. Commun.* **2013**, *4*, 1–8. [CrossRef]
31. Zhou, Z.; Zhu, L. STED microscopy based on axially symmetric polarized vortex beams. *Chin. Phys. B* **2016**, *25*, 030701. [CrossRef]
32. Freund, I.; Shvartsman, N.; Freilikher, V. Optical dislocation networks in highly random media. *Opt. Commun.* **1993**, *101*, 247–264. [CrossRef]
33. Roux, F.S. Coupling of noncanonical optical vortices. *J. Opt. Soc. Am. B* **2004**, *21*, 664–670. [CrossRef]
34. Maji, S.; Brundavanam, M.M. Controlled noncanonical vortices from higher-order fractional screw dislocations. *Opt. Lett.* **2017**, *42*, 2322–2325. [CrossRef] [PubMed]
35. Yin, C.; Kan, X.; Dai, H.; Shan, M.; Han, Q.; Cao, Z. Propagation dynamics of off-axis noncanonical vortices in a collimated Gaussian beam. *Chin. Phys. B* **2019**, *28*, 034205. [CrossRef]
36. Pang, X.; Xiao, W.; Zhang, H.; Feng, C.; Zhao, X.Y. X-type vortex and its effect on beam shaping. *J. Opt.* **2021**, *23*, 125604. [CrossRef]
37. Alonzo, C.A.; Rodrigo, P.J.; Glückstad, J. Helico-conical optical beams: A product of helical and conical phase fronts. *Opt. Express* **2005**, *13*, 1749–1760. [CrossRef] [PubMed]
38. Zhang, Y.; Li, P.; Liu, S.; Han, L.; Cheng, H.; Zhao, J. Manipulating spin-dependent splitting of vector abruptly autofocusing beam by encoding cosine-azimuthal variant phases. *Opt. Express* **2016**, *24*, 28409–28418. [CrossRef] [PubMed]
39. Ji, W.; Lee, C.H.; Chen, P.; Hu, W.; Ming, Y.; Zhang, L.; Lin, T.H.; Chigrinov, V.; Lu, Y.Q. Meta-q-plate for complex beam shaping. *Sci. Rep.* **2016**, *6*, 25528. [CrossRef] [PubMed]
40. Pang, X.; Visser, T.D.; Wolf, E. Phase anomaly and phase singularities of the field in the focal region of high-numerical aperture systems. *Opt. Commun.* **2011**, *284*, 5517–5522. [CrossRef]
41. Zhang, J.; Pang, X.; Ding, J. Wavefront spacing and Gouy phase in strongly focused fields: The role of polarization. *J. Opt. Soc. Am. A* **2017**, *34*, 1132–1138. [CrossRef]
42. Padgett, M.J.; Allen, L. The Poynting vector in Laguerre-Gaussian laser modes. *Opt. Commun.* **1995**, *121*, 36–40. [CrossRef]
43. Qiu, S.; Ren, Y.; Liu, T.; Liu, Z.; Wang, C.; Ding, Y.; Sha, Q.; Wu, H. Directly observing the skew angle of a Poynting vector in an OAM carrying beam via angular diffraction. *Opt. Lett.* **2021**, *46*, 3484–3487. [CrossRef] [PubMed]
44. Toyoda, K.; Miyamoto, K.; Aoki, N.; Morita, R.; Omatsu, T. Using optical vortex to control the chirality of twisted metal nanostructures. *Nano Lett.* **2012**, *12*, 3645–3649. Available online: https://pubs.acs.org/doi/10.1021/nl301347j (accessed on 23 November 2022). [CrossRef] [PubMed]
45. Syubaev, S.; Zhizhchenko, A.; Vitrik, O.; Porfirev, A.; Fomchenkov, S.; Khonina, S.; Kudryashov, S.; Kuchmizhak, A. Chirality of laser-printed plasmonic nanoneedles tunable by tailoring spiral-shape pulses. *Appl. Surf. Sci.* **2019**, *470*, 526–534. [CrossRef]

photonics

Communication

Poincaré Beams at the Tight Focus: Inseparability, Radial Spin Hall Effect, and Reverse Energy Flow

Victor V. Kotlyar [1,2], Sergey S. Stafeev [1,2,*], Vladislav D. Zaitsev [1,2] and Alexey M. Telegin [2]

1 Image Processing Systems Institute of the RAS—Branch of FSRC "Crystallography & Photonics" of the RAS, 151 Molodogvardeyskaya St., 443001 Samara, Russia
2 Samara National Research University, 34 Moskovskoe Shosse, 443086 Samara, Russia
* Correspondence: sergey.stafeev@gmail.com

Abstract: The family of Poincaré beams has three parameters, including two real-valued angular parameters, which specify a definite polarization state on the Poincaré sphere, and a third integer parameter n specifying the beam singularity order. We show theoretically and through a numerical simulation that, while being inseparable and not allowing for the separation of polarization and orbital degrees of freedom in the general case, the Poincaré beams display remarkable properties when tightly focused. We find that at $n = 2$, a reverse energy flow occurs near the optical axis, which is mathematically expressed as the negative projection of the Poynting vector. We also reveal that given certain parameters of the Poincaré beams, the energy flow rotates around the optical axis due to spin–orbital conversion. We also reveal a radial optical Hall effect that occurs at the tight focus of Poincaré beams, when the on-axis components of the spin angular momentum vector have different signs on certain different-radius circles centered at the focal spot center.

Keywords: Poincaré beams; sharp focus; inseparability; radial spin Hall effect; spin angular momentum; orbital angular momentum; reverse energy flow

Citation: Kotlyar, V.V.; Stafeev, S.S.; Zaitsev, V.D.; Telegin, A.M. Poincaré Beams at the Tight Focus: Inseparability, Radial Spin Hall Effect, and Reverse Energy Flow. *Photonics* **2022**, *9*, 969. https://doi.org/10.3390/photonics9120969

Received: 18 November 2022
Accepted: 8 December 2022
Published: 11 December 2022

1. Introduction

Among the recipients of the 2022 Nobel Prize in physics, A. Zeilinger was recognized for research in quantum entanglement between photons. By way of illustration, his work [1] has amassed 6000+ citations. As far as photons are concerned, (hybrid) quantum entanglement (in terms of both polarization and angular momentum) implies that two photons are in superposition, and can be described by the expression [2]:

$$|\psi\rangle = a|H\rangle|l\rangle + e^{i\delta}\sqrt{1-a^2}|V\rangle|-l\rangle,$$

where H and V stand for the horizontal and vertical polarization vectors, l is the orbital angular momentum (OAM) of photon or topological charge, a is the weight coefficient (probability density of detecting the photon state), and δ is the phase delay. At $a = 1/\sqrt{2}$, the state $|\psi\rangle$ becomes fully entangled. It turns out that similar entanglement may be found in classical physics and refers to a light field described by the on-axis superposition of two vortex laser beams with different polarization states and different topological charges. Physically, a classical entangled state is when the light field is impossible to describe as a direct product $|P\rangle|OAM\rangle$, with no way to simultaneously determine a particular polarization state $|P\rangle$ and a particular OAM state $|OAM\rangle$ of the light field of interest. Hence, the light field described by the aforementioned superposition is ill-defined. In this work, we discuss an example of ill-defined light fields in the form of a wide class Poincaré beams [3,4], which include linearly and circularly polarized vortex laser beams and cylindrical vector beams [5]. Below, we show that given specific parameters, remarkable optical effects occur at the tight focus of such beams, including a reverse on-axis energy flow [6,7] and an optical Hall effect [8]. The optical Hall effect is divided into spin (SHE) [9–13], orbital [14], and spin–orbital [15] Hall effects. An optical Hall effect occurs due to the conservation of angular momentum of light and spin–orbit interaction. As a rule, the optical Hall effect occurs because of the light–medium interaction, for instance when light reflects at a metasurface [16]. What we are more interested in is studying the SHE that occurs without the involvement of medium and microparticles, but by tightly focusing laser beams propagating in free space (see review [17]). For instance, an off-axis focal spot has been shown to be generated by passing a tightly focused vortex laser beam through an aperture [18]. In this work, analyzing the free-space propagation of Poincaré beams, we show that given specific parameters, there is a transverse energy flow at the tight focus, rotating about the optical axis clockwise or anticlockwise. We also reveal that at the tight focus, a radial spin Hall effect occurs, when on-axis spin angular momentum (SAM) projections (that may be looked considered as 'spin') have different signs on different-radius circles centered at the optical axis. That is, on the different-radius circles, vectors of elliptic polarization rotate oppositely.

2. Inseparability of Vector and Spatial Degrees of Freedom

In a previous work [19], we derived expressions to describe an intensity pattern from Poincaré beams at the tight focus. In this work, our purpose is to derive expressions for Poynting vector projections and an on-axis projection of the SAM at the tight focus of the Poincaré beam. However, beforehand it makes sense to discuss the topic of inseparability (entanglement) of polarization and spatial degrees of freedom of the Poincaré beam.

The light beams for which the polarization state is described by unit vectors on a Poincaré sphere are described by a Jones vector given by [3,4]:

$$\mathbf{E}_P(\varphi) = \frac{1}{\sqrt{2}} \begin{pmatrix} ae^{-in\varphi} + be^{in\varphi} \\ iae^{-in\varphi} - ibe^{in\varphi} \end{pmatrix} \tag{1}$$

where $a = \cos(\theta/2)e^{-i\psi/2}$, $b = \sin(\theta/2)e^{i\psi/2}$, $a^2 + b^2 = 1$, θ and ψ are the polar and azimuthal angles on the sphere, and n is an integer number that describes the topological charge of the optical vortex or the order of cylindrical (azimuthal or radial) polarization. Beam (1) can be represented as the superposition of two beams where one has right-handed and the other has left-handed polarization:

$$\frac{1}{\sqrt{2}} \begin{pmatrix} ae^{-in\varphi} + be^{in\varphi} \\ iae^{-in\varphi} - ibe^{in\varphi} \end{pmatrix} = \frac{ae^{-in\varphi}}{\sqrt{2}} \begin{pmatrix} 1 \\ i \end{pmatrix} + \frac{be^{in\varphi}}{\sqrt{2}} \begin{pmatrix} 1 \\ -i \end{pmatrix} = ae^{-in\varphi}|R\rangle + be^{in\varphi}|L\rangle, \tag{2}$$

where $|R\rangle$ and $|L\rangle$, respectively, denote Jones vectors for the right- and left-handed polarization. Considering that $a^2 + b^2 = 1$, Equation (2) can be rearranged as:

$$\mathbf{E}_p(\varphi) = \sqrt{\gamma}e^{-in\varphi}|R\rangle + \sqrt{1-\gamma}e^{in\varphi}|L\rangle, \quad \gamma = a^2. \tag{3}$$

The two quantities are said to be inseparable [20–22] when it not possible to represent the vector field as the direct product of the spatial scalar function by a Jones vector that describes the polarization state of the field. In other words, the vector field of inseparable quantities is not able to be represented as $U(r,\varphi,z)|P\rangle$, where $U(r,\varphi,z)$ is a scalar function that describes the solution of a paraxial Helmholtz equation and $|P\rangle$ is the state of transverse polarization of a paraxial light beam.

The entanglement notion is a classical analog of quantum entanglement [1,2,20–22]. The degree of inseparability is defined based on Neumann's entropy *Ent*, which, in a simple case of Equation (3), is given by:

$$Ent(E_p) = -[\gamma \log_2(\gamma) + (1-\gamma)\log_2(1-\gamma)] \tag{4}$$

Entropy in Equation (4) lies in the range $0 < Ent(E_p) < 1$. If entropy is zero, the vector field is fully separable; if entropy equals unity, the vector field is fully (maximally) inseparable. At $\gamma = 1/2$, the Poincaré beam is maximally inseparable. In this case, $a = b = 1/\sqrt{2}$, which means that the Poincaré beam with maximal inseparability is identical to an n-th-order cylindrical vector beam:

$$\mathbf{E}_P(\varphi; a = b = 1/\sqrt{2}) = \begin{pmatrix} \cos(n\varphi) \\ \sin(n\varphi) \end{pmatrix}. \tag{5}$$

At other values of a and b, the Poincaré beams in Equation (1) are partially inseparable, and if one of the parameters equals zero (either $a = 0$ or $b = 0$), the Poincaré beams become separable circularly polarized vortex beams. Hereinafter, we analyze what happens with beam (1) at the tight focus.

We note that exact solutions of a Helmholtz equation in spherical coordinates, as well as all six projections of the electric and magnetic fields for non-paraxial laser modes, were deduced in Ref. [23]. A linear combination of such beams produces hybrid vector beams that may be looked upon as an example of beams with partial inseparability.

3. Flow Energy at the Tight Focus of the Poincaré Beam

Using Richards–Wolf formalism [24], projections of the electric and magnetic fields at the tight focus of the Poincaré beam (1) can be derived in the form:

$$\begin{aligned}
E_x &= \frac{i^{n-1}}{\sqrt{2}}\left[\left(be^{in\varphi} + ae^{\ in\varphi}\right)I_{0,n} + \left(be^{i(n-2)\varphi} + ue^{-i(n-2)\varphi}\right)I_{2,n-2}\right], \\
E_y &= \frac{i^{n}}{\sqrt{2}}\left[\left(ae^{-in\varphi} - be^{in\varphi}\right)I_{0,n} - \left(ae^{-i(n-2)\varphi} - be^{i(n-2)\varphi}\right)I_{2,n-2}\right], \\
E_z &= \sqrt{2}i^{n}\left(be^{i(n-1)\varphi} + ae^{-i(n-1)\varphi}\right)I_{1,n-1}, \\
H_x &= \frac{i^{n}}{\sqrt{2}}\left[\left(be^{in\varphi} - ae^{-in\varphi}\right)I_{0,n} + \left(be^{i(n-2)\varphi} - ae^{-i(n-2)\varphi}\right)I_{2,n-2}\right], \\
H_y &= \frac{i^{n-1}}{\sqrt{2}}\left[\left(be^{in\varphi} + ae^{-in\varphi}\right)I_{0,n} - \left(be^{i(n-2)\varphi} + ae^{-i(n-2)\varphi}\right)I_{2,n-2}\right], \\
H_z &= \sqrt{2}i^{n+1}\left(be^{i(n-1)\varphi} - ae^{-i(n-1)\varphi}\right)I_{1,n-1}.
\end{aligned} \tag{6}$$

In Equation (6), the notation $I_{\nu,\mu}$ stands for integrals that depend just on the radial variable r:

$$I_{\nu,\mu} = \left(\frac{4\pi f}{\lambda}\right)\int_0^{\theta_0} \sin^{\nu+1}\left(\frac{\theta}{2}\right)\cos^{3-\nu}\left(\frac{\theta}{2}\right)\cos^{1/2}(\theta)A(\theta)e^{ikz\cos\theta}J_\mu(x)d\theta, \tag{7}$$

where f is the focal length of an aplanatic optical system (an ideal spherical lens), λ is the wavelength, $NA = \sin\theta_0$ is the numerical aperture, $J_\mu(x)$ is the Bessel function of the first kind and μth order, $x = kr\sin\theta$, and (r, φ, z) are the cylindrical coordinates. The function $A(\theta)$ can be in the form of Gaussian or Bessel–Gaussian function or a constant (plane wave).

Using Equation (6), we can write a relationship for the intensity distribution across the focal spot:

$$\begin{aligned}
I(r,\varphi) &= |E_x|^2 + |E_y|^2 + |E_z|^2 \\
&= I_{0,n}^2 + I_{2,n-2}^2 + 2I_{1,n-1}^2 + 2\sin\theta\cos(2(n-1)\varphi + \psi)\left(I_{0,n}I_{2,n-2} + 2I_{1,n-1}^2\right).
\end{aligned} \tag{8}$$

From Equation (8), it is seen that at $\sin\theta = 0$, the intensity pattern at the focus is radially symmetric and independent of the polar angle φ. Specifically, at $n = 0, 1, 2$, the focal spot forms a circle, but at $n > 2$ or $n < 0$, the focal spot forms a ring. At $\sin\theta \neq 0$, the focal spot is devoid of radial symmetry, depending on the polar angle φ and index n in Equation (8).

By using relationships for the field projections in Equation (6), we obtain Poynting vector projections (energy flow density) in the form [24]:

$$\mathbf{P} = \frac{c}{2\pi}\mathrm{Re}(\mathbf{E}* \times \mathbf{H}) \tag{9}$$

where **E** and **H** are the electric and magnetic field vectors, * denotes complex conjugation, × indicates the vector product, and c is the speed of light in vacuum. Hereinafter, the constant $c/(2\pi)$ is neglected. Substituting the relationships for electromagnetic field projections at the focus in Equation (6) into Equation (9) yields:

$$\begin{aligned} P_x &= 2\left(|a|^2 - |b|^2\right)\sin(\varphi) I_{1,n-1}(I_{0,n} + I_{2,n-2}), \\ P_y &= -2\left(|a|^2 - |b|^2\right)\cos(\varphi) I_{1,n-1}(I_{0,n} + I_{2,n-2}), \\ P_z &= I_{0,n}^2 - I_{2,n-2}^2. \end{aligned} \quad (10)$$

In view of the relationships for the parameters a and b from Equation (1), the Poynting vector projections of Equation (10) at the focus can be rewritten as:

$$\begin{aligned} P_x &= 2\cos(\theta)\sin(\varphi) I_{1,n-1}(I_{0,n} + I_{2,n-2}), \\ P_y &= -2\cos(\theta)\cos(\varphi) I_{1,n-1}(I_{0,n} + I_{2,n-2}), \\ P_z &= I_{0,n}^2 - I_{2,n-2}^2. \end{aligned} \quad (11)$$

In Equation (11), the angle θ is a parameter that defines a point on the Poincaré sphere of polarization states. The transverse Poynting vector projections in Equation (11) can be converted from the Cartesian to the polar coordinates, with Equation (11) transformed into:

$$\begin{aligned} P_r &= 0, \\ P_\varphi &= -\cos(\theta) I_{1,n-1}(I_{0,n} + I_{2,n-2}), \\ P_z &= I_{0,n}^2 - I_{2,n-2}^2. \end{aligned} \quad (12)$$

An analysis of Equation (12) affords the following conclusions. With the on-axis Poynting vector component depending only on the radial coordinate, the distribution of the energy flow propagating along the optical axis becomes circularly symmetric for any parameters (a, b) or (θ, ψ). Interestingly, at $n = 2$, it follows from Equation (12) that near the optical axis in the focal plane, a reverse energy flow occurs:

$$\begin{aligned} P_{z,2}(r) &= -I_{2,0}^2 + I_{0,2}^2, \\ P_{z,2}(r=0) &= -I_{2,0}^2 < 0. \end{aligned} \quad (13)$$

The second equality in Equation (13) stems from the fact that among the integrals entering Equation (7), only those with zero second index ($\mu = 0$) remain non-zero on the optical axis ($r = 0$). This is because only the zero-order Bessel function is non-zero at the origin: $J_0(0) = 1$. Equation (13) suggests that an on-axis reverse flow will occur when tightly focusing a multitude of initial optical fields, including a second-order optical vortex with left-handed circular polarization, $\exp(i2\varphi)(1, -i)^T$, where T denotes vector transposition, a minus second-order optical vortex with right-handed circular polarization $\exp(-i2\varphi)(1,i)^T$, a second-order radially polarized vector field $(\cos(2\varphi), \sin(2\varphi))^T$, a second-order azimuthally polarized optical field, $(-\sin(2\varphi), \cos(2\varphi))^T$, and so on. Previously, the reverse energy flow at the tight focus was discussed in Refs. [6,7].

From Equation (12), it also follows that at $\cos\theta \neq 0$, the near-axis transverse energy flow at the focus is rotating clockwise at $\cos\theta > 0$ and anticlockwise at $\cos\theta < 0$.

At $\cos\theta = 0$, the energy flow at the focal plane is laminar (the transverse Poynting vector component is zero) and not rotating. Given that $\cos\theta \neq 0$ (meaning that elliptic polarization is found in the source field), the transverse energy flow at the tight focus of Poincaré beams will rotate around the optical axis at any topological charge n due to spin–orbital conversion. At $\cos\theta > 0$ and $n = 0$, the energy flow will rotate anticlockwise and clockwise at $n > 0$.

4. Spin Angular Momentum at the Tight Focus of Poincaré Beams

In this section, we derive the spin angular momentum (SAM) of the Poincaré beam (1) at the focus. The spin density vector is calculated using a familiar formula [25]:

$$\mathbf{S} = \frac{1}{8\pi\omega}\mathrm{Im}(\mathbf{E} * \times \mathbf{E}) \tag{14}$$

where ω is the cyclic frequency of light. Hereinafter, the constant $1/(8\pi\omega)$ is neglected. From (14), the on-axis SAM component (without regard for the constant) is seen to coincide with the third Stokes vector component, s_3:

$$s_3 = S_z = 2\mathrm{Im}(E_x^* E_y) \tag{15}$$

Thus, we infer that it is the on-axis SAM component that indicates the field at the focus is circularly or elliptically polarized. Substituting the electric field projections (6) into (15) yields:

$$S_z = \left(|a|^2 - |b|^2\right)\left[I_{0,n}^2 - I_{2,n-2}^2\right] \tag{16}$$

For the source Poincaré field (1), the on-axis SAM component is given by:

$$S_{z,in} = |a|^2 - |b|^2 = \cos\theta. \tag{17}$$

From Equations (16) and (17), the initial homogeneous spin density, defined by $\cos\theta$, is seen at the focus to become inhomogeneous, as well as changing sign and taking a zero value at certain radii. The radius-dependent separation of regions with the opposite sign of spin (16) may be interpreted as a radial spin Hall effect for all n. If, in the initial plane, the spin density is zero ($\cos\theta = 0$), then at the focus the on-axis spin density component of Equation (16) will also be zero. If the original spin in (17) is positive ($\cos\theta > 0$), i.e., the initial polarization vector rotates anticlockwise (Equation (12), then, given $n = 0$, the transverse energy flow at the focus will also rotate anticlockwise thanks to the spin–orbital conversion.

5. Orbital Angular Momentum at the Tight Focus of Poincaré Beams

Now, let us analyze the behavior of the angular momentum at the tight focus of a Poincaré beam (1). The angular momentum is given by [26]:

$$\mathbf{J} = \frac{1}{2c}\mathrm{Re}(\mathbf{r} \times (\mathbf{E}^* \times \mathbf{H})) = \frac{2\pi}{c^2}(\mathbf{r} \times \mathbf{P}) \tag{18}$$

where $\mathbf{r}$ is the radius vector and c is the speed of light in vacuum. Dropping the constant $2\pi/c^2$ in Equation (18) and considering Equation (12), the on-axis projection of the angular momentum vector (18) takes the form:

$$J_z = rP_\varphi = -r\cos(\theta) I_{1,n-1}(I_{0,n} + I_{2,n-2}). \tag{19}$$

From Equation (19), the angular momentum is seen to always be zero on the optical axis at $r = 0$, because the "arm" is zero. The angular momentum is the sum of the SAM and the orbital angular momentum (OAM) [26]:

$$\mathbf{J} = \mathbf{S} + \mathbf{L} = \frac{1}{8\pi\omega}\mathrm{Im}(\mathbf{E}^* \times \mathbf{E}) + \frac{1}{8\pi\omega}\sum_{p=x,y,z}\mathrm{Im}\left(E_p^*(\mathbf{r} \times \nabla)E_p\right) \tag{20}$$

Dropping the constant $1/(8\pi\omega)$ from Equation (20) and taking into account Equations (16) and (19), the relation for the on-axis component L_z of the OAM density takes the form:

$$L_z = -\cos(\theta)\left[rI_{1,n-1}(I_{0,n} + I_{2,n-2}) + I_{0,n}^2 - I_{2,n-2}^2\right]. \tag{21}$$

Equation (21) suggests that if the source Poincaré beam has zero spin ($\cos\theta = 0$), the OAM is zero at the focus ($L_z = 0$). If, however, the source Poincaré beam has non-zero spin ($\cos\theta \neq 0$), both the angular momentum (19) and OAM (21) occur at the focus.

6. Numerical Modeling

The numerical simulation in this section is based on calculating the Debye integral using a Richards–Wolf method [24]. The focusing is assumed to be conducted with an aplanatic objective with NA = 0.95 for a wavelength of $\lambda = 0.532$ nm. Figure 1 depicts intensity patterns $I = |E_x|^2 + |E_y|^2 + |E_z|^2$ at the tight focus of a Poincaré beam at the parameters $\theta = \pi/4$, $\psi = \pi/4$, $n = 0$ (non-vortex elliptically polarized beam, Figure 1a); $\theta = 0$, $\psi = \pi/2$, $n = 0$ (non-vortex right-handed circularly polarized beam, Figure 1b); $\theta = \pi/4$, $\psi = \pi/4$, $n = 1$ (elliptically polarized vortex, Figure 1c); and $\theta = 0$, $\psi = 0$, $n = -1$ (right-handed circularly polarized vortex, Figure 1d).

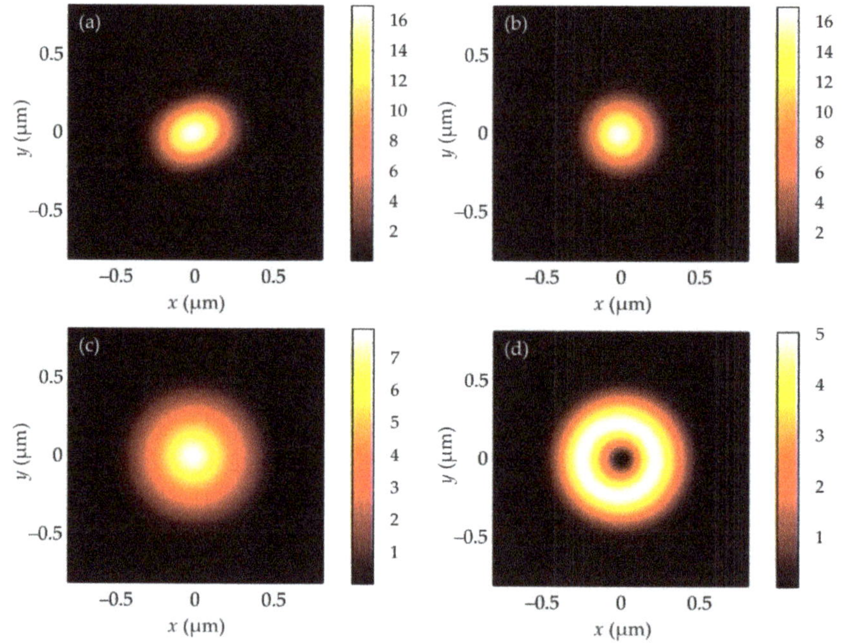

Figure 1. Intensity patterns at the focus for different Poincaré beams of Equation (1): $\theta = \pi/4$, $\psi = \pi/4$, $n = 0$ (**a**); $\theta = 0$, $\psi = \pi/2$, $n = 0$ (**b**); $\theta = \pi/4$, $\psi = \pi/4$, $n = 1$ (**c**); $\theta = 0$, $\psi = 0$, $n = -1$ (**d**).

From Figure 1, beams with elliptical (a), circular (b), and near-radial (c) polarization are seen to produce an elliptic or circular focal spot, whereas a vortex beam with circular polarization (d) produces an annular focal spot. Such types of intensity patterns agree well with Equation (8).

Figure 2 depicts transverse Poynting vector components P_x (a,d) and P_y (b,e) at the foci of two out of the four beams shown in Figure 1. The energy flow in Figure 1a for the beam with the parameters $\theta = \pi/4$, $\psi = \pi/4$, $n = 0$ is identical to those for the beams with $\theta = \pi/4$, $\psi = \pi/4$, $n = 1$ (c) and $\theta = 0$, $\psi = 0$, $n = -1$ (d). From Figure 2a,b, the transverse energy flows at the focus are seen to be rotating anticlockwise for these three beams. Meanwhile, the transverse energy flow produced by the beam with $\pi/4$, $\psi = \pi/4$, $n = 1$ (Figure 2d,e) will experience clockwise rotation. This may be concluded from Equation (12), because at $n = 0, -1$ the angular projection of the near-axis energy flow ($kr < 1$) is positive: $P_\varphi = \cos(\theta) I_{1,1}(I_{0,0} + I_{2,2}) > 0$, $P_\varphi = \cos(\theta) I_{1,2}(I_{0,1} + I_{2,3}) > 0$, becoming negative at

$n = 1$: $P_\varphi = -\cos(\theta) I_{1,0}(I_{0,1} - I_{2,1}) < 0$. The only situation when the energy flow does not rotate is at $\cos\theta = 0$.

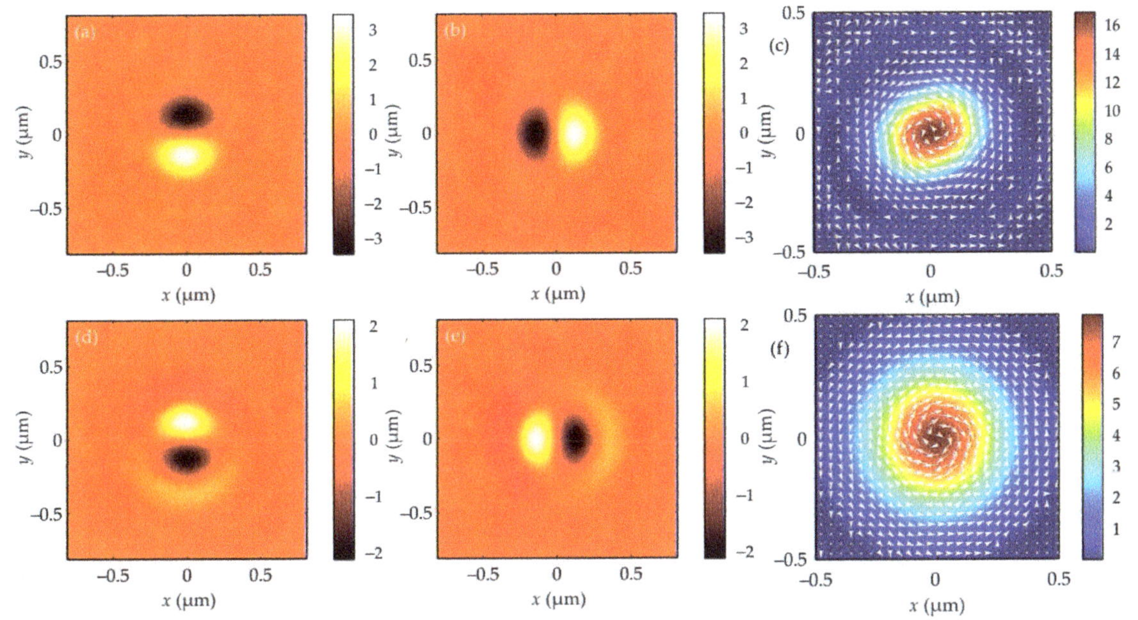

Figure 2. Patterns of the transverse energy flow (Poynting vector projections) at the tight focus: P_x (**a**,**d**), P_y (**b**,**e**) and the direction (arrows) of rotation of the transverse energy flow against the intensity background (**c**,**f**) for two different Poincaré beams: $\theta = \pi/4$, $\psi = \pi/4$, $n = 0$ (**a**,**b**,**c**); $\theta = \pi/4$, $\psi = \pi/4$, $n = 1$ (**d**,**e**,**f**).

Figure 3 depicts patterns of the on-axis SAM vector component (16) at the foci of two beams with the parameters: $\theta = \pi/4$, $\psi = \pi/4$, $n = 0$ (a,b); $\theta = \pi/4$, $\psi = \pi/4$, $n = 1$ (c,d). We note that the beam with the parameters $\theta = 0$, $\psi = \pi/2$, $n = 0$ produces a SAM pattern similar to that in Figure 3a, with the beam at $\theta = 0$, $\psi = 0$, $n = -1$ producing a pattern of the on-axis SAM projection analogous to that in Figure 3c. This conclusion follows from Equation (16), because at $n = 0$, the on-axis SAM projection does not take zero values, $S_z(r = 0) = \cos\theta\left[I_{0,0}^2 - I_{2,2}^2\right] \neq 0$; meanwhile at $n = 1$, everywhere on the optical axis the SAM projection is zero, $S_z(r = 0) = \cos\theta\left[I_{0,1}^2 - I_{2,1}^2\right] = 0$, as is the case at $n = -1$, $S_z(r = 0) = \cos\theta\left[I_{0,1}^2 - I_{2,3}^2\right] = 0$.

Figure 3b,d,f show sections along the x-axis of the axial projection of the SAM vector. It can be seen from Figure 3b,d that at some distance from the optical axis, the SAM projection becomes negative (see enlarged fragments). In Figure 3f, negative SAM values appear near the optical axis. Negative values of the longitudinal projection of the SAM take place in those places of the focus plane where the axial energy flow is negative, since it follows from (12) and (16) that:

$$S_z = P_z \cos\theta \quad (22)$$

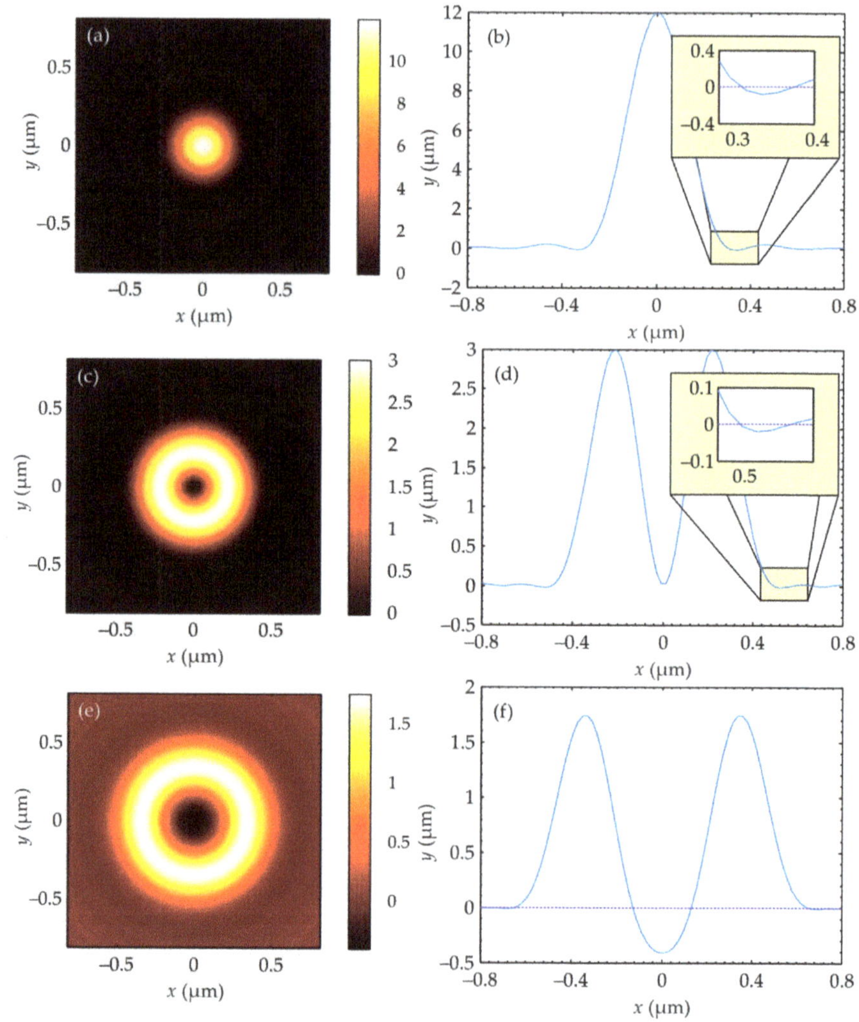

Figure 3. Distributions of the axial projection of the SAM (S_z) and their cross-sections along the x-axis for three different Poincaré beams: $\theta = \pi/4$, $\psi = \pi/4$, $n = 0$ (**a**,**b**); $\theta = \pi/4$, $\psi = \pi/4$, $n = 1$ (**c**,**d**); $\theta = \pi/4$, $\psi = \pi/4$, $n = 2$ (**e**,**f**).

7. Conclusions

In this work, we obtained the following new results. Using a relationship for von Neumann entropy it has been shown that the family of Poincare beams represents a classical entangled (nonseparable) state of light. The entanglement has been shown to become maximal when the Poincaré beams are reduced to cylindrical n-th-order vector beams, Equation (5). Classical entanglement, or nonseparability, of a light beam implies that polarization and orbital degrees of freedom cannot be separated. However, at certain parameters, the Poincaré beams are fully separable, in which case they are reduced to left- or right-handed circularly polarized vortex beams. The Poincaré beams show remarkable properties at the tight focus. Namely, at $n = 2$, a near-axis reverse energy flow has been found to occur when the on-axis Poynting vector component is negative, Equation (13). Given certain parameters of the Poincaré beam ($\cos\theta \neq 0$), the transverse energy flow has been found to rotate around the optical axis at any topological charge n thanks to spin–orbital conversion, Equation (12). More specifically, the energy flow rotates anticlockwise at $\cos\theta > 0$ and $n = 0$, rotating clockwise at $n > 0$. The condition $\cos\theta \neq 0$ means that in the source plane, the on-axis SAM component of Poincaré beams is non-zero, Equation (17). In this case, at the focus, a non-zero on-axis SAM component with alternating sign is also observed, Equation (16), that can take zero values on certain radii. Radius-dependent separation of focal regions with alternating sign of spin density (16) can be interpreted as a radial spin Hall effect for all n. Finally, it has been revealed that thanks to spin–orbital conversion, a non-zero on-axis component of the OAM vector occurs at the focus of Poincaré beams, Equation (21).

In conclusion, it is worth noting that the optical spin Hall effect, which manifests itself as the propagation of left- and right-hand elliptically polarized light fields at different angles to the optical axis (Figure 3), is closely related to the optical Magnus effect [27]. The latter occurs upon propagation of (i) circularly polarized optical beams in a multimode fiber [27] or (ii) circularly polarized optical vortices in a uniaxial crystal [28]. The optical Magnus effect consists in the propagation of left- and right-hand circularly polarized optical beams at different angles in the inhomogeneous or anisotropic medium and is caused by geometrical Berry phase anisotropy [29].

Finally, we note that Poincaré beams can be generated with the aid of spatial light modulators (SLM) [30], metasurfaces [31], and q-plates in a laser cavity [32].

Author Contributions: Conceptualization, V.V.K.; methodology, V.V.K. and S.S.S.; software, S.S.S., V.D.Z. and A.M.T.; validation, V.V.K.; formal analysis, V.V.K.; investigation, V.V.K., S.S.S., V.D.Z. and A.M.T.; resources, S.S.S.; data curation, S.S.S.; writing—original draft preparation, V.V.K.; writing—review and editing, V.V.K.; visualization, S.S.S. and V.D.Z.; supervision, V.V.K.; project administration, V.V.K.; funding acquisition, V.V.K. All authors have read and agreed to the published version of the manuscript.

Funding: This work was funded by the Russian Science Foundation under Project No. 22-12-00137.

Institutional Review Board Statement: Not applicable.

Informed Consent Statement: Not applicable.

Data Availability Statement: Not applicable.

Conflicts of Interest: The authors declare no conflict of interest.

References

1. Bouwmeester, D.; Pan, J.W.; Mattle, K.; Eibl, M.; Weinfurter, H.; Zeilinger, A. Experimental quantum teleportation. *Philos. Trans. R. Soc. A Math. Phys. Eng. Sci.* **1998**, *356*, 1733–1737. [CrossRef]
2. Fickler, R.; Campbell, G.; Buchler, B.; Lam, P.K.; Zeilinger, A. Quantum entanglement of angular momentum states with quantum numbers up to 10,010. *Proc. Natl. Acad. Sci. USA* **2016**, *113*, 13642–13647. [CrossRef] [PubMed]
3. Beckley, A.M.; Brown, T.G.; Alonso, M.A. Full Poincaré beams. *Opt. Express* **2010**, *18*, 10777. [CrossRef] [PubMed]
4. Chen, S.; Zhou, X.; Liu, Y.; Ling, X.; Luo, H.; Wen, S. Generation of arbitrary cylindrical vector beams on the higher order Poincaré sphere. *Opt. Lett.* **2014**, *39*, 5274–5276. [CrossRef]
5. Zhan, Q. Cylindrical vector beams: From mathematical concepts to applications. *Adv. Opt. Photonics* **2009**, *1*, 1–57. [CrossRef]
6. Kotlyar, V.V.; Kovalev, A.A.; Nalimov, A.G. Energy density and energy flux in the focus of an optical vortex: Reverse flux of light energy. *Opt. Lett.* **2018**, *43*, 2921–2924. [CrossRef]
7. Kotlyar, V.V.; Stafeev, S.S.; Kovalev, A.A. Reverse and toroidal flux of light fields with both phase and polarization higher-order singularities in the sharp focus area. *Opt. Express* **2019**, *27*, 16689–16702. [CrossRef]
8. Leyder, C.; Romanelli, M.; Karr, J.P.; Giacobino, E.; Liew, T.C.H.; Glazov, M.M.; Kavokin, A.V.; Malpuech, G.; Bramati, A. Observation of the optical spin hall effect. *Nat. Phys.* **2007**, *3*, 628–631. [CrossRef]
9. Yin, X.; Ye, Z.; Rho, J.; Wang, Y.; Zhang, X. Photonic spin hall effect at metasurfaces. *Science* **2013**, *339*, 1405–1407. [CrossRef]
10. Nath Baitha, M.; Kim, K. All angle polarization-independent photonic spin Hall effect. *Opt. Laser Technol.* **2022**, *156*, 108458. [CrossRef]
11. Li, S.-M.; Chen, J. Spin Hall effect of reflected light from an air-glass interface around the Brewster's angle. *Appl. Phys. Lett.* **2012**, *100*, 071109. [CrossRef]

12. Roy, B.; Ghosh, N.; Banerjee, A.; Gupta, S.D.; Roy, S. Manifestations of geometric phase and enhanced spin Hall shifts in an optical trap. *New J. Phys.* **2014**, *16*, 083037. [CrossRef]
13. Kumar, R.N.; Yatish; Gupta, S.D.; Ghosh, N.; Banerjee, A. Probing the rotational spin-Hall effect in a structured Gaussian beam. *Phys. Rev. A* **2022**, *105*, 023503. [CrossRef]
14. Zhang, J.; Zhou, X.-X.; Ling, X.-H.; Chen, S.-Z.; Luo, H.-L.; Wen, S.-C. Orbit-orbit interaction and photonic orbital Hall effect in reflection of a light beam. *Chin. Phys. B* **2014**, *23*, 064215. [CrossRef]
15. He, Y.; Xie, Z.; Yang, B.; Chen, X.; Liu, J.; Ye, H.; Zhou, X.; Li, Y.; Chen, S.; Fan, D. Controllable photonic spin Hall effect with phase function construction. *Photonics Res.* **2020**, *8*, 963. [CrossRef]
16. Fu, S.; Guo, C.; Liu, G.; Li, Y.; Yin, H.; Li, Z.; Chen, Z. Spin-orbit optical hall effect. *Phys. Rev. Lett.* **2019**, *123*, 243904. [CrossRef]
17. Ling, X.; Zhou, X.; Huang, K.; Liu, Y.; Qiu, C.-W.; Luo, H.; Wen, S. Recent advances in the spin Hall effect of light. *Rep. Prog. Phys.* **2017**, *80*, 066401. [CrossRef]
18. Ling, X.; Yi, X.; Zhou, X.; Liu, Y.; Shu, W.; Luo, H.; Wen, S. Realization of tunable spin-dependent splitting in intrinsic photonic spin Hall effect. *Appl. Phys. Lett.* **2014**, *105*, 151101. [CrossRef]
19. Kotlyar, V.V.; Kovalev, A.A.; Stafeev, S.S.; Zaitsev, V.D. Index of the polarization singularity of Poincare beams. *Bull. Russ. Acad. Sci. Phys.* **2022**, *86*, 1158–1163. [CrossRef]
20. Borges, C.V.S.; Hor-Meyll, M.; Huguenin, J.A.O.; Khoury, A.Z. Bell-like inequality for the spin-orbit separability of a laser beam. *Phys. Rev. A* **2010**, *82*, 033833. [CrossRef]
21. Otte, E.; Rosales-Guzmán, C.; Ndagano, B.; Denz, C.; Forbes, A. Entanglement beating in free space through spin–orbit coupling. *Light Sci. Appl.* **2018**, *7*, 18009. [CrossRef] [PubMed]
22. McLaren, M.; Konrad, T.; Forbes, A. Measuring the nonseparability of vector vortex beams. *Phys. Rev. A* **2015**, *92*, 023833. [CrossRef]
23. Volyar, A.V.; Shvedov, V.G.; Fadeeva, T.A. Structure of a nonparaxial Gaussian beam near the focus: III stability, eigenmodes, and vortices. *Opt. Spectr.* **2001**, *91*, 235–245. [CrossRef]
24. Richards, B.; Wolf, E. Electromagnetic diffraction in optical systems. II. Structure of the image field in an aplanatic system. *Proc. R. Soc. A Math. Phys. Eng. Sci.* **1959**, *253*, 358–379. [CrossRef]
25. Bliokh, K.Y.; Ostrovskaya, E.A.; Alonso, M.A.; Rodríguez-Herrera, O.G.; Lara, D.; Dainty, C. Spin-to-orbital angular momentum conversion in focusing, scattering, and imaging systems. *Opt. Express* **2011**, *19*, 26132. [CrossRef]
26. Bliokh, K.Y.; Bekshaev, A.Y.; Nori, F. Extraordinary momentum and spin in evanescent waves. *Nat. Commun.* **2014**, *5*, 3300. [CrossRef]
27. Dooghin, A.V.; Kundikova, N.D.; Liberman, V.S.; Zel'Dovich, B.Y. Optical Magnus effect. *Phys. Rev. A* **1992**, *45*, 8204. [CrossRef]
28. Fadeyeva, T.A.; Rubass, A.F.; Volyar, A.V. Transverse shift of a high-order paraxial vortex-beam induced by a homogeneous anisotropic medium. *Phys. Rev. A* **2009**, *79*, 053815. [CrossRef]
29. Bliokh, K.Y.; Bliokh, Y.P. Optical Magnus effect as a consequence of Berry phase anisotropy. *JEPT Lett.* **2004**, *79*, 519–522. [CrossRef]
30. Gu, Z.; Yin, D.; Gu, F.; Zhang, Y.; Nie, S.; Feng, S.; Ma, J.; Yuan, C. Generation of concentric perfect Poincaré beams. *Sci. Rep.* **2019**, *9*, 15301. [CrossRef]
31. Liu, M.; Huo, P.; Zhu, W.; Zhang, C.; Zhang, S.; Song, M.; Zhang, S.; Zhou, Q.; Chen, L.; Lezec, H.J.; et al. Broadband generation of perfect Poincaré beams via dielectric spin-multiplexed metasurface. *Nat. Commun.* **2021**, *12*, 2230. [CrossRef] [PubMed]
32. Naidoo, D.; Roux, F.S.; Dudley, A.; Litvin, I.; Piccirillo, B.; Marrucci, L.; Forbes, A. Controlled generation of higher-order Poincaré sphere beams from a laser. *Nat. Photonics* **2016**, *10*, 327–332. [CrossRef]

photonics

Communication

Tight Focusing Properties of Ring Pearcey Beams with a Cross Phase

Chaoqun Yu [1,†], Zhimin He [1,†], Cheng Huang [1], Fuchang Chen [1], Jun Zeng [1], Yan Li [1], Yongtao Zhang [1], Jixiong Pu [1,2] and Huichuan Lin [1,*]

1 College of Physics and Information Engineering, Minnan Normal University, Zhangzhou 363000, China
2 Fujian Provincial Key Laboratory of Light Propagation and Transformation, College of Information Science & Engineering, Huaqiao University, Xiamen 361021, China
* Correspondence: lhc1810@mnnu.edu.cn; Tel.: +86-18-8763-30655
† These authors contributed equally to this work.

Abstract: We theoretically investigated the properties of tightly focused ring Pearcey beams with a cross phase (CPRPB). The expressions of the distributions of both electric field and magnetic field in the focal region of an objective were first derived from the vectorial Debye theory, and then numerical calculations were carried out to obtain the focused intensity distribution and the Poynting vector of CPRPB near the focus. Numerical calculations indicate that as CPRPB is focused on an objective of high numerical aperture (NA), two nonuniform self-focusing spots occur at both sides of the geometrical focus of the objective symmetrically, and the angle between their directions is 90 degrees. The stronger is the strength of cross-phase modulation, the flatter are the ellipses of the self-focusing spots, and the smaller is the intensity at the geometrical focus of the objective. Numerical calculations also demonstrate that the optical gradient force produced by tightly focused CPRPB in the focal region can be manipulated in magnitude and in direction by tuning the strength of cross-phase modulation. Due to these properties of tightly focused CPRPB, they might find applications in the manipulation of micro- and nanoparticles and so on.

Keywords: cross phase; Pearcey beams; tight focusing; self-focusing

Citation: Yu, C.; He, Z.; Huang, C.; Chen, F.; Zeng, J.; Li, Y.; Zhang, Y.; Pu, J.; Lin, H. Tight Focusing Properties of Ring Pearcey Beams with a Cross Phase. *Photonics* **2022**, *9*, 964. https://doi.org/10.3390/photonics9120964

Received: 27 October 2022
Accepted: 7 December 2022
Published: 10 December 2022

1. Introduction

Taking advantage of focusing energy at a point along the propagation of light beams, self-focusing beams have found applications in many scenarios. In laser medicine, for instance, self-focusing beams maintain low power during propagation, avoiding the damage of tissues outside the region of interest [1]. Considerable optical gradient force can be produced by the self-focusing beams to trap or manipulate the micro- and nanoscale particles accurately [2,3]. Therefore, self-focusing beams always attract the keen interests of researchers [4–9]. Ring Pearcey beams were first reported by Chen et al., in 2018, and it was indicated that ring Pearcey beams belonged to a family of self-focusing beams. Compared with other self-focusing beams, ring Pearcey beams are characterized by their higher peak intensity contrast, shorter focal length, and no oscillatory energy loss beyond focus [10]. Cross phase is a unique phase structure. By cross-phase modulation, interconversion between the Hermite–Gaussian (HG) modes and rotation of the partially coherent Gaussian-Schell model beams can be realized [11–13]. Cross-phase modulation can also be employed to generate, to measure, and to manipulate high-order optical vortices [14,15]. To investigate the propagation of modulated laser beams, cross-phase modulation was utilized by Liu et al. in 2021 to modulate the ring Pearcey beams, and the ring Pearcey beams with a cross phase (CPRPB) was developed. It was found that CPRPB's focal length, direction, rotation angle, and symmetry of their self-focusing spots can be flexibly changed by cross-phase modulation [6].

Subwavelength focal spots can be achieved by focusing the laser beams with an objective of high numerical aperture (NA). Laser beams can also be shaped by modulating the polarization property, phase distribution, and coherent characteristics of the incident beams. These beam-shaping schemes are desirable for many applications, such as optical data storage, microscope, particle capture, and material processing. Thus, many research efforts have been made to study the tight-focusing properties of the laser beams of different types in these years [16–24]. Considering the self-focusing property and unique phase structure of CPRPB, it will be a significant work to explore its tight-focusing properties Based on the Richards-Wolf vector diffraction theory, in this paper, the electric field and intensity distributions of CPRPB tightly focused on an objective of high NA in the focal region were derived. The gradient forces imposed on micro- and nanoparticles by the focused beams were computed, and the influence of cross-phase on the self-focusing spots and optical gradient forces was discussed.

2. Theoretical Analysis

The electric field in the source plane of CPRPB can be expressed as

$$\begin{aligned} E(x,y,z=0) = {} & \mathrm{Pe}(\gamma,-r/p) \\ & \times \exp\Big\{-\alpha r^2 + i\mu[x\cos\delta - y\sin\delta] \times [x\sin\delta + y\cos\delta]\Big\} \end{aligned} \tag{1}$$

where $r = \sqrt{x^2+y^2}$ is the modulus in the polar coordinate system of the source plane, $\mathrm{Pe}(\gamma, -r/p)$ is the Pearcey function, γ and p are the factors that affect the spatial distribution of the light intensity, α is the attenuation factor that guarantees finite beam energy, μ is a real coefficient that determines the intensity of cross-phase modulation, and δ is the azimuth parameter that determines the autofocus orientation [12].

As illustrated in Figure 1, CPRPB is tightly focused on an objective lens of high NA. According to the vectorial Debye theory [25], the field distribution in the region of the objective's geometrical focus can be expressed as

$$\begin{aligned} E(\rho,\psi,z) = {} & \frac{iE_0}{\lambda}\int_0^{\beta}\int_0^{2\pi} \exp[-ikz\cos\theta - ik\rho\sin\theta\cos(\psi-\varphi)] \\ & \times P(\varphi,\theta)K(\varphi,\theta)\sin\theta\sqrt{\cos\theta}\,d\varphi d\theta \end{aligned} \tag{2}$$

where θ is the angle of refraction, λ is the wavelength of the beam, k is the wavenumber, $\beta = \sin^{-1}$ (NA) is the maximum angle of refraction that is decided by the numerical aperture of the objective, E_0 is a constant here representing the amplitude of the incident beam and the variables, and ρ, ψ, and z are the cylindrical coordinates in the focal region. Assume that the incident CPRPB are the linearly polarized beams, $k(\varphi, \theta)$, in Equation (2):

$$K(\varphi,\theta) = \begin{bmatrix} (\cos\theta + \sin^2\varphi(1-\cos\theta))\boldsymbol{e_x} \\ \cos\varphi\sin\varphi(\cos\theta - 1)\boldsymbol{e_y} \\ \cos\varphi\sin\theta\boldsymbol{e_z} \end{bmatrix} \tag{3}$$

where $\boldsymbol{e_x}$, $\boldsymbol{e_y}$, and $\boldsymbol{e_z}$ are the unit vectors along the x, y, and z axes in the Cartesian coordinate system of the focal region. By projecting the incident CPRPB to the aperture of the objective, the apodization function of CPRPB at the objective can be obtained from Equation (1) as

$$\begin{aligned} P(\varphi,\theta) = {} & \mathrm{Pe}(\gamma,-f\sin\theta/p)\exp\Big(-\alpha(f\sin\theta)^2\Big) \\ & \times \exp\{i\mu f\sin\theta[\cos\varphi\cos\delta - \sin\varphi\sin\delta] \\ & \times f\sin\theta[\cos\varphi\sin\delta + \sin\varphi\cos\delta]\} \end{aligned} \tag{4}$$

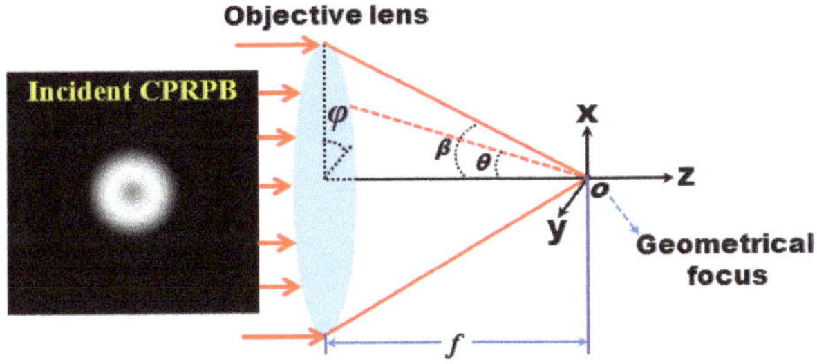

Figure 1. Intensity distribution of incident linearly polarized CPRPB (left) and schematic of a tight-focusing setup (right).

Put $k(\varphi, \theta)$ in Equation (3) and $P(\varphi, \theta)$ in Equation (4) into Equation (2) and set $\delta = 0$. Three polarized components of the electric field distribution of tightly focused CPRPB in the focal region of the objective of high NA are achieved in Cartesian coordinates as

$$\begin{aligned} E_x(\rho,\psi,z) = & \frac{\pi E_0}{2\lambda}\int_0^{\beta} \mathrm{Pe}(\gamma, -f\sin\theta/p)\exp\left[-\alpha f^2\sin^2\theta - ikz\cos\theta\right] \\ & \times \sin\theta \sum_{l=-\infty}^{\infty} i^{-l} J_l\left(\frac{1}{2}\mu f^2 \sin^2\theta\right)\exp\left(-il\frac{\pi}{2}\right)\exp(2il\psi) \\ & +\{2J_{2l}(k\rho\sin\theta)(\cos\theta+1)\, J_{2l+2}(k\rho\sin\theta)\exp(2i\psi)(1-\cos\theta) \\ & +J_{2l-2}(k\rho\sin\theta)\exp(-2i\psi)(1-\cos\theta)\}d\theta \end{aligned} \quad (5)$$

$$\begin{aligned} E_y(\rho,\psi,z) = & \frac{\pi E_0(1-\cos\theta)}{2\lambda}\int_0^{\beta} \mathrm{Pe}(\gamma, -f\sin\theta/p)\exp\left[-\alpha f^2\sin^2\theta - ikz\cos\theta\right] \\ & \times \sin\theta \sum_{l=-\infty}^{\infty} i^{-(l+1)} J_l\left(\frac{1}{2}\mu f^2 \sin^2\theta\right)\exp\left(-il\frac{\pi}{2}\right)\exp(2il\psi) \\ & \times\{J_{2l+2}(k\rho\sin\theta)\exp(2i\psi) - J_{2l-2}(k\rho\sin\theta)\exp(-2i\psi)\}d\theta \end{aligned} \quad (6)$$

$$\begin{aligned} E_z(\rho,\psi,z) = & \frac{-\pi E_0}{\lambda}\int_0^{\beta} \mathrm{Pe}(\gamma, -f\sin\theta/p)\exp\left[-\alpha f^2\sin^2\theta - ikz\cos\theta\right] \\ & \times \sin^2\theta \sum_{l=-\infty}^{\infty} i^{-(l+1)} J_l\left(\frac{1}{2}\mu f^2 \sin^2\theta\right)\exp\left(-il\frac{\pi}{2}\right)\exp(2il\psi) \\ & \times\{J_{2l+1}(k\rho\sin\theta)\exp(i\psi) + J_{2l-1}(k\rho\sin\theta)\exp(-i\psi)\}d\theta \end{aligned} \quad (7)$$

With Equations (5)–(7), it is feasible to study the properties of CPRPB focused on an objective of high NA. In order to illustrate the effect of cross-phase modulation on the properties of tightly focused CPRPB more explicitly, numerical calculations were carried out utilizing the derived equations above, concentrating on the influence of the intensity factor of cross-phase modulation, μ, on the focused beams.

3. Results and Discussions

Based on the field distribution equations derived previously, the intensity distributions in the focal plane of the objective were first computed and are illustrated in Figure 2. Figure 2a indicates the x-polarization component, Figure 2b indicates the z-polarization component (longitudinal field), and Figure 2c indicates the y-polarization component. It could be found out from these figures that CPRPB was depolarized as focused on an objective of high NA. The intensity of the y-polarization component is about one thousandth of the intensity of the x-polarization component, while the intensity of the z-polarization component is about one millionth of that of the x-polarization component. As shown in Figure 2c, the y-polarization component appears as a doughnut-shaped ring. There exists a phase singularity at the center of the ring, illustrated by the color inset in the upper-left corner of Figure 2c, which was obtained by computing the phase distribution of this component. Figure 2d is the total intensity distribution of three components. Since the intensity of the *x*-polarization component far outweighs that of the *y*- and *z*-polarization components, the total intensity distribution seems similar to the intensity distribution of the *x*-polarization component.

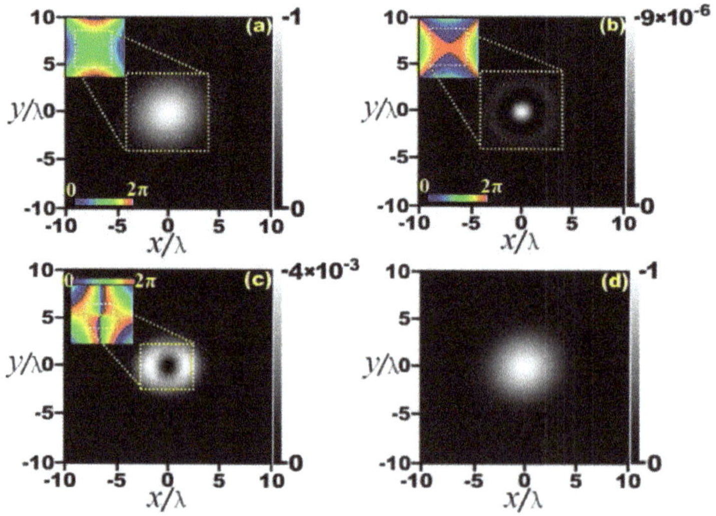

Figure 2. Intensity distributions in focal plane, (**a**) *x*-polarization component, (**b**) *z*-polarization component, (**c**) *y*-polarization component, (**d**) total intensity. Parameters for calculation are chosen as $\lambda = 633$ nm, $\alpha = 3$, NA = 0.9, $f = 3.4$ cm, $\mu = 10$ mm^{-2}, and $E_0 = 1$.

Figure 3 indicates the intensity distributions of tightly focused CPRPB in the ρ-z planes around the focus of the objective, in which (a), (b), (c), and (d) are the intensity distributions in the planes of $\psi = 0$, $\psi = \pi/4$, $\psi = \pi/2$, and $\psi = 3\pi/4$, respectively. It can be concluded from the figures that the focused beam is axially asymmetrical in the region around the focus, and the maxima are not located at the geometrical focus of the objective, as CPRPB is tightly focused.

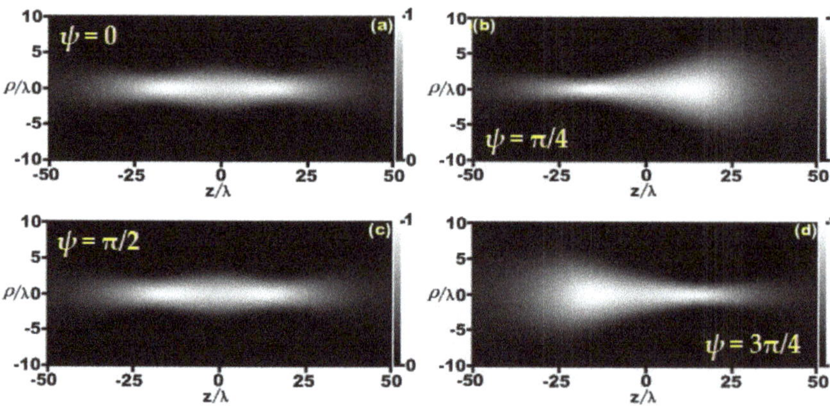

Figure 3. Intensity distributions in ρ-z planes around the focus. (**a**) $\psi = 0$, (**b**) $\psi = \pi/4$, (**c**) $\psi = \pi/2$, and (**d**) $\psi = 3\pi/4$. Parameters for calculation are chosen as $\lambda = 633$ nm, $\alpha = 3$, NA = 0.9, $f = 3.4$ cm, $\mu = 10$ mm^{-2}, and $E_0 = 1$.

In order to find out the reason for the asymmetry of the focused beam and deviation of the maxima from the focus, intensity distributions of tightly focused CPRPB in the cross sections around the focal plane were computed and are listed in Figure 4 according to the sequence of beam propagation. The subdiagrams, (a)–(g), are the intensity distributions in the planes of (a) $z = -34\lambda$, $z = -17\lambda$, $z = -7\lambda$, $z = 0$, $z = 7\lambda$, $z = 17\lambda$, and $z = 34\lambda$, respectively. The plane of $z = 0$ is the focal plane as well. Numerical results show that the self-focusing phenomenon occurs twice as CPRPB tightly focused on the objective. One self-focusing is prior to the focus, and the other is behind the focus. The angle between their directions is 90 degrees. Therefore, it can be concluded that the beam rotates as it travels through the focal region. Two self-focusings separated at both sides of the geometrical focus of the objective and rotation of the beam lead to the asymmetry of intensity distribution. In addition, Figure 4b shows that the major radius of the elliptical beam profile has an inclination angle of 135° from the x-axis, which is in agreement with the experimental result shown in Reference [6]. Figure 4h plots the ratio of the major radius to the minor radius of the elliptical beam profile versus its position in the optical axis. It can be seen clearly from the curve that the beam's cross section deforms from an elliptical shape to a circular shape, and then back to an ellipse during propagation. Elliptical Gaussian beam takes a similar characteristic of asymmetric self-focusing during propagation; therefore, it might have the same manner of beam deformation as it is tightly focused. Asymmetrical intensity distribution and rotation of the beam in the focal region results in rotational intensity gradient distribution, and thus an optical torque, which might play an important role in optical manipulation.

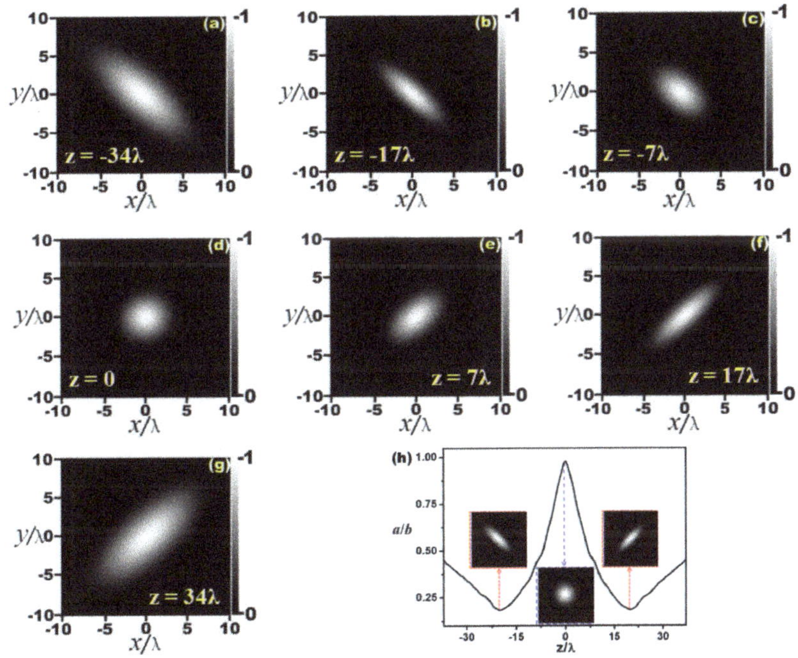

Figure 4. Intensity distributions in different cross sections around the focus, (**a**) $z = -34\lambda$, (**b**) $z = -17\lambda$, (**c**) $z = -7\lambda$, (**d**) $z = 0$, (**e**) $z = 7\lambda$, (**f**) $z = 17\lambda$, and (**g**) $z = 34\lambda$; (**h**) the ratio of the major radius to the minor radius of the elliptical beam profile versus its position in the optical axis. Parameters for calculation are chosen as $\lambda = 633$ nm, $\alpha = 3$, NA = 0.9, $f = 3.4$ cm, $\mu = 10$ mm^{-2}, and $E_0 = 1$.

It was reported that cross-phase modulation altered the symmetry of the intensity distributions of ring Pearcey beams, which could have affected their optical trapping capability. Therefore, it is essential to figure out the effect of the cross-phase modulation on the properties of tightly focused CPRPB. Figure 5 illustrates the intensity distributions of the self-focusing spots of tightly focused CPRPB modulated with different cross-phase modulation intensities. According to the ratio of the major radius to the minor radius of the beam profiles (a/b) in the subdiagrams in Figure 5, it can be seen that the stronger is the cross-phase modulation, the flatter is the ellipse of the beam profile. This conclusion is in good agreement with the practical results illustrated in Reference [6], which proves that the theoretical derivation is correct. With the intensity modulation factor, μ, reduced to 5 mm^{-2}, the beam profile is pretty close to a circle. Provided that the self-focusing spot of a ring Pearcey beam is exactly a circle, the beam does not rotate in the focal region as it is tightly focused. In this case, there exists no optical torque when it is employed as optical tweezers. In brief, in the manipulation of micro- and nanoparticles by tightly focused CPRPB, the optical torque in the focal region can be tuned by altering the intensity of the cross-phase modulation.

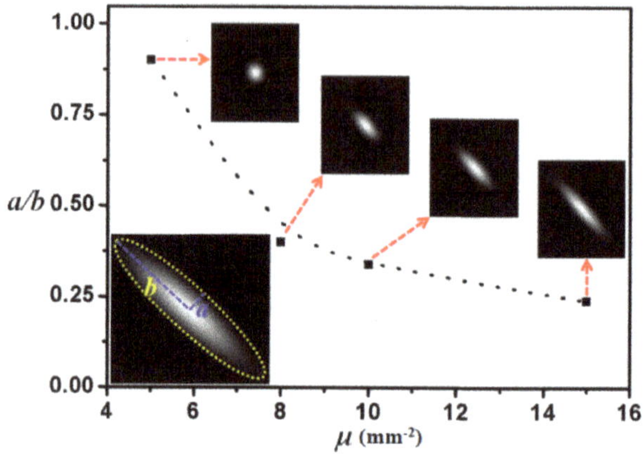

Figure 5. Influence of the cross-phase modulation on the intensity distribution of self-focusing spots. Parameters for calculation are chosen as λ = 633 nm, α = 3, NA = 0.9, f = 3.4 cm, and E_0 = 1.

As discussed previously, the maximum intensity of tightly focused CPRPB in the focal region is not located at the geometrical focus of the objective due to their self-focusing property, and cross-phase modulation also affects the intensity distributions of their self-focusing spots. In the optical trapping of micro- and nanoparticles, particles are usually trapped at the point of intensity maximum by the optical gradient force, where the optical gradient force is equal to zero [26]. According to different intensities of cross-phase modulation, the intensity distributions along the optical axis were calculated and are plotted in Figure 6. With relatively mild modulation intensity (μ = 5 mm^{-2}), the intensity of the beam focused on an objective of high NA peaks at the geometrical focus of the objective. As modulation intensity increases, the intensity distribution profile takes a saddle shape in which the intensity maxima shift towards both sides of the geometrical focus, while the saddle point located at the focus moves down. When the intensity factor of cross-phase modulation, μ, equals 18 mm^{-2}, the intensity of the tightly focused beam at the geometrical focus of the objective falls to zero.

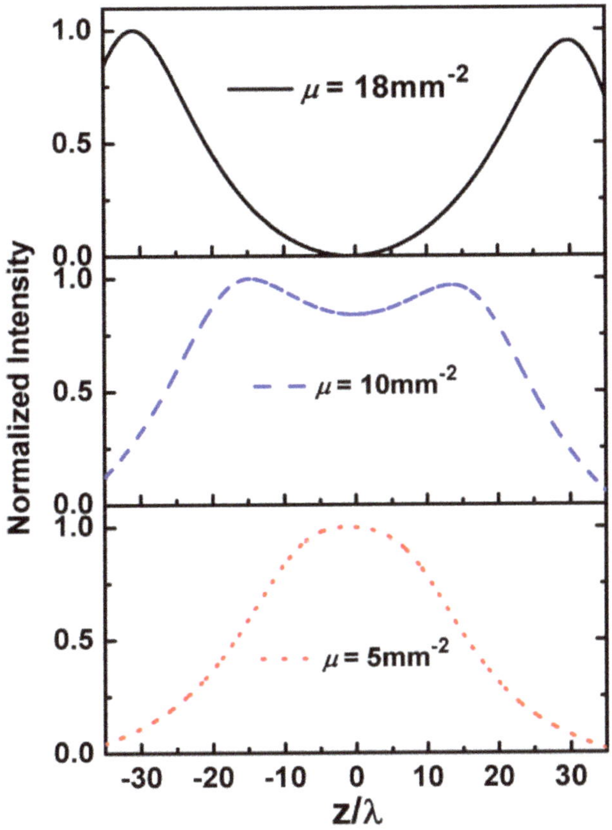

Figure 6. Influence of cross-phase modulation on the intensity distribution along the optical axis. Parameters for calculation are chosen as λ = 633 nm, α = 3, NA = 0.9, f = 3.4 cm, and E_0 = 1.

It can be concluded from the calculations above that the cross-phase modulation plays a key role in the intensity distribution of the ring Pearcey beams in the focal region. In the study of how the change of intensity distribution affects the optical trapping of micro- and nanoparticles, the gradient forces produced by tightly focused CPRPB on the particles were calculated through Equation (8). For simplicity, it was assumed that the radii of the particles trapped were far smaller than the wavelength of the focused beam.

$$\mathbf{F}_g(\rho, \psi, z) = 2\pi a^3 \frac{n_2}{c} \frac{m^2 - 1}{m^2 + 2} \nabla I(\rho, \psi, z) \tag{8}$$

In Equation (8), $I(\rho, \psi, z)$ is the intensity distribution of CPRPB in the focal region, a is the radius of the trapped particle, n_2 is the refractive index of the particle, n_1 is the refractive index of the background medium, $m = n_2/n_1$, and c is the speed of light in vacuum. The formula of optical gradient force of Equation (8) implies that the intensity gradients are zero at the intensity maxima, and thus the gradient forces produced by the beam at those points are nil as well. As the refractive index of a particle is greater than the refractive index of the background medium, that is, $n_2 > n_1$, the particle would have been trapped at the positions of intensity maxima, which are usually located at the optical axis of the beam. The gradient forces along the optical axis of the tightly focused CPRPB with reference to different intensities of cross-phase modulation were calculated and are plotted in Figure 7. Numerical results explicitly show that there exists a balance point of zero gradient force formed at the geometrical focus of the objective regardless of the modulation intensity. However, the gradient force along the axis varies acutely with the increase in cross-phase modulation intensity. With μ equaling 5 mm^{-2}, the gradient force is positive on the left side of the focus while negative on the right side. When μ rises to 10 mm^{-2}, the profile of gradient force is flipped vertically. Meanwhile, another two balance points of zero gradient force appear at distances of $\pm 15\lambda$ from the geometrical focus in the optical axis. As the modulation intensity increases further, two new balance points move farther away from the focus, and the profile of gradient force is flipped vertically again. Besides intensity distribution and gradient force, the Poynting vector that describes the energy flows of a beam would be affected by the cross-phase modulation as well. Based on the studies in [27–29], it is meaningful to study the Poynting vector of tightly focused CPRPB.

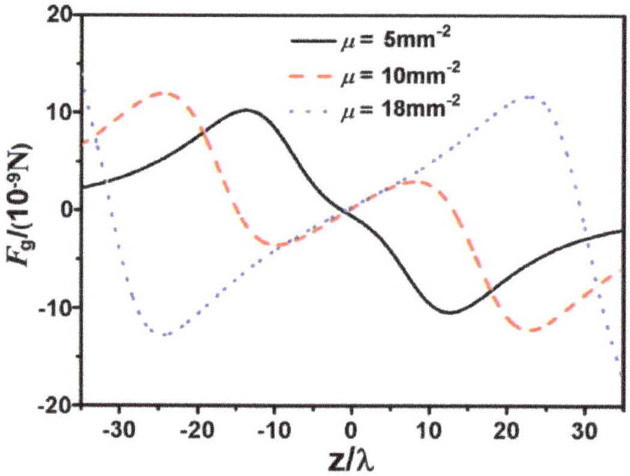

Figure 7. Influence of cross-phase modulation on the gradient force along the optical axis. Parameters for calculation are chosen as λ = 633 nm, α = 3, NA = 0.9, f = 3.4 cm, n_2 = 1.5, n_1 = 1.3, and a = 20 nm, $E_0 = \sqrt{100(\mathrm{mW/um^2})}$.

The Poynting vector is another important characteristic of a beam indicating the beam's energy flows. The Poynting vector can be expressed as $S = E \times H$, in which E and H are the electric and magnetic field vectors, respectively. In order to study the Poynting vector, both E and H have to be solved. The magnetic field distribution of the linearly polarized CPRPB in the focal region is expressed as

$$\begin{aligned} B(\rho,\psi,z) = {} & \frac{iB_0}{\lambda}\int_0^{\beta}\int_0^{2\pi}\exp[-ikz\cos\theta - ik\rho\sin\theta\cos(\psi-\varphi)] \\ & \times P(\varphi,\theta)\begin{Bmatrix} \sin\varphi\cos\varphi(\cos\theta-1)e_x \\ \left[\cos\theta+\cos^2\varphi(1-\cos\theta)\right]e_y \\ \sin\varphi\sin\theta e_z \end{Bmatrix}\sin\theta\sqrt{\cos\theta}d\varphi d\theta \end{aligned} \quad (9)$$

By solving Equation (9), three polarized components of the magnetic field distribution of the tightly focused linearly polarized CPRPB in the focal region are achieved in Cartesian coordinates as

$$\begin{aligned} B_x(\rho,\psi,z) = {} & \frac{\pi B_0(1-\cos\theta)}{2\lambda}\int_0^{\beta}\mathrm{Pe}(\gamma,-f\sin\theta/p)\exp\left[-\alpha f^2\sin^2\theta - ikz\cos\theta\right] \\ & \times\sin\theta\sum_{l=-\infty}^{\infty} i^{-(l+1)}J_l\left(\frac{1}{2}\mu f^2\sin^2\theta\right)\exp\left(-il\frac{\pi}{2}\right)\exp(2il\psi) \\ & \times\{J_{2l+2}(k\rho\sin\theta)\exp(2i\psi) - J_{2l-2}(k\rho\sin\theta)\exp(-2i\psi)\}d\theta \end{aligned} \quad (10)$$

$$B_y(\rho,\psi,z) = \frac{\pi B_0}{2\lambda}\int_0^{\beta} \mathrm{Pe}(\gamma, -f\sin\theta/p)\exp\left[-\alpha f^2\sin^2\theta - ikz\cos\theta\right]$$
$$\times \sin\theta \sum_{l=-\infty}^{\infty} i^{-l} J_l\left(\frac{1}{2}\mu f^2\sin^2\theta\right)\exp\left(-il\frac{\pi}{2}\right)\exp(2il\psi) \quad (11)$$
$$+\{2J_{2l}(k\rho\sin\theta)(\cos\theta+1)\, J_{2l+2}(k\rho\sin\theta)\exp(2i\psi)(\cos\theta-1)$$
$$+J_{2l-2}(k\rho\sin\theta)\exp(-2i\psi)(\cos\theta-1)\}d\theta$$

$$B_z(\rho,\psi,z) = \frac{-\pi B_0}{\lambda}\int_0^{\beta} \mathrm{Pe}(\gamma, -f\sin\theta/p)\exp\left[-\alpha f^2\sin^2\theta - ikz\cos\theta\right]$$
$$\times \sin^2\theta \sum_{l=-\infty}^{\infty} i^{-l} J_l\left(\frac{1}{2}\mu f^2\sin^2\theta\right)\exp\left(-il\frac{\pi}{2}\right)\exp(2il\psi) \quad (12)$$
$$\times\{J_{2l+1}(k\rho\sin\theta)\exp(i\psi) - J_{2l-1}(k\rho\sin\theta)\exp(-i\psi)\}d\theta$$

Three polarized components of the electric filed distribution are shown in Equations (5)–(7). Numerical calculations were then applied to these equations, and the distributions of the Poynting vector of the tightly focused CPRPB were achieved, as shown in Figure 8.

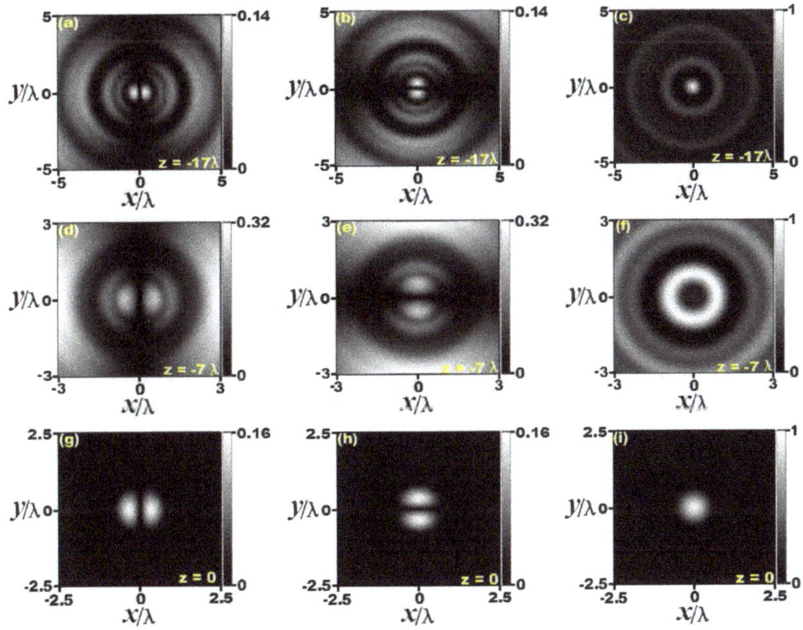

Figure 8. Poynting vector distributions in different cross sections around the focus. (**a**,**d**,**g**) are S_x; (**b**,**e**,**h**) are S_y; and (**c**,**f**,**i**) are S_z. Parameters for calculation are chosen as λ = 633 nm, α = 3, NA = 0.9, f = 3.4 cm, μ = 10 mm^{-2}, and $E_0 = B_0 = 1$.

It can been found from Figure 8 that there exist lateral energy flows (S_x and S_y) in the focal region, besides the energy flow along the optical axis (S_z). However, the ratios of S_x and S_y to S_z vary with the position. At the self-focusing position, that is, z = -17λ, the ratios of S_x and S_y to S_z are 0.14, while at the positions of z = -7λ and z = 0, the ratios are 0.32 and 0.16, respectively.

4. Concluding Remarks

The properties of tightly focused CPRPB were presented in this paper, focusing on the influences of the cross-phase modulation intensity on the intensity distribution of the focused beam and on the optical gradient force produced along the optical axis of the beam. Numerical results show that nonuniform self-focusing phenomena occur at both sides of the geometrical focus of the objective, and the angle between the directions of two self-focusing spots is 90 degrees, as CPRPB is tightly focused on an objective of high NA. The intensity of cross-phase modulation plays a key role in the properties of tightly focused beams. The stronger is the modulation strength, the flatter are the ellipses of the self-focusing spots beside the focal plane and the smaller is the intensity at the focus of the objective. The study of gradient forces along the optical axis in different intensities of cross-phase modulation demonstrates that the optical gradient force produced by tightly focused CPRPB in the focal region can be manipulated in magnitude and in direction by tuning the modulation intensity. Due to these properties of tightly focused CPRPB, they might find use in the manipulation of micro- and nanoparticles.

Funding: This research was funded by National Natural Science Foundation of China (NSFC), grant numbers 61975072 and 12174173; This research was funded by Natural Science Foundation of Fujian province, grant numbers 2022H0023, 2022J02047 and 2022G02006; This research was funded by the Young Core Instructor from the Education Department of Fujian Province, grant numbers JT180295 and JAT190393.

Institutional Review Board Statement: Not applicable.

Informed Consent Statement: Not applicable.

Conflicts of Interest: No conflict of interest exits in this manuscript, and manuscript is approved by all authors.

References

1. Ioannis, C.; Efremidis, N.; Christodoulides, D. Pre-engineered abruptly autofocusing beams. *Opt. Lett.* **2011**, *36*, 1890–1892.
2. Jiang, Y.; Huang, K.; Lu, X. Radiation force of abruptly autofocusing Airy beams on a Rayleigh particle. *Opt. Express* **2013**, *21*, 24413–24421. [CrossRef] [PubMed]
3. Lu, W.; Sun, X.; Chen, H.; Liu, S.; Lin, Z. Abruptly autofocusing property and optical manipulation of circular Airy beams. *Phys. Rev. A* **2019**, *99*, 013817. [CrossRef]
4. Efremidis, N.K.; Christodoulides, D.N. Abruptly autofocusing waves. *Opt. Lett.* **2010**, *35*, 4045–4047. [CrossRef]
5. Manousidaki, M.; Papazoglou, D.; Farsari, M.; Tzortzakis, S. Abruptly autofocusing beams enable advanced multiscale photo-polymerization. *Optica* **2016**, *3*, 525–530. [CrossRef]
6. Xin, L.; Li, Z.; Monfared, Y.E.; Liang, C.; Wang, F.; Hoenders, B.J.; Cai, Y.; Ma, P. Flexible autofocusing properties of ring Pearcey beams by means of a cross phase. *Opt. Lett.* **2021**, *46*, 70–73. [CrossRef]
7. Ding, C.; Korotkova, O.; Horoshko, D.; Zhao, Z.; Pan, L. Evolution of Spatiotemporal Intensity of Partially Coherent Pulsed Beams with Spatial Cosine-Gaussian and Temporal Laguerre–Gaussian Correlations in Still, Pure Water. *Photonics* **2021**, *8*, 102. [CrossRef]
8. Chen, K.H.; Qiu, H.X.; Wu, Y.; Lin, Z.; Huang, H.; Shui, L.; Deng, D.; Liu, H.; Chen, Z. Generation and control of dynamically tunable circular Pearcey beams with annular spiral-zone phase. *Sci. China Phys. Mech. Astron.* **2021**, *64*, 104211. [CrossRef]
9. Xu, D.; Mo, Z.; Jiang, J.; Huang, H.; Wei, Q.; Wu, Y.; Wang, X.; Liang, Z.; Yang, H.; Chen, H.; et al. Guiding particles along arbitrary trajectories by circular Pearcey-like vortex beams. *Phys. Rev. A* **2022**, *106*, 013509. [CrossRef]
10. Chen, X.; Deng, D.; Zhuang, J.; Peng, X.; Li, D.; Zhang, L.; Zhao, F.; Yang, X.; Liu, H.; Wang, G. Focusing properties of circle Pearcey beams. *Opt. Lett.* **2018**, *43*, 3626–3629. [CrossRef]
11. Liang, G.; Wang, Q. Controllable conversion between Hermite Gaussian and Laguerre Gaussian modes due to cross phase. *Opt. Express* **2019**, *27*, 10684–10691. [CrossRef]
12. Wan, L.; Zhao, D. Controllable rotating Gaussian Schell-model beams. *Opt. Lett.* **2019**, *44*, 735–738. [CrossRef]
13. Liu, Z.; Zhao, D. Experimental generation of a kind of reversal rotating beams. *Opt. Express* **2020**, *28*, 2884–2894. [CrossRef] [PubMed]
14. Shen, D.; Zhao, D. Measuring the topological charge of optical vortices with a twisting phase. *Opt. Lett.* **2019**, *44*, 2334–2337. [CrossRef] [PubMed]
15. Ren, Y.; Wang, C.; Liu, T.; Wang, Z.; Yin, C.; Qiu, S.; Li, Z.; Wu, H. Polygonal shaping and multi-singularity manipulation of optical vortices via high-order cross-phase. *Opt. Express* **2020**, *28*, 26257–26266. [CrossRef] [PubMed]
16. Gu, B.; Wu, J.L.; Pan, Y.; Cui, Y. Achievement of needle-like focus by engineering radial-variant vector fields. *Opt. Express* **2013**, *21*, 30444–30452. [CrossRef]
17. Weng, X.; Gao, X.; Guo, H.; Zhuang, S. Creation of tunable multiple 3D dark spots with cylindrical vector beam. *Appl. Opt.* **2014**, *53*, 2470–2476. [CrossRef]

18. Man, Z.; Min, C.; Du, L.; Zhang, Y.; Zhu, S.; Yuan, X. Sub-wavelength sized transversely polarized optical needle with exceptionally suppressed side-lobes. *Opt. Express* **2016**, *24*, 874–882. [CrossRef]
19. Chen, Z.; Zhao, D. 4Pi focusing of spatially modulated radially polarized vortex beams. *Opt. Lett.* **2012**, *37*, 1286–1288. [CrossRef]
20. Wang, X.; Gong, L.; Zhu, Z.; Gu, B.; Zhan, Q. Creation of identical multiple focal spots with three-dimensional arbitrary shifting. *Opt. Express* **2017**, *25*, 17737–17745. [CrossRef]
21. Ouyang, X. Tight-focusing properties of the chirp modulation linearly polarized circular Airy Gaussian vortex beam. *Results Phys.* **2022**, *32*, 105119. [CrossRef]
22. Chen, Z.; Hua, L.; Pu, J. Tight focusing of light beams: Effect of polarization, phase, and coherence. *Prog. Optics. Elsevier* **2012**, *57*, 219–260.
23. Pu, J.; Zhang, Z. Tight focusing of spirally polarized vortex beams. *Opt. Laser Technol.* **2010**, *42*, 186–191. [CrossRef]
24. Lin, H.; Zhou, X.; Chen, Z.; Sasaki, O.; Li, Y.; Pu, J. Tight focusing properties of a circular partially coherent Gaussian beam. *JOSA A* **2018**, *35*, 1974–1980. [CrossRef] [PubMed]
25. Richards, B.; Wolf, E. Electromagnetic diffraction in optical systems. II. Structure of the image field in an aplanatic system. *Proc. R. Soc. Lond. A Math. Phys. Eng. Sciences. R. Soc.* **1959**, *253*, 358–379.
26. Kotlyar, V.V.; Nalimov, A.G. Analytical expression for radiation forces on a dielectric cylinder illuminated by a cylindrical Gaussian beam. *Opt. Express* **2006**, *14*, 6316–6321. [CrossRef]
27. Kotlyar, V.V.; Nalimov, A.G.; Stafeev, S.S. Focusing a vortex laser beam with polarization conversion. *Photonics* **2021**, *8*, 480. [CrossRef]
28. Lavanya, M.; Thiruarul, D.; Rajesh, K.B.; Mahadevan, G.; Velauthapillai, D.; Jaroszewicz, Z. Energy flux density for higher-order cylindrical vector vortex beam tightly focused through a dielectric interface. *J. Opt.* **2021**, *50*, 548–558. [CrossRef]
29. Kotlyar, V.V.; Stafeev, S.S.; Nalimov, A.G. Energy backflow in the focus of a light beam with phase or polarization singularity. *Phys. Rev. A* **2019**, *99*, 033840. [CrossRef]

photonics

Article

Evolution Properties of a Partially Coherent Twisted Laguerre-Gaussian Pulsed Beam Propagating through Anisotropic Atmospheric Turbulence

Ying Xu [1], Yonggen Xu [1,*] and Tiejun Wang [2,3]

[1] Department of Physics, School of Science, Xihua University, Chengdu 610039, China
[2] State Key Laboratory of High Field Laser Physics, Shanghai Institute of Optics and Fine Mechanics, CAS Center for Excellence in Ultra-Intense Laser Science, Chinese Academy of Sciences, Shanghai 201800, China
[3] Center of Materials Science and Optoelectronics Engineering, University of Chinese Academy of Sciences, Beijing 100049, China
* Correspondence: xuyonggen06@126.com

Abstract: Analytical expressions for the cross-spectral density matrix of a partially coherent twisted Laguerre-Gaussian pulsed (PCTLGP) beam in anisotropic atmospheric turbulence are derived based on the extended Huygens–Fresnel principle. Numerical results indicate that the atmospheric turbulence induces the degeneration of the spectral intensity distribution of the PCTLGP beam, and the PCTLGP beam also shows different evolution properties on propagation in weaker turbulence and stronger turbulence. The PCTLGP beam with a negative twisted factor exhibits an advantage over the Laguerre-Gaussian pulsed beam for reducing the atmospheric turbulence-induced degeneration, and this advantage is further strengthened with increasing the topological charge, mode order and absolute value of the twisted factor. In addition, we also find that the pulse duration will affect the spectral intensity of the PCTLGP beam in turbulence. This kind of beam will show potential application value in free-space optical communications and remote sensing.

Keywords: PCTLGP beam; twist phase; pulse duration; spectral intensity; anisotropic atmospheric turbulence

Citation: Xu, Y.; Xu, Y.; Wang, T. Evolution Properties of a Partially Coherent Twisted Laguerre-Gaussian Pulsed Beam Propagating through Anisotropic Atmospheric Turbulence. *Photonics* **2022**, *9*, 707. https://doi.org/10.3390/photonics9100707

Received: 14 September 2022
Accepted: 27 September 2022
Published: 28 September 2022

1. Introduction

Optical pulse has attracted enormous attention since the concept of the Gaussian Schell-Model pulsed (GSMP) beam was introduced by Paakkonen et al. [1,2]. Subsequently, the generation and propagation of various pulsed beams have been extensively studied [3–9]. In recent years, the partially coherent (PC) beams with a vortex phase carrying orbital angular momentum (OAM) have attracted considerable interest due to their various applications in free-space optical communications (FSOC), optical tweezers, astronomy [9–16], etc. The partially coherent Laguerre-Gaussian (PCLG) beam is a classical vortex beam that shows more obvious advantages in reducing the turbulence-induced scintillation [13–15]. A large number of studies have indicated that PCLG beams can be applied to the optical manipulation of Bose–Einstein condensates, optical imaging, biomedicine, quantum information coding and other fields [4,13–20]. Based on the merits of the OAM mode and the pulsed beam, the propagation of Laguerre-Gaussian pulsed (LGP) beams in the different channels have been investigated widely [4,21–23]. For examples, Martínez-Matos et al. studied the spatial pulse shaping of ultrashort Laguerre-Gaussian beams at the focal plane [21]. The time-varying of the pulse Laguerre-Gaussian beam propagating in free space has been analyzed by Gao et al. [22]. Li et al. investigated the probability distribution of OAM modes of the ultrashort LGP beam propagating through oceanic turbulence [23].

Many studies have been conducted on the propagation of PC beams through a turbulent atmosphere [6,7,9,13–15,20]. Researches have demonstrated that the anisotropic atmospheric turbulence will seriously affect the characteristics of the beam in practical applications [7,13,15]. Therefore, reducing the influence of turbulence on beam propagation is necessary. The modulation of the vortex phase and twist phase is one of the important methods to effectively suppress the influence of turbulence [24–30]. Peng et al. introduced a PC twisted beam, named the twisted Laguerre-Gaussian Schell-model beam, which carries both the vortex phase and twist phase. The results show that the influence of atmospheric turbulence on the propagation characteristics for this beam can be effectively reduced by adjusting the twisted factor, topological charge and mode order [13,16,28–30].

However, to the best of our knowledge, the evolution properties of a partially coherent twisted Laguerre-Gaussian pulsed (PCTLGP) beam propagating through anisotropic atmospheric turbulence are still unreported so far. In this paper, we focus on the evolution of the spectral intensity distribution for the PCTLGP beam in anisotropic non-Kolmogorov turbulence. Analytical expressions for the cross-spectral density matrix (CSDM) of the PCTLGP beam propagating through the turbulence are derived, and numerical results are illustrated in detail. The results have reference value for the application of beams in FSOC and other fields.

2. Theoretical Formulation

In the source plane ($z = 0$), the statistical properties of the partially coherent Laguerre-Gaussian pulsed (PCLGP) beam at points $\boldsymbol{r}_1 = (r_{1x}, r_{1y})$ and $\boldsymbol{r}_2 = (r_{2x}, r_{2y})$, and at different instant times t_1 and t_2, can be characterized by a two-point two-time mutual coherent function matrix (MCFM), which is given by [7,9,13,15,20]:

$$\begin{aligned}
\Gamma^{(0)}(\boldsymbol{r}_1,\boldsymbol{r}_2,0;t_1,t_2) &= \begin{bmatrix} \Gamma^{(0)}_{xx}(\boldsymbol{r}_1,\boldsymbol{r}_2,0;t_1,t_2) & \Gamma^{(0)}_{xy}(\boldsymbol{r}_1,\boldsymbol{r}_2,0;t_1,t_2) \\ \Gamma^{(0)}_{yx}(\boldsymbol{r}_1,\boldsymbol{r}_2,0;t_1,t_2) & \Gamma^{(0)}_{yy}(\boldsymbol{r}_1,\boldsymbol{r}_2,0;t_1,t_2) \end{bmatrix} \\
&= \frac{1}{\left(2^{2n+l}n!\right)^2}\sum_{a_1=0}^{n}\sum_{b_1=0}^{l}\sum_{p_1=0}^{n}\sum_{q_1=0}^{l}\left(i^{b_1}\right)^* i^{q_1}\binom{n}{a_1}\binom{l}{b_1}\binom{n}{p_1}\binom{l}{q_1} \\
&\times H_{l+2a_1-b_1}\left(\frac{\sqrt{2}r_{1x}}{\sigma}\right)H_{l+2p_1-q_1}\left(\frac{\sqrt{2}r_{2x}}{\sigma}\right)H_{2n-2a_1+b_1}\left(\frac{\sqrt{2}r_{1y}}{\sigma}\right)H_{2n-2p_1+q_1}\left(\frac{\sqrt{2}r_{2y}}{\sigma}\right) \\
&\times \exp\left[-\frac{r_1^2+r_2^2}{\sigma^2}-\frac{(\boldsymbol{r}_1-\boldsymbol{r}_2)^2}{2\delta^2}\right]\exp\left[-\frac{t_1^2+t_2^2}{2T_0^2}-\frac{(t_1-t_2)^2}{2T_c^2}\right] \\
&\times \exp[-i\omega_0(t_2-t_1)],
\end{aligned} \tag{1}$$

where $H(\cdot)$ being the Hermite polynomial, n and l denote the mode order and the topological charge, respectively. Their meanings are consistent throughout the paper. $\binom{n}{a_1}$, $\binom{l}{b_1}$, $\binom{n}{p_1}$ and $\binom{l}{q_1}$ are the binomial coefficients. σ and δ describe the waist width and the spatial coherence length, respectively. T_0 and T_c are the pulse duration and the temporal coherence length, respectively. ω_0 is the central angular frequency of the pulse. "*" describes the complex conjugate.

According to the Fourier-transform [9]

$$W_{\alpha\beta}(\boldsymbol{r}_1,\boldsymbol{r}_2,0;\omega_1,\omega_2) = \iint \Gamma^{(0)}_{\alpha\beta}(\boldsymbol{r}_1,\boldsymbol{r}_2,0;t_1,t_2)\exp[-i(\omega_2 t_2-\omega_1 t_1)]dt_1 dt_2 \quad (\alpha,\beta = x,y), \tag{2}$$

the two-point two-frequency CSDM of an PCLGP beam at $z = 0$ plane can be expressed as [7,13,20]:

$$\begin{aligned} W_{\alpha\beta}(\boldsymbol{r}_1,\boldsymbol{r}_2,0;\omega_1,\omega_2) &= \tfrac{T_0}{2\pi\Omega_0}\exp\left[-\tfrac{r_1^2+r_2^2}{\sigma^2}-\tfrac{(r_1-r_2)^2}{2\delta^2}\right] \\ &\times\exp\left[-\tfrac{(\omega_1-\omega_0)^2+(\omega_2-\omega_0)^2}{2\Omega_0^2}-\tfrac{(\omega_1-\omega_2)^2}{2\Omega_c^2}\right] \\ &\times\tfrac{1}{(2^{2n+l}n!)^2}\sum_{a_1=0}^{n}\sum_{b_1=0}^{l}\sum_{p_1=0}^{n}\sum_{q_1=0}^{l}(i^{b_1})^* i^{q_1}\binom{n}{a_1}\binom{l}{b_1}\binom{n}{p_1}\binom{l}{q_1} \\ &\times H_{l+2a_1-b_1}\left(\tfrac{\sqrt{2}r_{1x}}{\sigma}\right)H_{l+2p_1-q_1}\left(\tfrac{\sqrt{2}r_{2x}}{\sigma}\right)H_{2n-2a_1+b_1}\left(\tfrac{\sqrt{2}r_{1y}}{\sigma}\right)H_{2n-2p_1+q_1}\left(\tfrac{\sqrt{2}r_{2y}}{\sigma}\right) \end{aligned} \tag{3}$$

where $\Omega_0 = \sqrt{\frac{1}{T_0^2}+\frac{2}{T_c^2}}$ and $\Omega_c = \frac{\sqrt{T_0^2+T_c^2}}{T_0^2}$ denote the spectral width and the spectral coherent width of the pulse, respectively.

According to the extended Huygens–Fresnel principle, the two-point two-frequency CSDM at the arbitrary two points $\boldsymbol{\rho}_1 = (x_1, y_1)$ and $\boldsymbol{\rho}_2 = (x_2, y_2)$ in the received plane ($z = L$) can be expressed as [7,9]:

$$\begin{aligned} W_{\alpha\beta}(\boldsymbol{\rho}_1,\boldsymbol{\rho}_2,L;\omega_1,\omega_2) &= \tfrac{\omega_1\omega_2}{(2\pi cL)^2}\exp\left[-\tfrac{i(\omega_2-\omega_1)L}{c}\right]\int\iiint d^2\boldsymbol{r}_1 d^2\boldsymbol{r}_2 \\ &\times W_{\alpha\beta}(\boldsymbol{r}_1,\boldsymbol{r}_2,0;\omega_1,\omega_2)\langle\exp[\psi_1+\psi_2^*]\rangle \\ &\times\exp\left[-\tfrac{i\omega_1}{2cL}(\boldsymbol{r}_1-\boldsymbol{\rho}_1)^2+\tfrac{i\omega_2}{2cL}(\boldsymbol{r}_2-\boldsymbol{\rho}_2)^2\right], \end{aligned} \tag{4}$$

where L is the propagation distance, and c is the speed of light in a vacuum. $\langle\cdot\rangle$ denotes the ensemble average, and the turbulent factor $\langle\exp[\psi_1 + \psi_2^*]\rangle$ is given by [3,13,15,20]:

$$\langle\exp[\psi_1+\psi_2^*]\rangle \approx \exp\left[-\frac{(\boldsymbol{\rho}_1-\boldsymbol{\rho}_2)^2+(\boldsymbol{\rho}_1-\boldsymbol{\rho}_2)(\boldsymbol{r}_1-\boldsymbol{r}_2)+(\boldsymbol{r}_1-\boldsymbol{r}_2)^2}{\rho_0^2}\right], \tag{5}$$

where ρ_0 denotes the coherence length of a spherical wave propagating in turbulence and is given by [3,13,16]:

$$\rho_0 = \left[\frac{\pi^2k^2L}{3}\int_0^\infty \kappa^3\Phi_n(\kappa)d\kappa\right]^{-1/2}, \tag{6}$$

where $k = 2\pi/\lambda$, is the wave number, with λ as the wavelength. $\Phi_n(\boldsymbol{\kappa})$ represents the spatial power spectrum of the refractive-index fluctuations of anisotropic atmospheric turbulence with $\boldsymbol{\kappa} = (\kappa_x, \kappa_y, \kappa_z)$ denoting the spatial wavenumber, and is given by [7,13]:

$$\begin{aligned} \Phi_n(\boldsymbol{\kappa}) &= \frac{A(\alpha)\widetilde{C}_n^2\xi_x\xi_y}{(\xi_x^2\kappa_x^2+\xi_y^2\kappa_y^2+\kappa_z^2+\kappa_0^2)^{\alpha/2}}\exp\left(-\frac{\xi_x^2\kappa_x^2+\xi_y^2\kappa_y^2+\kappa_z^2}{\kappa_m^2}\right) \\ &(3<\alpha<4), \end{aligned} \tag{7}$$

where α is the spectral power law, $\widetilde{C}_n^2$ is the generalized refractive structure constant with units $\mathrm{m}^{3-\alpha}$. ξ_x and ξ_y are the anisotropic factors that can describe the asymmetry of turbulence cells. $A(\alpha) = \cos(\alpha\pi/2)\Gamma(\alpha-1)/(4\pi^2)$, $c(\alpha) = [2\pi A(\alpha)\Gamma(5/2-\alpha/2)/3]^{1/(\alpha-5)}$ with $\Gamma(\cdot)$ represents the Gamma function. κ_x, κ_y and κ_z are three components of the $\boldsymbol{\kappa}$ along the x, y and z directions. $\kappa_m = c(\alpha)/l_0$ and $\kappa_0 = 2\pi/L_0$. l_0 and L_0 are the turbulence inner and outer scales, respectively.

After the twist phase modulation is applied to PCLGP beam, with the setting $\omega_1 = \omega_2 = \omega$, the CSDM of a PCTLGP beam in the received plane can be obtained [7,9,13]:

$$\begin{aligned}
W_{\alpha\beta}(\boldsymbol{\rho}_1, \boldsymbol{\rho}_2, L; \omega) &= \frac{\omega^2}{\left(2^{2n+l}\pi cLn!\right)^2} \frac{T_0}{8\pi\Omega_0} \exp\left[-\frac{(\omega-\omega_0)^2}{\Omega_0^2}\right] \\
&\times \int \iiint d^2\mathbf{r}_1 d^2\mathbf{r}_2 \sum_{a_1=0}^{n} \sum_{b_1=0}^{l} \sum_{p_1=0}^{n} \sum_{q_1=0}^{l} \left(i^{b_1}\right)^* i^{q_1} \binom{n}{a_1}\binom{l}{b_1}\binom{n}{p_1}\binom{l}{q_1} \\
&\times H_{l+2a_1-b_1}\left(\frac{\sqrt{2}r_{1x}}{\sigma}\right) H_{l+2p_1-q_1}\left(\frac{\sqrt{2}r_{2x}}{\sigma}\right) H_{2n-2a_1+b_1}\left(\frac{\sqrt{2}r_{1y}}{\sigma}\right) H_{2n-2p_1+q_1}\left(\frac{\sqrt{2}r_{2y}}{\sigma}\right) \\
&\times \exp\left[-ik\eta\left(r_{1x}r_{2y} - r_{2x}r_{1y}\right)\right] \exp\left[-\frac{r_1^2+r_2^2}{\sigma^2} - \frac{(\mathbf{r}_1-\mathbf{r}_2)^2}{2\delta^2}\right] \\
&\times \langle \exp[\psi_1 + \psi_2^*] \rangle \exp\left[-\frac{i\omega}{2cL}(\mathbf{r}_1 - \boldsymbol{\rho}_1)^2 + \frac{i\omega}{2cL}(\mathbf{r}_2 - \boldsymbol{\rho}_2)^2\right],
\end{aligned} \tag{8}$$

where η denotes the twisted factor.

Substituting Equations (5)–(7) into Equation (8), we can obtain the expression of the CSDM for a PCTLGP beam propagating through the anisotropic atmospheric turbulence

$$\begin{aligned}
W_{\alpha\beta}(\boldsymbol{\rho}_1, \boldsymbol{\rho}_2, L; \omega) &= \frac{W_1 W_2 T_0}{8\pi\Omega_0} \frac{\omega^2}{\left(2^{2n+l}cLn!\right)^2} \exp\left[-\frac{(\omega-\omega_0)^2}{\Omega_0^2}\right] \\
&\times \exp\left[-\frac{i\omega}{2cL}\left(\boldsymbol{\rho}_1^2 - \boldsymbol{\rho}_2^2\right)\right] \exp\left[-\frac{1}{\rho_0^2}(\boldsymbol{\rho}_1 - \boldsymbol{\rho}_2)^2\right] \\
&\times \sum_{a_1=0}^{n} \sum_{b_1=0}^{l} \sum_{p_1=0}^{n} \sum_{q_1=0}^{l} \sum_{a_2=0}^{l+2a_1-b_1} \sum_{b_2=0}^{l+2a_1-b_1-a_2} \sum_{p_2=0}^{[a_2/2]} \sum_{q_2=0}^{[(l+2p_1-q_1)/2]} \sum_{a_3=0}^{l+2p_1-q_1+a_2-2p_2-2q_2} \sum_{b_3=0}^{a_3} \sum_{p_3=0}^{[b_3/2]} \sum_{q_3=0}^{[(2n-2a_1+b_1)/2]} \\
&\times \sum_{a_4=0}^{2n-2a_1+b_1+b_3-2p_3-2q_3} \sum_{b_4=0}^{[b_2/2]} \sum_{p_4=0}^{[(a_3-b_3)/2]} \sum_{q_4=0}^{[a_4/2]} \sum_{q_5=0}^{[(2n-2p_1+q_1)/2]} \left(i^{b_1}\right)^* i^{q_1} \Delta_1\Delta_2\Delta_3\Delta_4 \\
&\times \left(\frac{\omega\eta}{c}\right)^{b_2+a_3-2b_4-2p_3-2p_4} H_{l+2a_1-a_2-b_1-b_2}\left(\frac{\sqrt{2}u_{1x}}{\sqrt{A^2\sigma^2-2A}}\right) H_{l+a_2-a_3+2p_1-q_1-2p_2-2q_2}\left(\frac{\sqrt{2}iE_2}{2\sqrt{E_1}}\right) \\
&\times H_{2n-2a_1-a_4+b_1+b_3-2p_3-2q_3}\left(\frac{\sqrt{2}iM_2}{2\sqrt{E_3}}\right) H_{2n+a_3+a_4+b_2-b_3-2b_4-2p_1+q_1-2p_4-2q_4-2q_5}\left(\frac{iM_4}{2\sqrt{M_3}}\right)
\end{aligned} \tag{9}$$

with

$$\left\{\begin{aligned}
& A = \frac{1}{\sigma^2} + \frac{1}{2\delta^2} + \frac{1}{\rho_0^2} + \frac{i\omega}{2cL}, B = \frac{1}{\delta^2} + \frac{2}{\rho_0^2}, u_{1x} = \frac{i\omega x_1}{cL} - \frac{x_1-x_2}{\rho_0^2}, \\
& u_{1y} = \frac{i\omega y_1}{cL} - \frac{y_1-y_2}{\rho_0^2}, u_{2x} = -\frac{i\omega x_2}{cL} + \frac{x_1-x_2}{\rho_0^2}, u_{2y} = -\frac{i\omega y_2}{cL} + \frac{y_1-y_2}{\rho_0^2}, \\
& E_1 = -\frac{B^2}{4A} + A^*, E_2 = \frac{u_{1x}B}{2A} + u_{2x}, E_3 = \frac{\omega^2\eta^2}{4c^2E_1} + A, \\
& M_1 = \frac{\omega^2\eta^2 B}{4c^2AE_1} + B, M_2 = \frac{i\omega\eta E_2}{2cE_1} + u_{1y}, M_3 = \frac{\omega^2\eta^2}{4c^2A} + \frac{\omega^2\eta^2B^2}{16c^2A^2E_1} - \frac{M_1^2}{4E_3} + A^*, \\
& M_4 = -\frac{i\omega\eta u_{1x}}{2cA} - \frac{i\omega\eta BE_2}{4cAE_1} + \frac{M_1M_2}{2E_3} + u_{2y}, \\
& W_1 = \exp\left(\frac{u_{1x}^2}{4A}\right)\exp\left(\frac{E_2^2}{4E_1}\right)\exp\left(\frac{M_2^2}{4E_3}\right)\exp\left(\frac{M_4^2}{4M_3}\right), \\
& W_2 = \left(\frac{\sqrt{A\sigma^2-2}}{2i\sigma^2}\right)^l \left(\frac{1}{\sqrt{AE_1}}\right)^{l+1} \left(\frac{2}{\sigma^4}\right)^n \left(\frac{1}{\sqrt{E_3M_3}}\right)^{2n+1}, \\
& \Delta_1 = \binom{n}{a_1}\binom{l}{b_1}\binom{n}{p_1}\binom{l}{q_1}\binom{a_3}{b_3}\binom{l+2a_1-b_1}{a_2}\binom{l+2a_1-b_1-a_2}{b_2} \\
& \times \binom{l+2p_1-q_1+a_2-2p_2-2q_2}{a_3}\binom{2n-2a_1+b_1+b_3-2p_3-2q_3}{a_4}, \\
& \Delta_2 = \frac{a_2!a_4!b_2!b_3!}{b_4!p_2!q_2!p_3!q_3!p_4!q_4!q_5!(a_2-2p_2)!(a_4-2q_4)!(b_2-2b_4)!(b_3-2p_3)!} \\
& \times \frac{(a_3-b_3)!(2n-2p_1+q_1)!(l+2p_1-q_1)!(2n-2a_1+b_1)!}{(a_3-b_3-2p_4)!(2n-2p_1+q_1-2q_5)!(l+2p_1-q_1-2q_2)!(2n-2a_1+b_1-2q_3)!}, \\
& \Delta_3 = \left(\frac{2E_3-\sigma^2AE_3}{4A}\right)^{a_1}\left(\frac{2\sqrt{A}}{i\sqrt{\sigma^2AE_3-2E_3}}\right)^{b_1}\left(\frac{M_3}{2E_1}\right)^{p_1}\left(\sqrt{\frac{2E_1}{M_3}}\right)^{q_1} \\
& \times \left(\frac{B}{i\sqrt{A^2E_1\sigma^2-2AE_1}}\right)^{a_2}\left(-\frac{\sqrt{2}}{\sqrt{\sigma^2A^2M_3-2AM_3}}\right)^{b_2}\left(\frac{2A^2E_1\sigma^2-4AE_1}{B^2}\right)^{p_2}\left(\sigma^2E_1\right)^{q_2}, \\
& \Delta_4 = \left(\frac{B}{2iA\sqrt{2E_1M_3}}\right)^{a_3}\left(-\frac{A\sqrt{2M_3}}{B\sqrt{E_3}}\right)^{b_3}(2E_1E_3)^{p_3}\left(\sigma^2E_3\right)^{q_3} \\
& \times \left(\frac{M_1}{\sqrt{2E_3M_3}}\right)^{a_4}\left(-\frac{\sigma^2A^2M_3-2AM_3}{2}\right)^{b_4}\left(\frac{4A^2E_1M_3}{B^2}\right)^{p_4}\left(-\frac{2E_3M_3}{M_1^2}\right)^{q_4}\left(\frac{\sigma^2M_3}{2}\right)^{q_5}.
\end{aligned}\right. \tag{10}$$

In the above derivations, we have used the following formulas [13,20]:

$$H_n(x+y) = \frac{1}{2^{n/2}} \sum_{m=0}^{n} \binom{n}{m} H_m\left(\sqrt{2}x\right) H_{n-m}\left(\sqrt{2}y\right), \tag{11}$$

$$\int_{-\infty}^{\infty} x^n \exp\left[-(x-\gamma)^2\right] dx = (2i)^{-n} \sqrt{\pi} H_n(i\gamma), \tag{12}$$

$$H_n(x) = \sum_{m=0}^{[n/2]} (-1)^m \frac{n!}{m!(n-2m)!} (2x)^{n-2m}, \tag{13}$$

$$\int_{-\infty}^{\infty} \exp\left[-(x-y)^2\right] H_n(mx) dx = \sqrt{\pi}\left(1-m^2\right)^{n/2} H_n\left[\frac{my}{(1-m^2)^{1/2}}\right]. \tag{14}$$

The spectral density along the observation plane at the "coincident point" $\rho_1 = \rho_2 = \rho = (x, y)$ is given by the formula [7,9,13]:

$$S(\rho, L; \omega) = W_{xx}(\rho, \rho, L; \omega) + W_{yy}(\rho, \rho, L; \omega). \tag{15}$$

According to Equations (9), (10) and (15), the spectral density of a PCTLGP beam is mainly determined by the twisted factor η, topological charge l, mode order n, pulse duration T_0 and turbulence parameters.

3. Numerical Simulation and Discussion

In this section, we numerically calculate the spectral intensity of the PCTLGP beam propagating through the anisotropic non-Kolmogorov turbulence by using the above analyses. The parameters in the following calculations were set as λ = 1550 nm, σ = 20 mm, δ = 10 mm, ω_0 = 2.9788 rad/fs, $\omega = 1.02\omega_0$, T_0 = 100 fs, T_c = 50 fs, α = 3.4, L_0 = 50 m, l_0 = 10 mm; the other parameters were specified in the figure captions.

To understand the twist phase induced evolution of the spectral intensity distribution on propagation, Figure 1 shows both the 3D-normalized spectral intensity distribution of the PCLGP beam (η = 0 m^{-1}) and the PCTLGP beam ($\eta = -5 \times 10^{-4}$ m^{-1}) at several propagation distances in anisotropic non-Kolmogorov turbulence. In Figure 1(a1,b1), one finds that the initial beam profiles of the PCLGP beam and the PCTLGP beam show a multi-ring dark hollow beam profile. From Figure 1(a1–a8), the evolution process of the PCLGP beam can be seen: the spectral intensity in the dark center region gradually increases with the increase in the propagation distance, and the dark ring gradually disappears, and the outer ring gradually degenerates, and finally the beam spot evolves into a Gaussian shape. Similarly, for a PCTLGP beam (see Figure 1(b1–b8)), the spectral intensity distribution gradually evolves to a flat-top distribution, and eventually converges into a Gaussian-like distribution. This is consistent with the conclusions in Refs. [15,31–33]: The Kurtosis parameters of the arbitrary beams will converge to a constant 2 as the propagation distance $L \to \infty$, which means that arbitrary beams will evolve into a Gaussian shape in atmospheric turbulence with the increase in L. Compared to Figure 1a,b, it can be found that the PCLGP beam evolves into Gaussian distribution more quickly than the PCTLGP beam during transmission, which indicates that the twist phase plays an important role in resistance to the turbulence-induced degeneration.

Next, we focus on the influence of the topological charge on the evolution properties during propagation. Figure 2 shows the normalized spectral intensity distribution of the PCTLGP beam propagating through the anisotropic non-Kolmogorov turbulence ($\widetilde{C}_n^2 = 1 \times 10^{-17}$ m$^{3-\alpha}$) at several propagation distances for the different topological charges. Figure 2a,c,d presents that the normalized spectral intensity distribution of the PCTLGP beam gradually evolves from a dark hollow distribution to a flat-topped distribution when $n = 0$, and finally converges into a Gaussian-like distribution. Figure 2b displays the normalized spectral intensity distribution of the PCTLGP beam when $n \neq 0$, and the results shown agree with those discussed in Figure 1. Importantly, we find that the evolution of the PCTLGP beam can be accelerated with the decrease in l, indicating that the PCTLGP beam with a larger l has a stronger ability to resist the degeneration caused by the turbulent atmosphere. In Ref. [14], the partially coherent radially polarized vortex beam studied by Yu et.al. also showed a similar phenomenon. Furthermore, it can be seen that the progress of the beam evolving into a Gaussian-like shape will not be accelerated with the increase in the propagation distance when it propagates for a long distance ($L > 3$ km), which is caused by the too weak turbulence, but the beam spot is larger with the increase in L.

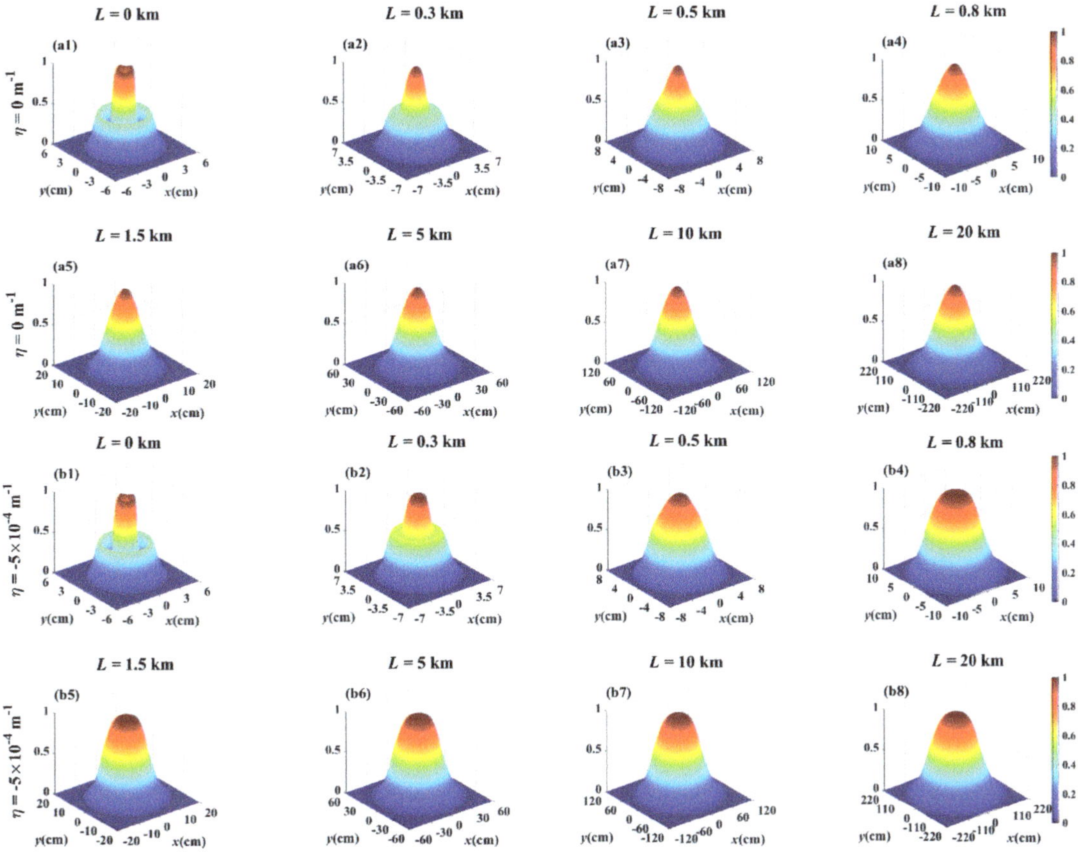

Figure 1. Normalized spectral intensity distribution of (**a1–a8**) the PCLGP beam and (**b1–b8**) the PCTLGP beam at several propagation distances in anisotropic turbulence ($\xi_x = \xi_y = 3$, $l = 1$, $n = 1$, $\widetilde{C}_n^2 = 1 \times 10^{-14}$ $m^{3-\alpha}$).

For comparison, Figure 3 depicts the normalized spectral intensity distribution of the PCTLGP beam propagating through the anisotropic non-Kolmogorov turbulence ($\widetilde{C}_n^2 = 1 \times 10^{-13}$ $m^{3-\alpha}$) at the same propagation distances and the topological charges with Figure 2. Figure 3a–d shows the progress of the normalized spectral intensity distribution for the PCTLGP beam gradually evolves to a Gaussian-like distribution under different mode orders and topological charges. It can be easily found that the degeneration of the PCTLGP beam is accelerated by the stronger turbulence by comparing Figures 2 and 3. It is also found that the PCTLGP beam shows different evolution properties when propagating in weaker turbulence and stronger turbulence. The degradation progress of the beam is accelerated with increasing L when propagating for $L > 3$ km in stronger atmospheric turbulence, and it eventually converges to a Gaussian-like distribution, while the weakly turbulent atmosphere has little effect on the progress of the beam evolving into a Gaussian-like shape, and only the beam spot becomes lager with the increase in L.

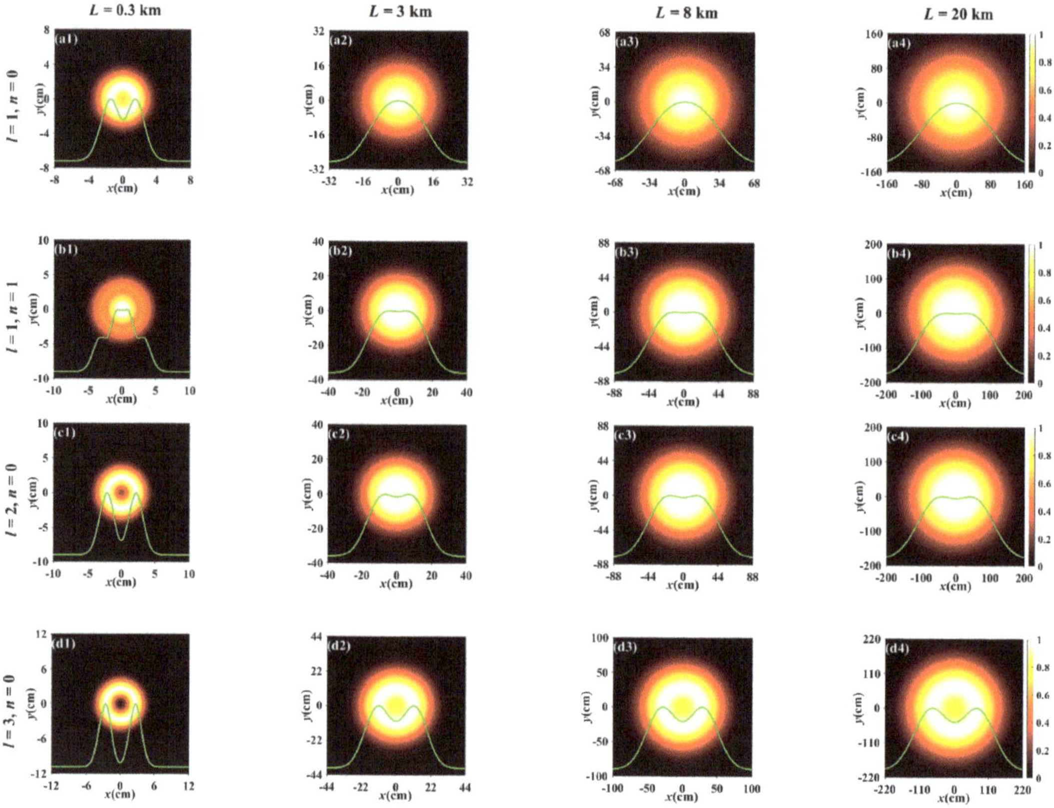

Figure 2. Normalized spectral intensity distribution of the PCTLGP beam at several propagation distances in anisotropic turbulence for different mode orders and topological charges ($\xi_x = \xi_y = 3$, $\widetilde{C}_n^2 = 1 \times 10^{-17}\ \mathrm{m}^{3-\alpha}$, $\eta = -5 \times 10^{-4}\ \mathrm{m}^{-1}$).

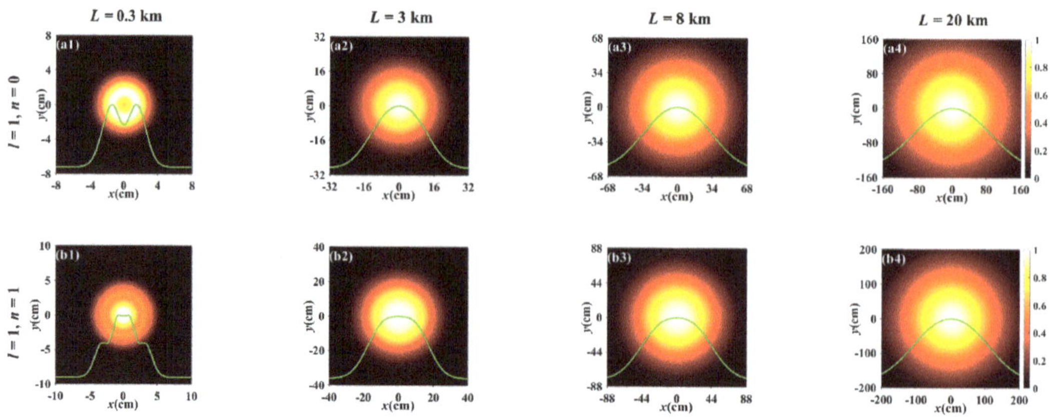

Figure 3. *Cont.*

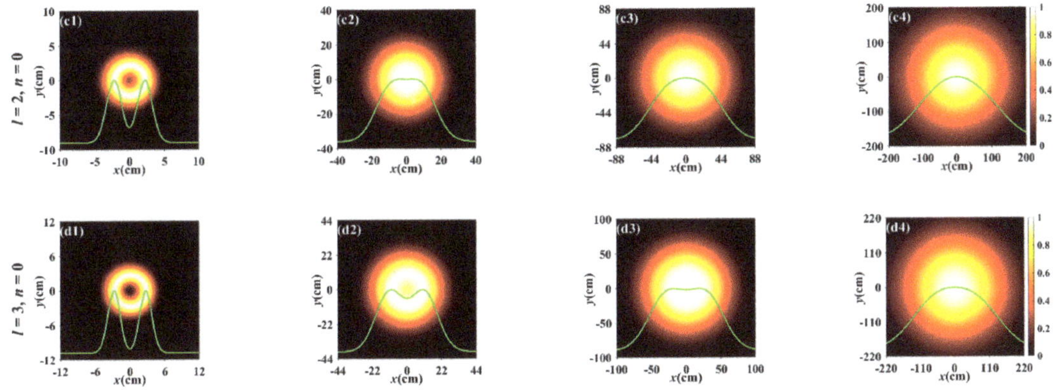

Figure 3. Normalized spectral intensity distribution of the PCTLGP beam at several propagation distances in anisotropic turbulence for different mode orders and topological charges ($\xi_x = \xi_y = 3$, $\widetilde{C}_n^2 = 1 \times 10^{-13}$ m$^{3-\alpha}$, $\eta = -5 \times 10^{-4}$ m^{-1}).

To explore the influence of the twisted factor on the evolution of the beam, Figure 4 studies the normalized spectral intensity distribution of the PCTLGP beam with $\eta = -1 \times 10^{-3}$ m^{-1} propagating through the weaker turbulence ($\widetilde{C}_n^2 = 1 \times 10^{-17}$ m$^{3-\alpha}$). Figure 4a–d illustrates the evolution of the PCTLGP beam in the turbulent atmosphere, and it can be seen that the degeneration of the PCTLGP beam will be suppressed with the increase in l. Compared to Figures 2 and 4, it can be found that the PCTLGP beam with the smaller η has a stronger ability to maintain the dark hollow distribution when propagating in the weaker turbulence, meaning that the PCTLGP beam has an advantage over the PCLGP beam for resisting the degeneration caused by turbulence, and the advantage is further enhanced with decreasing η.

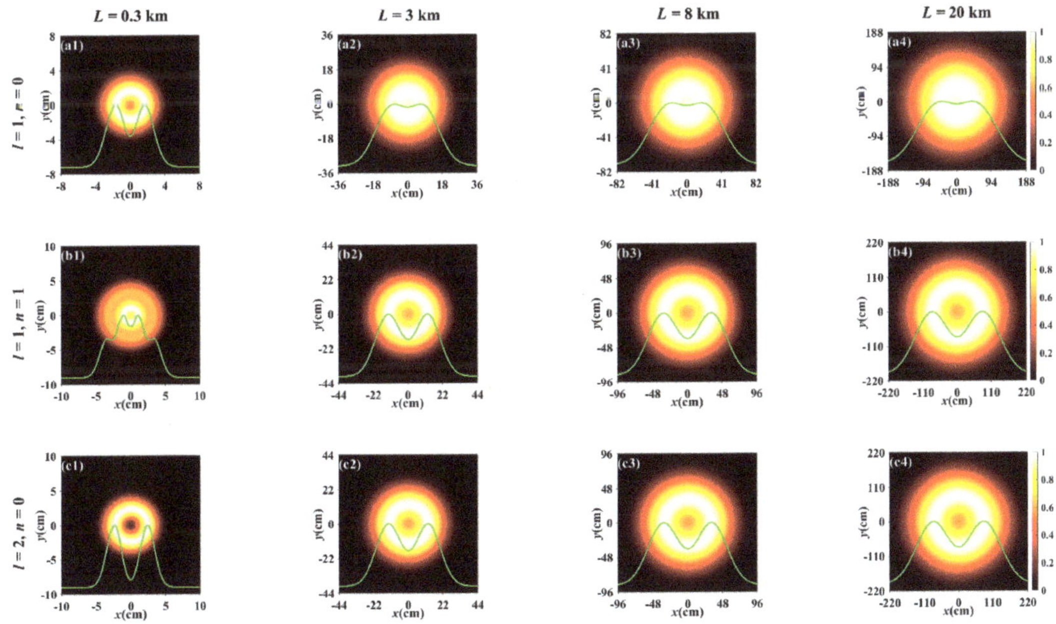

Figure 4. *Cont.*

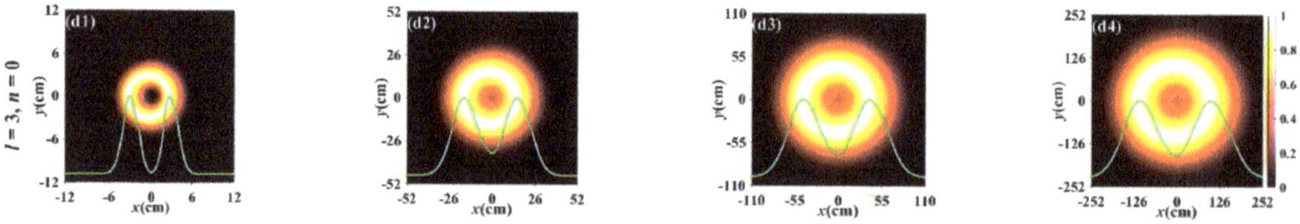

Figure 4. Normalized spectral intensity distribution of the PCTLGP beam at several propagation distances in anisotropic turbulence for different mode orders and topological charges ($\xi_x = \xi_y = 3$, $\widetilde{C}_n^2 = 1 \times 10^{-17}$ m$^{3-\alpha}$, $\eta = -1 \times 10^{-3}$ m^{-1}).

Figure 5 illustrates the evolution of the normalized spectral intensity distribution for a PCTLGP beam with $\eta = -1 \times 10^{-3}$ m^{-1} propagating through the stronger turbulence ($\widetilde{C}_n^2 = 1 \times 10^{-13}$ m$^{3-\alpha}$). Figure 5a–d depicts the degeneration of the PCTLGP beam with different l and n in the turbulent atmosphere. Compared to Figures 3 and 5, it can be seen that the above conclusions obtained from comparing Figures 2 and 4 are also applicable to the stronger turbulence, that is to say, whether in stronger turbulence or weaker turbulence, the PCTLGP beam with a smaller η can more effectively resist the influence of the turbulence on the degeneration. Next, we focus on the effect of the joint modulation of the twist factor and the topological charge on the beam evolution.

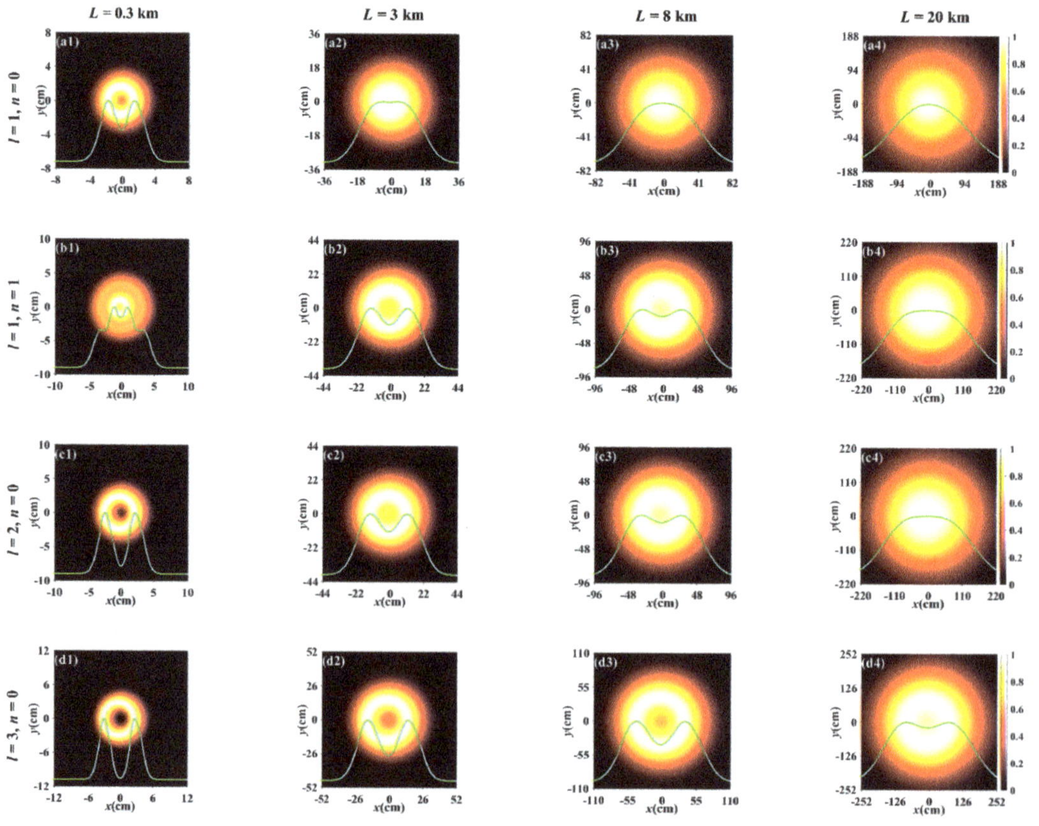

Figure 5. Normalized spectral intensity distribution of the PCTLGP beam at several propagation distances in anisotropic turbulence for different mode orders and topological charges ($\xi_x = \xi_y = 3$, $\widetilde{C}_n^2 = 1 \times 10^{-13}$ m$^{3-\alpha}$, $\eta = -1 \times 10^{-3}$ m^{-1}).

Figure 6 gives the evolution of the normalized spectral intensity distribution of the PCTLGP beam on propagation through anisotropic non-Kolmogorov turbulent atmosphere for the different topological charges and twisted factors. Figure 6a–c studies the influence of the sign of the twisted factor on the beam evolution, and the results present that the process of the beam evolving into a Gaussian shape is accelerated by modulating the sign of twisted factor is "+". Conversely, the evolution process is suppressed when the sign is "−". In addition, Figure 6c,d shows that the values of l and η will affect the degeneration of the PCTLGP beam, and one finds that the beam degenerates more slowly when both l and $|\eta|$ are larger, which means that the PCTLGP beam with a negative twisted factor shows an advantage in resisting the turbulence-induced degeneration, and this advantage is further enhanced with increasing l and $|\eta|$.

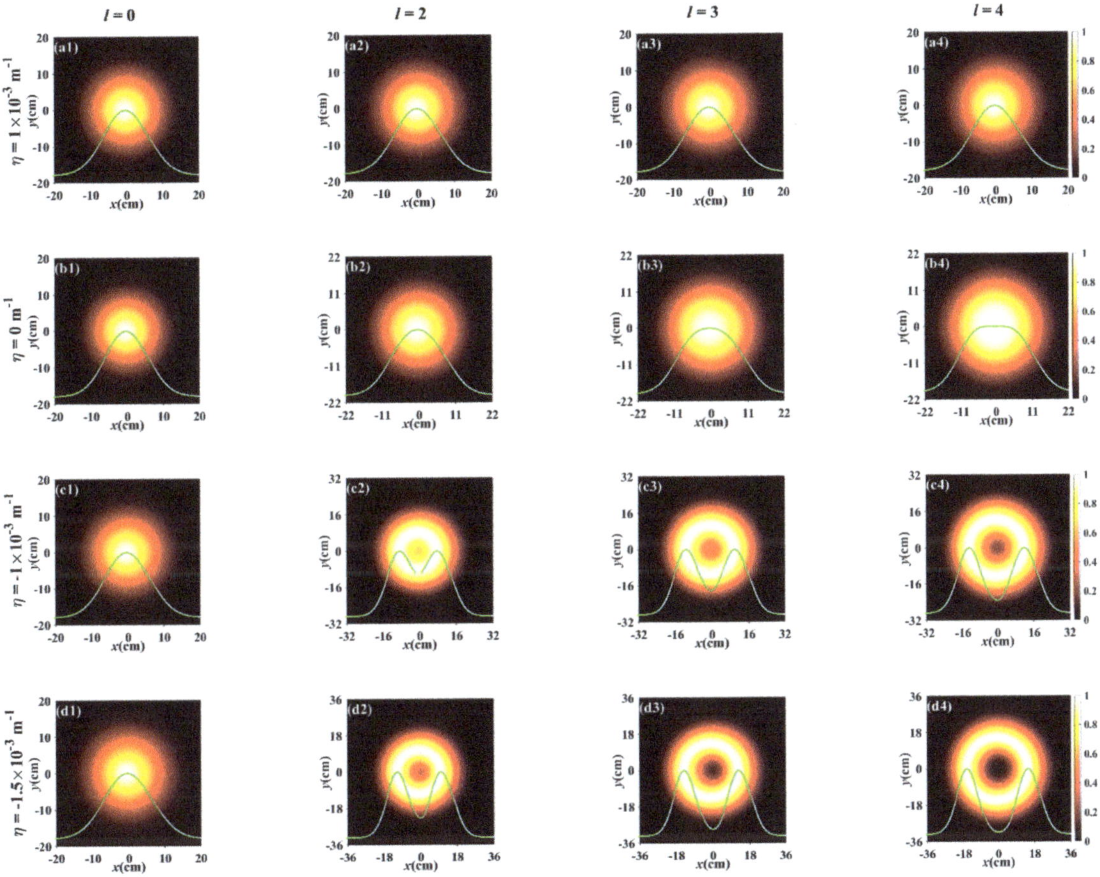

Figure 6. Normalized spectral intensity distribution of the PCTLGP beam propagating through the anisotropic turbulence for different twisted factors and topological charges ($\tilde{C}_n^2 = 5 \times 10^{-14}$ m$^{3-\alpha}$, $\xi_x = \xi_y = 2$, $L = 2$ km, $n = 0$).

For comparison, we introduce the normalized spectral intensity $S(T_0) = S(\rho, L; \omega)/S_{\max}(\rho, L; \omega)$, where $S_{\max}(\rho, L; \omega)$ is the maximum spectral intensity at the points $\rho = (x, y)$ when the pulse duration is 100 fs. Figure 7 illustrates the $S(T_0)$ of the PCTLGP beam propagating for 0.3 km in anisotropic non-Kolmogorov turbulent atmosphere under the different pulse durations. Obviously, Figure 7 shows that the overall $S(T_0)$ increases with increasing T_0. Figure 7a–c illustrates the influence of the twisted factor, the topological charge and the mode order on $S(T_0)$, respectively. It can be found that the ability of the

PCTLGP beam to maintain the dark hollow shape is enhanced with increasing l, and the dark ring of the beam is more obvious when n is larger. Moreover, the PCTLGP beam degrades more slowly when the sign of the twisted factor is "−". Therefore, we can draw a conclusion: the PCTLGP beam with a negative twisted factor can more effectively resist the influence of turbulence, and this resistance ability is further enhanced with increasing l, n.

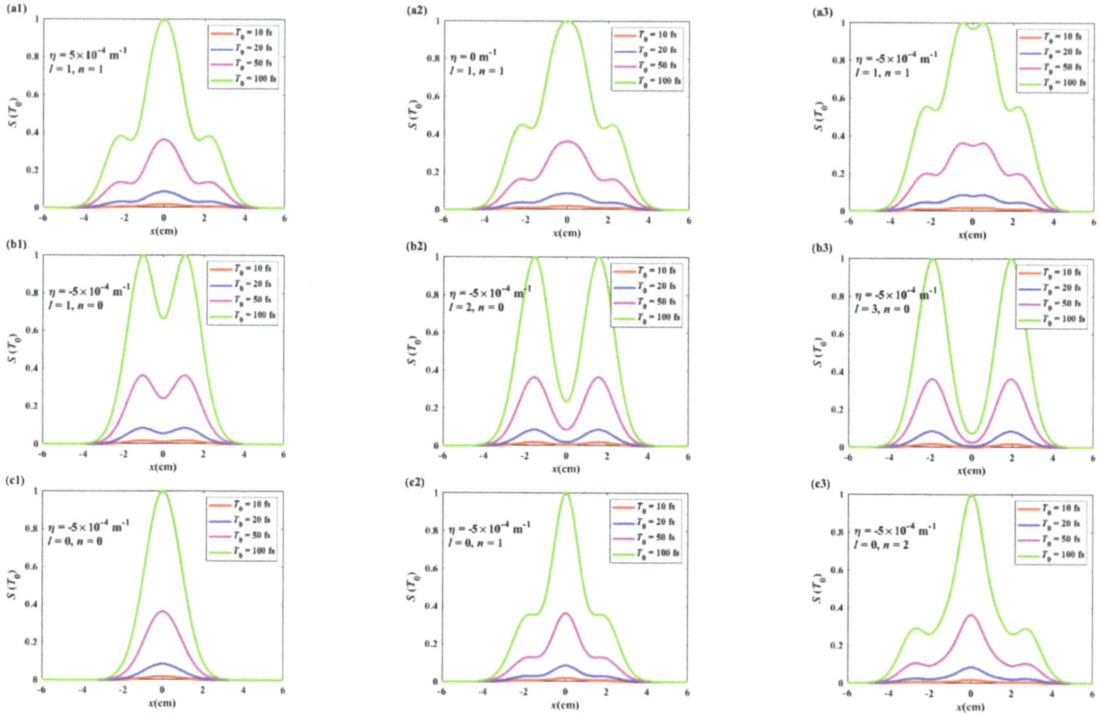

Figure 7. The relationship between the normalized spectral intensity with pulse duration of the PCTLGP beam propagating through the anisotropic turbulence for different (**a1–a3**) twisted factors, (**b1–b3**) topological charges, (**c1–c3**) mode orders ($\xi_x = \xi_y = 3$, $L = 0.3$ km, $\widetilde{C}_n^2 = 1 \times 10^{-14}$ $m^{3-\alpha}$).

4. Conclusions

In this paper, we have derived the analytical expressions for the diagonal elements of the CSDM for a PCTLGP beam in anisotropic turbulent atmosphere. The evolution properties of the PCTLGP beam on propagation are numerically discussed in detail. It is found that the initial beam profile of the PCTLGP beam ($n \neq 0$) shows a multi-ring dark hollow beam profile, and it will degenerate into a flat-top distribution with the increase in the propagation distance and eventually evolves into a Gaussian-like distribution in the far field. In addition, the influence of the twisted factor η, topological charge l, mode order n, pulse duration T_0 and turbulence parameters on the spectral intensity of the PCTLGP beam propagating through the turbulent atmosphere have been studied, and we find the most important parameters affecting the spectral intensity are the twisted factor η and topological charge l. Numerical results show that the PCTLGP beam with a negative twisted factor shows an advantage over the PCLGP beam in resisting the turbulence-induced degeneration, and this advantage is further strengthened with increasing l and $|\eta|$. Moreover, the ability of the PCTLGP beam to resist turbulence also increases with the increase in n, and the pulse duration will affect the value of the spectral intensity, which increases with increasing T_0. It is also found that the PCTLGP beam shows different evolution properties when propagating in weaker turbulence and stronger turbulence.

When the PCTLGP beam propagates for long distances in a weakly turbulent atmosphere, the progress of the beam evolving into a Gaussian-like shape will not be accelerated with the increase in the propagation distance, and only the beam spot will become lager. However, the degradation progress of the beam is accelerated with increasing L when propagating for long distances in stronger atmospheric turbulence, and it eventually converges to the Gaussian-like distribution. Compared with the general PCLGP beam, the PCTLGP beam shows some advantages. Therefore, the PCTLGP beam will be useful in free-space optical communications and remote sensing.

Author Contributions: Writing—original draft, Y.X. (Ying Xu); writing—review and editing, Y.X. (Ying Xu), Y.X. (Yonggen Xu); supervision, Y.X. (Ying Xu), Y.X. (Yonggen Xu) and T.W. All authors have read and agreed to the published version of the manuscript.

Funding: This research was funded by Sichuan Provincial University Key Laboratory of Detection and Application of Space Effect in Southwest Sichuan (ZDXM202201003); Department of Science and Technology of Sichuan Province (2019YJ0470); NSAF (U2130123); International Partnership Program of Chinese Academy of Sciences (181231KYSB20200033); and Shanghai Science and Technology Program (21511105000 and 19ZR1464100).

Institutional Review Board Statement: Not applicable.

Informed Consent Statement: Not applicable.

Data Availability Statement: Exclude this statement.

Conflicts of Interest: The authors declare no conflict of interest.

References

1. Paakkonen, P.; Turunen, J.; Vahimaa, P.; Friberg, A.; Wyrowski, F. Partially coherent Gaussian pulses. *Opt. Commun.* **2002**, *204*, 53–58. [CrossRef]
2. Lin, Q.; Wang, L.; Zhu, S. Partially coherent light pulse and its propagation. *Opt. Commun.* **2003**, *219*, 65–70. [CrossRef]
3. Liu, D.; Wang, Y.; Wang, G.; Yin, H.; Wang, J. The influence of oceanic turbulence on the spectral properties of chirped Gaussian pulsed beam. *Opt. Laser Technol.* **2016**, *82*, 76–81. [CrossRef]
4. Ding, C.; Korotkova, O.; Horoshko, D.; Zhao, Z.; Pan, L. Evolution of Spatiotemporal Intensity of Partially Coherent Pulsed Beams with Spatial Cosine-Gaussian and Temporal Laguerre-Gaussian Correlations in Still, Pure Water. *Photonics* **2021**, *8*, 102. [CrossRef]
5. Wang, H.; Yan, X.; Feng, X.; Zhao, Z.; Pan, L. Intensity Evolution of Cosine-Gaussian-Correlated Schell-Model Pulse Scattered by a Medium. *Appl. Sci.* **2020**, *10*, 1825. [CrossRef]
6. Liu, D.; Luo, X.; Wang, G.; Wang, Y. Spectral and Coherence Properties of Spectrally Partially Coherent Gaussian Schell-model Pulsed Beams Propagating in Turbulent Atmosphere. *Curr. Opt. Photonics* **2017**, *1*, 271–277.
7. Li, Y.; Gao, M.; Lv, H.; Wang, L.; Ren, S.; Hu, X. Anisotropic turbulence-induced statistical properties evolution of a radially polarized partially coherent pulsed beam on propagation. *Opt. Commun.* **2020**, *466*, 125669. [CrossRef]
8. Wang, X.; Tang, J.; Wang, Y.; Liu, X.; Liang, C.; Zhao, L.; Hoenders, B.; Cai, Y.; Ma, P. Complex and phase screen methods for studying arbitrary genuine Schell-model partially coherent pulses in nonlinear media. *Opt. Express* **2022**, *30*, 24222–24231. [CrossRef]
9. Li, Y.; Gao, M.; Li, B. Vortex phase induced evolution of propagation properties of a spatiotemporally partially coherent partially polarized pulsed beam in atmospheric turbulence. *Opt. Commun.* **2022**, *518*, 128385. [CrossRef]
10. Zhao, J.; Wang, G.; Ma, X.; Zhong, H.; Yin, H.; Wang, Y.; Liu, D. Intensity and Coherence Characteristics of a Radial Phase-Locked Multi-Gaussian Schell-Model Vortex Beam Array in Atmospheric Turbulence. *Photonics* **2021**, *8*, 5. [CrossRef]
11. Guo, Z. Coherent Property Evolution of Partially Coherent Elliptical Vortex Beam Propagation Through Turbulence. *Laser Optoelectron. Prog.* **2021**, *58*, 0901001.
12. Liu, D.; Wang, Y.; Wang, G.; Yin, H. Propagation properties of a partially coherent anomalous hollow vortex beam in uniaxial crystal orthogonal to the optical axis. *J. Mod. Opt.* **2018**, *65*, 1442–1449. [CrossRef]
13. Xu, Y.; Zhao, L.; Yang, N.; Xu, Y. Propagation characteristics of partially coherent twisted Laguerre-Gaussian beam in atmospheric turbulence with anisotropy. *J. Mod. Opt.* **2022**, *69*, 200–209. [CrossRef]
14. Yu, J.; Huang, Y.; Wang, F.; Liu, X.; Gbur, R.; Cai, Y. Scintillation properties of a partially coherent vector beam with vortex phase in turbulent atmosphere. *Opt. Express* **2019**, *27*, 26676–26688. [CrossRef]
15. Zhao, L.; Xu, Y.; Dan, Y. Evolution properties of partially coherent radially polarized Laguerre-Gaussian vortex beams in an anisotropic turbulent atmosphere. *Opt. Express* **2021**, *29*, 34986–35002. [CrossRef]
16. Peng, X.; Liu, L.; Wang, F.; Popov, S.; Cai, Y. Twisted Laguerre-Gaussian Schell-model beam and its orbital angular moment. *Opt. Express* **2018**, *26*, 33956–33969. [CrossRef]

17. Zhong, W.; Yang, Z.; Belic, M.; Zhong, W. Two-dimensional asymmetric Laguerre-Gaussian diffraction-free beams. *Phys. Lett. A* **2022**, *423*, 127818. [CrossRef]
18. Zhong, W.P.; Zhong, W.; Belic, M.; Yang, Z. Controllable two-dimensional diffraction-free polygon beams. *Phys. Lett. A* **2022**, *432*, 128009. [CrossRef]
19. Pan, X.; Wu, J.; Li, Z.; Zhang, C.; Deng, C.; Zhang, Z.; Wen, H.; Gao, Q.; Yang, J.; Yi, Z.; et al. Laguerre-Gaussian mode purity of Gaussian vortex beams. *Optik* **2021**, *230*, 166320. [CrossRef]
20. Xu, Y.; Li, Y.; Zhao, X. Intensity and effective beam width of partially coherent Laguerre–Gaussian beams through a turbulent atmosphere. *J. Opt. Soc. Am. A* **2015**, *32*, 1623–1630. [CrossRef]
21. Martínez-Matos, Ó.; Vaveliuk, P.; Izquierdo, J.; Loriot, V. Femtosecond spatial pulse shaping at the focal plane. *Opt. Express* **2013**, *21*, 25010–25025. [CrossRef] [PubMed]
22. Liu, Y.; Gao, C. Study on the time-varying and propagating characteristics of ultrashort pulse Laguerre-Gaussian beam. *Opt. Express* **2010**, *18*, 12104–12110. [CrossRef] [PubMed]
23. Li, Y.; Zhang, Y.; Zhu, Y. Probability distribution of the orbital angular momentum mode of the ultrashort Laguerre-Gaussian pulsed beam propagation in oceanic turbulence. *Results Phys.* **2018**, *11*, 698–705. [CrossRef]
24. Liu, Z.; Wan, L.; Zhou, Y.; Zhang, Y.; Zhao, D. Progress on Studies of Beams Carrying Twist. *Photonics* **2021**, *8*, 92. [CrossRef]
25. Zhou, Y.; Zhao, D. Propagation properties of a twisted rectangular multi-Gaussian Schell-model beam in free space and oceanic turbulence. *Appl. Opt.* **2018**, *57*, 8978–8983. [CrossRef]
26. Cui, Y.; Wang, F.; Cai, Y. Propagation of a twist Gaussian-Schell model beam in non-Kolmogorov turbulence. *Opt. Commun.* **2014**, *324*, 108–113. [CrossRef]
27. Zhao, Y.; Liu, X.; Liang, C.; Liu, L.; Wang, F.; Cai, Y. Statistical Characteristics of a Twisted Anisotropic Gaussian Schell-Model Beam in Turbulent Ocean. *Photonics* **2020**, *7*, 37.
28. Liu, Y.; Lin, R.; Wang, F.; Cai, Y.; Yu, J. Propagation properties of Laguerre-Gaussian Schell-model beams with a twist phase. *J. Quant. Spectrosc. Radiat. Transf.* **2021**, *264*, 107556. [CrossRef]
29. Zhang, B.; Huang, H.; Xie, C.; Zhu, S.; Li, Z. Twisted rectangular Laguerre-Gaussian correlated sources in anisotropic turbulent atmosphere. *Opt. Commun.* **2020**, *459*, 125004. [CrossRef]
30. Madhu, V.; Ivan, J. Robustness of the twist parameter of Laguerre-Gaussian mode superpositions against atmospheric turbulence. *Phys. Rev. A* **2017**, *95*, 043836.
31. Xu, Y.; Dan, Y.; Yu, J.; Cai, Y. Kurtosis parameter K of arbitrary electromagnetic beams propagating through non-Kolmogorov turbulence. *J. Mod. Opt.* **2017**, *64*, 1976–1987. [CrossRef]
32. Xu, Y.; Xu, Y.; Wang, S.; Deng, X.; Liu, Y.; Wang, S. Kurtosis parameter and coupling coefficient of partially-coherent twisted Gaussian beam propagating through non-Kolmogorov atmospheric turbulence. *J. Russ. Laser Res.* **2022**, *43*, 509–519. [CrossRef]
33. Zhou, Y.; Cheng, K.; Zhu, B.; Yao, N.; Zhong, X. Beam spreading, kurtosis parameter and Strehl ratio of partially coherent cosh-Airy beams in atmospheric turbulence. *Optik* **2019**, *183*, 656–663. [CrossRef]

MDPI
St. Alban-Anlage 66
4052 Basel
Switzerland
www.mdpi.com

Photonics Editorial Office
E-mail: photonics@mdpi.com
www.mdpi.com/journal/photonics

www.ingramcontent.com/pod-product-compliance
Lightning Source LLC
Chambersburg PA
CBHW061550210926
51850CB00009B/2063
* 9 7 8 3 7 2 5 8 0 1 9 0 9 *